Springer Transactions in Civil and Environmental Engineering

More information about this series at http://www.springer.com/series/13593

G.L. Sivakumar Babu • Sireesh Saride
B. Munwar Basha
Editors

Sustainability Issues in Civil Engineering

Springer

Editors
G.L. Sivakumar Babu
Department of Civil Engineering
Indian Institute of Science Bangalore
Bengaluru, Karnataka
India

Sireesh Saride
Department of Civil Engineering
Indian Institute of Technology Hyderabad
Sangareddy, Telangana
India

B. Munwar Basha
Department of Civil Engineering
Indian Institute of Technology Hyderabad
Sangareddy, Telangana
India

ISSN 2363-7633 ISSN 2363-7641 (electronic)
Springer Transactions in Civil and Environmental Engineering
ISBN 978-981-10-1928-9 ISBN 978-981-10-1930-2 (eBook)
DOI 10.1007/978-981-10-1930-2

Library of Congress Control Number: 2016955959

Printed on acid-free paper

This Springer imprint is published by Springer Nature
The registered company is Springer Science+Business Media Singapore Pte Ltd.

Preface

Sustainability issues in civil engineering have been the topic of discussions and intense debate among civil engineering professionals and academics worldwide. Many universities across the globe are now offering courses on sustainability in civil engineering programs at both undergraduate and graduate levels. The topic is also at the core of many professional certification programs.

Sustainability issues become more pronounced in developing economies, which have high economic and infrastructural growth rates. This edited volume on *Sustainability Issues in Civil Engineering* is an outcome of deliberations and discussions in the "International Conference on Sustainable Civil Infrastructure" held on 17–18 October 2014 in Hyderabad, India. This international conference was organized by the American Society of Civil Engineers (ASCE)_ India Section in association with the Indian Institute of Technology (IIT), Hyderabad. At the conference, several keynote lectures were delivered, which have been incorporated as chapters in this book. In addition, a few invited experts outside of the conference have contributed chapters to make this volume complete.

Chapter 1 is on the usage of construction and demolition materials in road construction in which Prof. Arulrajan and his colleagues from Australia present the evaluation of suitability of recycled concrete aggregate (RCA), crushed brick (CB), reclaimed asphalt pavement (RAP), and waste rock (WR) as embankment fills, pavement base/subbase, and pipe bedding applications.

In Chap. 2, Prof. Filippo Practico from Italy presents an approach to estimate costs and benefits when one uses (1) pervious pavements, (2) continuous monitoring of pavement structural condition, (3) reinforcement to increase life expectancy, and (4) recycling in pavements.

Prof. Veeraraghavan and Ms. Nivedya from IIT Madras present the results of characterization of bituminous stabilized mixes (BSH) in Chap. 3. They also present the evaluation of sections which are constructed with cold in-place recycling (CIPR) technology in India.

In Chap. 4, Prof. Partha Chakraborty of IIT Kanpur presents the role of intelligent transportation systems in the context of sustainable transportation for Indian cities.

Prof. Animesh Das of IIT Kanpur presents economic sustainability considerations in asphalt pavement design and examines the role of alternative materials and technologies in Chap. 5.

In Chap. 6, Prof. Sireesh Saride and his colleagues from IIT Hyderabad present a sustainable design for rural roads in India using reclaimed asphalt materials.

Prof. Sarvesh Chandra of IIT Kanpur discusses sustainability issues in railway tracks in Chap. 7 and advocates the use of geosynthetics in railways.

Chapter 8 presents the use of system approach to identify and evaluate sustainability policy and practices in road construction industry. This article by Prof. Rajib B. Mallick demonstrates the ideas with an example of road construction and its impact and the use of recycling for achieving sustainable road construction.

Prof. Hemanta Hazarika and his colleagues from Kyushu and Kochi Universities, Japan, walk us through Great East Japan earthquake disasters and success stories of the innovative tire retaining wall, which survived during the earthquake. They propose a new concept of using waste tires behind seawalls to protect them from devastating tsunami impact forces in Chap. 9.

Prof. K. Rajagopal and Ms. Anjana of IIT Madras present the uses of reinforced piled embankments for sustainable infrastructure development in Chap. 10. They investigate the performance of these embankments in soft soil.

Chapter 11 presents the studies of Prof. Hoyos and Prof. Anand Puppala of the University of Arlington, USA, in modeling the unsaturated geomaterials for sustainable geotechnical designs.

Chapter 12 addresses the use of underground space for sustainable urban development in which Prof. T. G. Sitharam of IISc, Bangalore, presents the experiences in underground metro construction.

Prof. Murali Krishna Adapa and Mr. Bali Reddy S from IIT Guwahati demonstrate the advantages of adopting sand-tire chips mixture as a sustainable material in various geo-engineering applications in Chap. 13.

Chapters 14 and 15, contributed by Prof. G. L. Sivakumar Babu of IISc, Bangalore, address the sustainability issues in landfills and use the appropriate environmental, social, and economic indicators to analyze the cases considered.

Prof. Krishna Reddy and Ms. Bala Yamini Sadasivam of the University of Illinois, Chicago, USA, present approaches to select appropriate sustainable technologies for remediation of contaminated sites and illustrate them with case studies in Chap. 16.

Chapter 17 presents the development of a new disinfection system which has the potential to replace conventional systems, and the studies of Prof. Ligy Philip and her students from IIT Madras clearly establish the supremacy of the proposed system.

Chapter 18 presents the use of pre-engineered bamboo structures as a step toward sustainable construction. This article by Prof. Suresh Bhalla and his co-worker from IIT Delhi provides useful insights on the use of bamboo in housing construction.

Chapter 19 presents how the architectural engineering professionals incorporate sustainability concepts in practice. Mr. Anchuri and Prof. N. V. Ramana Rao illustrate the principles of sustainability in a consistent manner.

In Chap. 20, Mr. Chetan Hazaree and his colleagues present a case study of long-distance pumping of concrete and demonstrate how sustainable solutions were implemented in this unique case.

We wish to place on record our sincere thanks and appreciation to all the experts and contributors to this special volume. We wish to thank Springer for publishing this important contribution in the area of sustainable practices in civil engineering construction.

Bengaluru, Karnataka, India G.L. Sivakumar Babu
Sangareddy, Telangana, India Sireesh Saride
Sangareddy, Telangana, India B. Munwar Basha

Contents

About the Editors

G. L. Sivakumar Babu completed his Ph.D. (geotechnical engineering) in 1991 from the Indian Institute of Science, Bangalore, India, after earning his master's degree (soil mechanics and foundation engg.) in 1987 from Anna University, Madras, and B.Tech. (civil engineering) in 1983 from Sri Venkateswara University, Tirupati. He worked in the Central Road Research Institute and International Airports Authority of India, New Delhi. He joined the Indian Institute of Science in 1997, where his teaching and research activities include courses on geomechanics, soil investigations, and risk and reliability applications in civil engineering, pavement engineering, geosynthetics, and ground improvement.

He is the editor in chief of the *Indian Geotechnical Journal* and is an editorial board member of (1) the *Journal of Hazardous, Toxic, and Radioactive Waste* (ASCE), (2) the *International Journal of Geomechanics* (ASCE), (3) the *International Journal of Georisk: Assessment and Management of Risk for Engineered Systems and Geohazards* (Taylor & Francis Group), (4) the *International Journal of Systems Assurance Engineering and Management* (Springer Publications), (5) *the International Journal of Geosynthetics and Ground Engineering and Ground Improvement*, and (6) *the Journal of the Institution of Civil Engineers*, UK (from 2006 to 2013).

He is the governor of Region 10 of ASCE, the ex-president of ASCE (India Section), and a member of the Sectoral Innovation Council of the Ministry of Road Transport and Highways and the Highway Research Board (Identification, Monitoring, and Research Application) Committee of the Indian Roads Congress, New Delhi. He worked as a Humboldt fellow and visiting scientist in Germany from June 1999 to July 2000 and as a visiting scholar of Purdue University, Lafayette, USA, from February 1995 to February 1996. He is presently the chairman of the International Technical Committee (TC-302) on Forensic Geotechnical Engineering (FGE) of the International Society for Soil Mechanics and Geotechnical Engineering (ISSMGE) and served as its secretary from 2005 to 2009 and from 2009 to 2013. He was a member in the International Technical Committee (TC-32) on Risk Assessment in Geotechnical Engineering from 1997 to 2001 and from 2001 to 2005.

He is a fellow of ASCE, life member of the Indian Roads Congress, fellow of the Institution of Engineers India, fellow of the Indian Geotechnical Society, and fellow of the Association of Consulting Engineers (ACCE).

He has guided a number of students for research degrees (Ph.D. and M.Sc. (engg.)) and wrote a book on soil reinforcement and geosynthetics, has edited six books and proceedings, and has over 200 publications to his credit in journals and conferences of international repute. He has received several awards for his work such as best Ph.D. thesis in geotechnical engineering in India from the Indian Geotechnical Society, a Humboldt fellowship from Germany, a DST Boyscast fellowship, a Young Engineers Award from the Central Board of Irrigation and Power at New Delhi, and a few awards for best papers from ASCE and the Indian Geotechnical Society.

Sireesh Saride is currently an associate professor in the Department of Civil Engineering of the Indian Institute of Technology in Hyderabad. His expertise and research interests are in the area of pavement geotechnics, geosynthetics, and ground improvement. He earned his B.Tech (2000) from Jawaharlal Nehru Technological University, India, and his masters integrated with Ph.D. from the Indian Institute of Science, Bangalore, India, in 2006. He pursued his postdoctoral research at the University of Texas at Arlington. Sireesh's current research focuses on the dynamic behavior of geocell-reinforced base and subbase layers and improvement of rural roads using secondary/recycled materials. His research sponsorship comes from the Department of Science and Technology, the Ministry of Road Transportation and Highways, and Neyveli Lignite Corporation Ltd. and other private sector enterprises. Sireesh has published more than 100 referred journal and conference articles and technical reports. He is a currently serving as associate editor for the *Journal of Ground Improvement* (ICE) and the *Journal of Materials in Civil Engineering* (ASCE). He is also a guest editor for a special issue on "transportation geotechnics" of the *Indian Geotechnical Journal* (2015). He is a member of ASCE and ICE and life member of IGS and IRC. He served as a secretary for ASCE India Section (2012–2014). Dr. Sireesh received best paper awards from the Indian Geotechnical Society (2004, 2014) and ASCE GeoShanghai Conference (2010). He received the Young Scientist Award from DST of the government of India (2012).

B. Munwar Basha is currently an Assistant Professor in Civil Engineering Department at Indian Institute of Technology, Hyderabad, India. Earlier, he worked in IIT Delhi as Assistant Professor in Civil Engineering Department for 3 years from 2010 to 2013. He completed his Ph.D. from Indian Institute of Science Bangalore, India in August 2009, after earning his M.Tech from IIT Kanpur in 2004 and B.Tech (in Civil Engg) from JNTU college of Engineering, Anantapur, AP, India in June 2001. Further he worked a post-doctoral researcher at IISc Bangalore, India until 2010. His main research focus is on risk and reliability in geotechnical and geoenvironmental engineering, soil dynamics and earthquake resistant design of earth structures. He received Prof. N. S. Govinda Rao Gold Medal for the Best Ph.D thesis in the department of CE at IISc Bangalore for the year 2008–2009. He also received Vellanki Rama Rao Memorial Gold Medal Award for bagging University First Rank in B.Tech Civil Engineering Department of all constituent colleges of JNTU for the batch 1997–2001.

He received several awards for his work such as IGS – Dr B B Rai – Shri S N Gupta Biennial Award 2015 for the best paper on Earth and Earth Retaining Structures, DST Fast-Track Scheme for Young Scientist, Outstanding Young Faculty Fellow in the area of Engineering for the year 2011 by Kusuma Trust at IIT-Delhi, Excellent Paper Award in IACMAG, Best Paper Award in 3IYGEC, and Citation of his Doctoral thesis in Geotechnique journal. He has more than 80 publications in journals and conferences (24 publications in journals and 21 in ASCE Geotechnical Special Publications). He has been guiding a number of students for Ph.D degrees. He guided nine M.Tech students and five B.Tech projects. He has successfully completed a few sponsored research projects to the tune of 28.36 lakhs Rupees. He has carried out consultancy works for more than ten projects.

Part I
Sustainable Transportation Systems

Chapter 1
Sustainable Usage of Construction and Demolition Materials in Roads and Footpaths

Arul Arulrajah, Mahdi M. Disfani, and Suksun Horpibulsuk

Abstract The increase in generation of waste from construction activities along with significant increase in global population has led to increasing focus and research on reuse of waste material. In this paper, the application of construction and demolition (C&D) materials in road works is reviewed and discussed. C&D materials were evaluated by laboratory testing methods to assess their viability for reuse in roads and footpaths. Several unique field case studies where C&D materials have been used are also reported. C&D materials studied include recycled concrete aggregate (RCA), crushed brick (CB), reclaimed asphalt pavement (RAP), fine recycled glass (FRG), and waste rock (WR). C&D materials were found to be suitable for road and footpath applications such as embankment fills, pavement base/subbase, and pipe bedding applications.

Keywords Recycled materials • Demolition materials • Pavement • Footpath

1.1 Introduction

Waste materials are any type of material by-product of human and industrial activity that has no lasting value (Tam and Tam 2007). The disposal of solid waste is a major problem throughout the world. The sustainable usage of waste materials in engineering applications is of social and economic benefit. Due to the shortages of natural mineral resources and available land space and increasing waste disposal costs, recycling and reusing of solid wastes have become significant in recent years.

A. Arulrajah (✉)
Swinburne University of Technology, Melbourne, Australia
e-mail: aarulrajah@swin.edu.au

M.M. Disfani
The University of Melbourne, Melbourne, Australia

S. Horpibulsuk
Swinburne University of Technology, Melbourne, Australia

Suranaree University of Technology, Nakhon Ratchasima, Thailand

G.L. Sivakumar Babu et al. (eds.), *Sustainability Issues in Civil Engineering*,
Springer Transactions in Civil and Environmental Engineering,
DOI 10.1007/978-981-10-1930-2_1

Construction and demolition (C&D) materials are the excess or waste materials associated with the construction and demolition of buildings and structures, including concrete, brick, reclaimed asphalt, steel, timber, plastics, and other building materials and products (Sustainability Victoria 2010). Recycling and subsequent reuse of C&D materials will reduce the demand for scarce virgin natural resources and simultaneously reduce the quantity of this waste material destined for landfills (Arulrajah et al. 2014a; Disfani et al. 2011; Hoyos et al. 2011). The usage of C&D materials in civil engineering applications such as roads and footpaths is a viable and sustainable option to minimize the C&D waste while reducing the demand for scarce virgin quarried materials (Poon and Chan 2006; Tam and Tam 2007; Hoyos et al. 2011; Puppala et al. 2011; Arulrajah et al. 2013c).

In Australia, approximately 8.7 million tons of RCA, 1.3 million tons of CB, 3.3 million tons of WR, 1.0 million tons of FRG, and 1.2 million tons of RAP are stockpiled annually (Arulrajah et al. 2013c). The reuse of C&D materials would clearly provide substantial benefits in terms of reduced new material supply and waste disposal cost, increased sustainability, and reduced environmental impact (Sivakumar et al. 2004).

In this research, the geotechnical characteristics of five major categories of C&D materials and several of their blends have been characterized through an extensive series of geotechnical tests to address their usage in road pavement and footpath applications. Field test results from several unique case studies utilizing C&D materials are also presented. The properties of the C&D materials were tested and compared with typical road authority specified requirements for usage as a subbase material. Environmental assessments have also been carried out. The C&D materials studied in this research were recycled concrete aggregate (RCA), crushed brick (CB), reclaimed asphalt pavement (RAP), waste rock (WR), and fine recycled glass (FRG).

RCA is the by-product of C&D activities of concrete structures (Arulrajah et al. 2012a; Sustainability Victoria 2010). CB is a by-product of demolition activities of buildings and other structures (Arulrajah et al. 2012b). WR used in this study originates from "basalt floaters" or surface excavation rock (basalt) which commonly occurs near the surface to the west and north of Melbourne, Australia (Arulrajah et al. 2013c). RAP is the name given to spent asphalt that has been recycled during removal from roadways, which is done on a regular basis (Arulrajah et al. 2014b; Rahman et al. 2014). FRG is the by-product of crushing mixed color bottles and other glass products collected from both municipal and industrial waste streams (Landris 2007).

1.2 Laboratory Evaluation

The recycled CB, RCA, WR, and RAP used in this research had a maximum particle size of 20 mm. FRG has a maximum particle size of 4.75 mm and comprises sand size and a small percentage of silt size particles (Disfani et al. 2011).

Using sieve analysis results, the Unified Soil Classification System (ASTM 2010) was implemented to classify the recycled materials. Organic content of all the recycled material sources in this research was determined following ASTM (2007). pH values of the recycled materials were determined following the Standards Australia (1997a).

The test specimens for hydraulic conductivity tests were compacted with modified Proctor compaction effort, at optimum moisture content (OMC) to reach at least 98 % of maximum dry density (MDD). The falling head test method was chosen for all recycled aggregate with the exception of FRG, which was tested by the constant head method. CBR test specimens were prepared by applying modified compaction efforts to recycled aggregates mixed at the OMC obtained in compaction tests. A surcharge mass of 4.5 kg was placed on the surface of the compacted specimens, and then the samples were soaked in water for a period of 4 days. This is to simulate the confining effect of overlying pavement layers and also the likely worst case in-service scenario for a pavement (VicRoads 1998).

The consolidated drained (CD) static triaxial tests were performed with specimen dimensions of 100 by 200 mm (diameter by height) for all recycled material types except FRG which was tested with the dimensions of 50 by 100 mm. The test specimens were compacted to 98 % of MDD from modified compaction test in a split mold in eight layers. The shearing was performed under strain-controlled condition at the selected strain rate of 0.01 mm/min. Replicated samples were tested for the triaxial tests at the various stress levels.

Repeated load triaxial (RLT) tests were conducted to determine the resilient modulus and permanent deformation of the recycled materials. In this investigation, the RLT test was performed according to Austroads repeated load triaxial test method AG: PT/T053 (AustRoads 2000). The RLT testing consists of two phases of testing, permanent strain testing followed by resilient modulus testing. Permanent strain testing consists of three or four stages, each undertaken at different deviator stresses and a constant confining stress. The resilient modulus testing consists of 66 loading stages with 200 repetitions. In this test, the specimens were compacted to 98 % MDD based on modified compaction effort and tested at three target moisture contents of 70 %, 80 %, and 90 % of the OMC based on modified compaction effort, so as to simulate the dry-back process in the field. Replicated samples were tested for the RLT tests at each of the various moisture levels.

Total concentration (TC) and leachate analysis were carried out for a range of heavy metals on samples of C&D material. In the preparation of leachate, the method described in Australian Standard was followed, and slightly acidic leaching fluid (pH = 5) and alkaline leaching fluid (pH = 9.2) were used as leaching buffers (Standards Australia 1997b).

The particle size distributions of the five recycled materials are shown in Fig 1.1. The "after compaction" grading curves show that some breakdown has occurred during compaction especially for CB and RAP. However, all the recycled C&D materials, except for FRG, satisfied the guidelines for type 1 gradation C road base material according to ASTM D1241-07, except for slight deviations in the finer side

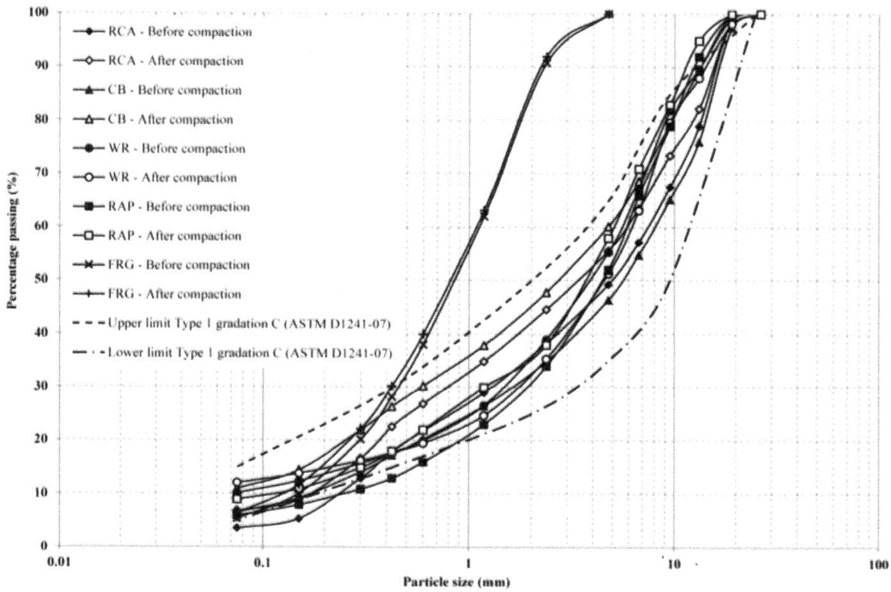

Fig. 1.1 Particle size distributions of the five recycled materials before and after compaction (Arulrajah et al. 2013c)

for some materials. The grain size distribution parameters are summarized in Table 1.1.

The RLT test results reported in Table 1.1 indicate that RCA, CB, and WR performed satisfactorily at 98 % MDD and at a target moisture content of 70 % of the OMC. RCA, CB, and WR materials showed sensitivity to moisture and produced higher limits of permanent strain and lower limits of resilient modulus, particularly at higher target moisture contents in the range of 80–90 % of the OMC. The performance of RCA, CB, and WR was found to be affected by increasing the target moisture contents and the density level. This is apparent particularly for CB which failed at the higher target moisture contents of 80–90 %. The results of permanent strain and resilient modulus for RAP and FRG could not be reported as these two materials possess very low cohesion values and their samples failed within a few cycles at a target moisture content of 60 % of the OMC.

TC and leachate analysis according to Australian Standard for Leaching Procedure (ASLP) with two buffer solutions (acidic and alkaline) were conducted on C&D materials. TC values of C&D samples were compared with EPA Victoria (Australia) requirements for fill material. The comparison implies that for all the contaminant constituents, TC values of C&D samples are far below the threshold.

According to the US EPA, a material is designated as a hazardous waste if any detected metal occurs at concentrations larger than 100 times the drinking water standard (Wartman et al. 2004). The ASLP values were found to be below the

Table 1.1 Geotechnical properties of recycled C&D materials (Arulrajah et al. 2013b)

Geotechnical parameters	RCA	CB	WR	RAP	FRG	Typical quarry material
D_{10} (mm)	0.24	0.18	0.075	0.24	0.16	–
D_{30} (mm)	1.3	1.7	1.5	1.9	0.45	–
D_{50} (mm)	5.0	5.6	3.9	4.5	0.85	–
D_{60} (mm)	7.5	8.0	5.6	5.9	1.2	–
C_u	31.2	44.4	74.7	25.6	7.5	–
C_c	0.9	2.0	5.4	2.5	1.5	–
Gravel content (%)	50.7	53.6	44.7	48.0	0.0	–
Sand content (%)	45.7	39.8	45.1	46.0	94.6	–
Fines content (%)	3.6	6.6	10.2	6.0	5.4	<10
USCS classification	GW	GW	SW	GW	SW	–
Particle density—coarse fraction (kN/m^3)	27.1	26.2	28.1	23.5	24.4	>19.62
Particle density—fine fraction (kN/m^3)	26.0	25.8	28.0	23.4	24.3	>19.62
Water absorption—coarse fraction (%)	4.7	6.2	3.3	2.2	1.0	<10
Water absorption—fine fraction (%)	9.8	6.9	4.7	2.4	1.8	<10
MDD (kN/m^3)—modified compaction	19.13	19.73	21.71	19.98	17.40	>17.5
OMC (%)—modified compaction	11.0	11.25	9.25	8.0	10.5	8–15
Organic content (%)	2.3	2.5	1.0	5.1	1.3	<5
pH	11.5	9.1	10.9	7.6	9.9	7–12
Hydraulic conductivity (m/s)	3.3×10^{-8}	4.5×10^{-9}	2.7×10^{-7}	3.5×10^{-7}	1.7×10^{-5}	$>1 \times 10^{-9}$
Flakiness index	11	14	19	23	–	<35
LA abrasion loss (%)	28	36	21	42	25	<40
CBR (%)	118–160	123–138	121–204	30–35	42–46	>80
Triaxial test (CD): apparent cohesion (kPa)	44	41	46	53	0	>35
Triaxial test (CD): friction angle (degree)	49	48	51	37	37	>35
Resilient modulus: target 90 % of the OMC	239–357	301–319	121–218	–	–	125–300
Resilient modulus: target 80 % of the OMC	487–729	303–361	202–274	–	–	150–300
Resilient modulus: target 70 % of the OMC	575–769	280–519	127–233	–	–	175–400

threshold of hazardous waste proving that they are not categorized as hazardous waste according to US EPA.

1.3 Field Case Study 1: Recycled Glass in Pavement Base

Nine sections of unbound granular base pavements, comprising of up to 30 % FRG in blends with RCA and waste rock (WR) in the pavement base, were constructed on the main haul road at a recycling site in Melbourne (Fig. 1.2). The design of these granular base pavements was based on the outcomes of the initial laboratory testing phase of this research. FRG/RCA and FRG/WR blends were found to satisfy the requirements of a pavement subbase material in the laboratory testing phase of this research; however, it was decided to use this material in the pavement base and assess its performance as a higher-quality pavement base material in the field trials. The 200 mm thick base layer comprised of FRG blends was placed and subsequently overlaid by a 50 mm thick glass asphalt (glassphalt) cover comprised of asphalt with 5 % glass content. The base material composition was with blends comprised of 10–30 % content of FRG/WR or FRG/RCA. Two control sections comprising 100 % of WR and RCA were also built. These aggregates are commonly accepted for usage in pavement base applications in Australia. Four sections were constructed with RCA with 10 %, 15 %, 20 %, and 30 % of FRG. Another three sections were constructed with WR with 10 %, 20 %, and 30 % of FRG. The 200 mm thick pavement base layer was constructed with various FRG/RCA or

Fig. 1.2 Laying of the base layer at one of the sections including a fraction of FRG (Arulrajah et al. 2013a)

FRG/WR blends in seven sections and with RCA and WR for the two remaining control sections.

For the assessment of the geotechnical performance of the recycled materials and their impact on base strength and stiffness, field testing was conducted at various locations after the placement of the pavement base layers. The field tests were undertaken 3 days after the placement of the subbase and base layers with a nuclear density gauge (NDG) and Clegg hammer (CH). It was expected that the field moisture conditions at the time of testing would be lower than the optimum moisture conditions at the time of compaction, as the materials were delivered within the recycling site and haulage time was 1–2 min.

The average field densities of the FRG-WR sections are noted to be higher than that of the FRG/RCA blends. The FRG/WR blends also had higher field density and laboratory density results than corresponding FRG-RCA blends with the same glass contents. The results indicate that average density ratios in individual sites varied in the range of 96–100 % MDD. The control sections (RCA, WR) achieved the highest density results in the field as compared to the FRG/RCA and FRG/WR sections. The field results indicate that WR is a higher-quality material than RCA. It was noted that the FRG30/WR70 blend containing 30 % recycled glass content produced the lowest density ratio compared to other FRG/WR blends with less glass additive content. Similarly, the FRG30/RCA70 blend containing 30 % recycled glass content also produced the lower density ratio compared to other FRG/RCA blends. From this finding, it can be concluded that a blend containing a recycled glass additive content of greater than 20 % would likely result in a lower field dry density being achieved.

The results from the Clegg hammer tests were analyzed to determine CBR values of the various pavement sections as well as to determine the strength ratios after field compaction. Figure 1.3 presents the Clegg hammer results for CBR for the various pavement base sections that have been transposed into the same figure for easy reference. Results of Clegg hammer tests meet the minimum soaked field CBR of 100 % for base materials in all sections except in FRG10/WR90, FRG10/RCA90, FRG20/RCA80, and FRG30/RCA70. Also, Clegg hammer results meet the specified minimum soaked field CBR of 80 % for subbase materials in all sections except over short stretches of FRG10/WR90, FRG20/RCA80, and FRG30/RCA70, for 10–20 m in which it was marginally below the specified requirements but is still deemed acceptable for haul roads.

The results seem to indicate that the FRG blends should be limited to pavement subbase applications and may not meet requirements for a pavement base material. The Clegg hammer results seem to also indicate variation in the recycled blends within each pavement section and between pavement sections. Both RCA and WR satisfied the requirements as a subbase material. Limited blends of 20 % FRG with coarse-sized recycled concrete aggregates (FRG20/RCA80) and waste rock aggregates (FRG20/WR80) appear to be the optimum limits of glass additives with recycled aggregates based on the field testing results.

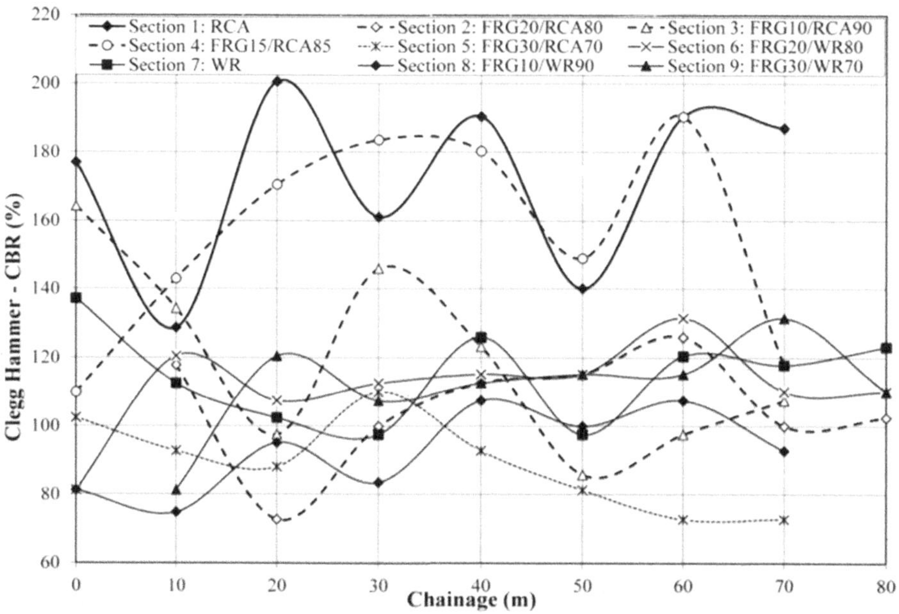

Fig. 1.3 Clegg hammer results for various combinations of C&D pavement base sections (Arulrajah et al. 2013a)

1.4 Field Case Study 2: Reclaimed Asphalt Pavement in Pavement Subbase

RAP was used as a subbase material in nine pavement sections for a haul road at a recycling site operator's facility. RAP was selected as it was available in large stockpiles at the recycling site, and there was an interest from various parties to evaluate the field performance of untreated RAP in subbase layer.

The earlier phase of laboratory evaluation of RAP indicated that it did not meet the local road authorities' requirements for usage in pavement subbase layers, particularly in terms of RLT and CBR requirements. Nevertheless, the field trial pavement constructed was for a private haul road in the recycling operator's site and as such did not have to meet the specified requirements of the local road authorities.

Based on the nuclear density results in construction of the subbase for the RAP pavement, the trial complied with target minimum moisture content requirement of 85 % OMC. The Clegg hammer results indicate that RAP did not meet the minimum soaked field CBR of 80 % for subbase materials in the various sections as advised by local road authorities. Based on the field and laboratory testing, RAP had insufficient strength requirements to meet the local road authority pavement subbase requirements.

1.5 Field Case Study 3: Recycled Glass in Footpaths

An asphalt footpath for shared use by pedestrians and cyclists was constructed using the outcomes of the laboratory testing phase of this research by a local government council in Melbourne. The shared path was comprised of a base layer of nominal 100 mm thickness, overlying a subgrade with a design soaked CBR greater than 3 %. The base layer was subsequently overlaid by a 30 mm thick asphalt cover. The asphalt footpath was constructed in three sections, with three different material blends of FRG/WR in the footpath base layer. The three trial sections constructed were 15 % FRG section (FRG15/WR85), 30 % FRG section (FRG30/WR70), and a control section comprising basaltic WR. Figure 1.4 shows the FRG15/WR85 and FRG30/WR70 sections after completion of field compaction of the base layer.

For the assessment of the construction variability of WR blended with FRG and their impact on base strength and stiffness, field testing was conducted at various locations along the pavement using a nuclear density gauge and Clegg hammer after the placement of the base layers. The tests were conducted 2 days after the placement of the base layers.

All the various material blends apart from FRG were found to plot within acceptable limits of the specified upper and lower limit requirements, except for some of the smaller fines at the lower particle sizes which were considered acceptable.

The variations between maximum and minimum dry density ratios along the chainage were found to be very small (i.e., maximum of 4 %). This indicates the field densities were achieved very consistently along the chainage. Clegg hammer test results were analyzed to determine CBR values of the various footpath sections

Fig. 1.4 A sections of shared path after field compaction of base layer (Arulrajah et al. 2013b)

as well as to determine the strength ratios based on a required minimum soaked field CBR of 28 % after field compaction.

Clegg hammer results suggested that finer grading (exceeding the fine grading limit) was found to significantly reduce the footpath base strength for FRG30/WR70. Limited blends of fine-sized recycled glass with coarse-sized rock aggregates, particularly FRG15/WR85, appear to be the optimum recycled material blend for a footpath base material.

1.6 Conclusions

This paper has reported on a comprehensive laboratory evaluation of the properties of C&D materials. In addition, several unique case studies are also reported in this paper, comprised of actual field implementation of recycled C&D materials in pavement bases, pavement subbases, and footpath base applications.

C&D materials were found to be suitable for road and footpath applications such as embankment fills, pavement base/subbase, and pipe bedding applications. The sustainable usage of C&D materials in sustainable civil engineering applications will result in a lower carbon footprint for our future roads, footpaths, and other civil engineering infrastructures.

Acknowledgments The third author is grateful to the financial support from the Thailand Research Fund (TRF) under the TRF Senior Research Scholar program, Grant No. RTA5680002.

References

Arulrajah A, Piratheepan J, Ali MMY, Bo MW (2012a) Geotechnical properties of re-cycled concrete aggregate in pavement sub-base applications. Geotech Test J ASTM 35(5):1–9

Arulrajah A, Piratheepan J, Bo MW, Sivakugan N (2012b) Geotechnical characteristics of recycled crushed brick blends for pavement sub-base applications. Can Geotech J 49 (7):796–811

Arulrajah A, Ali M, Disfani M, Horpibulsuk S (2013a) Recycled glass blends in pavement base/subbase applications: laboratory and field evaluation. J Mater Civ Eng 46:04014025. doi:10.1061/(ASCE)MT.1943-5533.0000966

Arulrajah A, Ali MMY, Disfani MM, Piratheepan J, Bo MW (2013b) Geotechnical performance of recycled glass-waste rock blends in footpath bases. J Mater Civ Eng ASCE 25(5):653–661

Arulrajah A, Piratheepan J, Disfani MM, Bo MW (2013c) Geotechnical and geoenvironmental properties of recycled construction and demolition materials in pavement subbase applications. J Mater Civ Eng ASCE 25(8):1077–1088

Arulrajah A, Disfani M, Horpibulsuk S, Suksiripattanapong C, Prongmanee N (2014a) Physical properties and shear strength responses of recycled construction and demolition materials in unbound pavement base/subbase applications. Constr Build Mater 58:245–257

Arulrajah A, Piratheepan J, Disfani MM (2014b) Reclaimed asphalt pavement/recycled concrete aggregate blends in pavement subbase applications: laboratory and field evaluation. J Mater Civ Eng ASCE 25(12):1920–1928

ASTM (2007) Standard test methods for moisture, ash, and organic matter of peat and other organic soils, ASTM D2974-07. American Society for Testing and Materials, West Conshohocken

ASTM (2010) Standard practice for classification of soils for engineering purposes (unified soil classification system), ASTM D2487-10. American Society for Testing and Materials, West Conshohocken

AustRoads (2000) Commentary to AG:PT/T053 – determination of permanent deformation and resilient modulus characteristics of unbound granular materials under drained conditions. AustRoads, Sydney

Disfani MM, Arulrajah A, Bo MW, Hankour R (2011) Recycled crushed glass in road work applications. Waste Manag 31(11):2341–2351

Hoyos LR, Puppala AJ, Ordonez CA (2011) Characterization of cement-fiber-treated reclaimed asphalt pavement aggregates: preliminary investigation. J Mater Civ Eng 23(7):977–989

Landris TL (2007) Recycled glass and dredged materials. US Army Corps of Engineers, Engineer Research and Development Center, Report No. ERDC TNDOER- T8, Mississippi, USA

Poon CS, Chan D (2006) Feasible use of recycled concrete aggregates and crushed clay brick as unbound road subbase. Constr Build Mater 20(8):578–585

Puppala AJ, Hoyos LR, Potturi AK (2011) Resilient moduli response of moderate-ly cement-treated reclaimed asphalt pavement aggregates. J Mater Civ Eng 23(7):990–998

Rahman MA, Arulrajah A, Piratheepan J, Bo MW, Imteaz M (2014) Resilient modulus and permanent deformation responses of geogrid-reinforced construction and demolition materials. J Mater Civ Eng ASCE 26(3):512–519

Sivakumar V, McKinley JD, Ferguson D (2004) Reuse of construction waste: performance under repeated loading. Proc Inst Civ Eng Geotech Eng 157(2):91–96

Standards Australia (1997a) Methods of testing soils for engineering purposes. Method 4.3.1: soil chemical tests-determination of the pH value of a soil-electrometric method. Standards Australia, AS 1289.4.3.1-1997, Homebush, NSW, Australia

Standards Australia (1997b) Wastes, sediments and contaminated soils, part 3: preparation of leachates-bottle leaching procedure. Standards Australia, AS 4439.3-1997, Homebush, NSW, Australia

Sustainability Victoria (2010) Victorian recycling industries annual report 2008–2009. ISSN 1836-9902, Melbourne, VIC, Australia

Tam VWY, Tam CM (2007) Crushed aggregate production from centralized combined and individual waste sources in Hong Kong. Constr Build Mater 21(4):879–886

VicRoads (1998) Guide to general requirements for unbound pavement materials, Technical bulletin 39. VicRoads, Melbourne, Australia. ISBN 0 7306 2162 6

Wartman J, Grubb DG, Nasim ASM (2004) Select engineering characteristics of crushed glass. J Mater Civ Eng 16(6):526–539

Chapter 2
A Few Dilemmas Pertaining Transportation Infrastructures and Their Sustainability

Filippo G. Pratico

Abstract This paper deals with transportation infrastructures. It focuses on the estimate of benefits and costs (agency, user and environmental externalities) deriving from the application of the following main concepts (dilemmas): (i) transition towards pervious pavements, (ii) 'hyper-'monitoring tendency (continuous monitoring of pavement structural condition), (iii) reinforcement tendency (increase of the expected life of a pavement due to reinforcement) and (iv) 'hyper-'recycling tendency (high percentage of recycling in road pavements). A method is set up. Consequences in terms of the present value of the assets costs are estimated.

Keywords Life cycle cost analysis • Transportation infrastructures • Porous European mixes • Recycling • Self-powered sensors

2.1 Objectives

The main objective of this study is to provide a theoretical framework to estimate the sustainability of a solution in the field of transportation infrastructures. A theoretical framework was set up for the quantitative assessment of the sustainability of a solution. Four main ideas (or tight spots) to improve the sustainability of a transportation infrastructure were discussed: (i) transition towards pervious pavements, (ii) 'hyper-'monitoring tendency (continuous monitoring of pavement structural condition), (iii) reinforcement (increase of the expected life of a pavement due to reinforcement) and (iv) 'hyper-'recycling tendency (high percentage of recycling in road pavements). The algorithms set up were finally applied and results discussed.

F.G. Pratico (✉)
University Mediterranea of Reggio Calabria, Reggio Calabria, Italy
e-mail: filippo.pratico@unirc.it

G.L. Sivakumar Babu et al. (eds.), *Sustainability Issues in Civil Engineering*,
Springer Transactions in Civil and Environmental Engineering,
DOI 10.1007/978-981-10-1930-2_2

2.2 Introduction and Discussion

The word sustainability is derived from the Latin sustinere (to hold). According to the Brundtland Commission of the United Nations (March 20, 1987), 'sustainable development is development that meets the needs of the present without compromising the ability of future generations to meet their own needs'. The Definition according to the 2005 World Summit is '. . .the reconciliation of environmental, social and economic demands – the 'three pillars' of sustainability (triple bottom line)'.

Among the sustainable initiatives/possibilities, the following can be listed: (i) vehicle technology (greener cars, fuel cell vehicles, etc.), (ii) travel behaviours, (iii) city planning, (iv) travel pathways, (v) innovative public transport means, (vi) pavement solutions (porous surfaces, recycling, sensor monitoring, reinforcement, etc.), (vii) reduction of traffic disruptions due to maintenance and rehabilitations and (viii) applications based on information and communication technologies (ICTs) for the transport sector. The remaining part of this introduction illustrates the following concepts: transition towards pervious friction courses (2.2.1), 'hyper-'monitoring (2.2.2), reinforcement (2.2.3) and 'hyper-'recycling (2.2.4). Analyses are carried out from the wider perspective of the formalization of a model able to take into account the consequences in terms of agency, user and externality costs.

2.2.1 Transition Towards Pervious Friction Courses

Porous European mixes (PEMs, air void content usually in excess of 20 %) act as a wearing course, usually 50 mm thick on impermeable base courses and have well-known points of strength:

1. Reduction of splash and spray
2. Mitigation of outdoor noise (high porosity, low flow resistivity)
3. Optimization of skid resistance at high speeds in wet conditions (high macrotexture)
4. Better water management:

   (4a) Regulation of water flows
   (4b) Higher water infiltration capacity/rate
   (4c) Higher water storage capacity
   (4d) Less deprived households at risk from flooding
   (4e) Reduced surface water runoff
   (4f) Reduced costs of property damage
   (4g) Reduced costs of grey infrastructure, e.g. dam construction
   (4h) Better regulation of water quality (drainage, irrigation and drought prevention, water purification), better water quality in aquatic ecosystems, improved biological indicators (Index of Biological Integrity), nitrogen

retention, nitrogen removal, population served by higher water quality and reduced wastewater treatment costs for domestic and commercial users.

In contrast, PEMs can have several weaknesses: clogging and variation of volumetric, noise, texture, friction and permeability performance over time.

It is noted that pervious pavements can yield higher wet friction and consequently lower accident ratios (Hall et al. 2009) and lower rolling and traffic noise (Praticò 2014). These facts can imply lower user costs.

2.2.2 'Hyper-'monitoring

Continuous monitoring of pavement structural condition (using self-powered sensors) with periodic high-speed inspections of the surface can enable to optimize cost, timing and intensity of maintenance and rehabilitation treatments, taking into account the actual and forecasted traffic and consequently minimizing disruptions due to construction and maintenance works. As for the self-powered systems for data gathering (smarter maintenance and rehabilitation), an energy harvesting stick can generate the power, with further, advanced applications (gathering and transmission of accident, speed, weather and pollution data, energy supply for road safety systems, gathering and transmission of administrative data, etc.). The self-powered systems for data gathering have a technology concept formulated readiness level (see Lajnef et al. 2013; Yun et al. 2014; Ceylan et al. 2013; Pearson et al. 2012).

Monitoring effect on agency, user and externality costs involves the consideration of the *reliability* of a pavement (probability that a pavement section designed using the process will perform satisfactorily for the anticipated traffic and environmental conditions for the design period). The following factors may impact the *design reliability*: materials, subgrade, traffic prediction accuracy, construction methods and environmental uncertainties (see AASHTO Guide 1993).

The aggregate standard deviation ($So =$ combined standard error of the traffic prediction and performance prediction) and the standard normal deviate (ZR, which corresponds to a given level of reliability) characterize the AASHTO reliability factor. Due to the improved timing of maintenance and rehabilitation operations, a decrease in agency costs (as well as user costs) is expected. Indeed, implementing a PMS takes resources, but it also has great rewards not only to the organization but also for the pavement users by reducing vehicle operating costs and travel time (Hudson and Haas 1994; Shahin et al. 2013; Smadi 2004; Wolters et al. 2011). As for *maintenance*, *sensors* and *energy harvesting*, several patents were set up (e.g. Pavement Management System, US Patent Application 20090319309, output power control circuit for a thermoelectric generator US 20130265023 A1). Anyhow, the following main issues still need to be addressed for a smarter maintenance and rehabilitation of multimodal transportation infrastructures:

(i) Sometimes survey-based asset management systems do not allow a timely and productive management of the infrastructure. This fact implies higher life cycle costs and higher environmental costs.
(ii) A better integration between ICT and infrastructure management is needed. Indeed, they both can address the overall aim of smarter transport infrastructures, by affecting users (information, time), agency (work zones, long life) and externality (carbon footprint of maintenance and rehabilitation occurrences) issues and costs.
(iii) When low-volume roads, pedestrian facilities, bicycle paths and low emission zones are considered, several energy harvesting systems, based on pressure-voltage transformation on a fixed path, may fail to achieve the best solution, and other systems can achieve viable results.

2.2.3 *Reinforcement*

As is well known, incorporating geosynthetic products into the pavement structure is a technique for reducing the severity and delaying the appearance of reflective cracking and reducing maintenance work and agency and user costs. Many methods and procedures can be listed (Romeo et al. 2014; Netherlands Pavements Consultants 1990; Perkins 2001; Chena et al. 2013; Button and Lytton 2007; Al-Qadi et al. 2000).

2.2.4 *'Hyper-'recycling*

As regards the high percentage of recycling (smarter design, smarter construction, smarter maintenance and rehabilitation), the technology readiness level is at the level of technology validated in lab (see, e.g. Praticò et al. 2013a, b). High-percentage recycling of porous asphalt mixes calls for further research in terms of their mechanical, functional and surface characteristics. Research carried out so far focused on single issues, while durability (expected life increase), friction and surface texture performance still call for further research. Note that the use as friction course of quasi-fully recycled (and porous) mixes still poses many questions and uncertainties (see, e.g. Malal et al. 2013; Kringos and Scarpas 2008; Dermot et al. 2012). Consequently, research and piloting are needed to demonstrate that premium and traditional properties of porous mixes can be regenerated in the recycled mixes.

Note that recycling more than 60 % of a porous asphalt concrete is an ambitious but feasible target. Indeed, previous projects demonstrated the feasibility (see, e.g. PRIN 2008 project – Italian Ministry of Research).

Note that, as for *recycling*, many patents can be considered (e.g. asphalt recycling; US Patent 4540287; Japanese Patent 1524756 dated October 12, 1989; Canadian Patent 2,102,090 dated February 15, 2000; Canadian Patent 2,131,429 dated November 11, 2003.). Furthermore, a number of practical applications and research projects can be listed (see, e.g. Hosokawa et al. 2014).

Finally it is noted that even the AASHTO guide (FHWA-SA-98-042) indicates that in essence there is no difference between hot recycled and virgin HMA material; however, the same report cautions that engineering judgement should always be applied for design of such mixes. This makes the comparison of the agency costs of recycled versus not recycled mixes quite challenging.

2.3 LCCA Modelling and Algorithms

2.3.1 LCCA

LCCA (life cycle cost analysis) is an engineering-economic analysis tool which compares the relative merits of competing project implementation alternatives. Minimizing the pavement life cycle costs (present worth value, PWV or *PV*, or equivalent uniform annual cost, EUAC) will increase the sustainability of the pavement system (see Douglas and Molenaar 2004; Walls and Smith 1998; Wilde et al. 1999; FHWA 2002).

The detailed analysis of the costs over the entire life cycle of the transportation infrastructure (LCCA, with respect to the zero option-traditional transportation facilities) can assess the decrease of agency (AC, e.g. maintenance and rehabilitation), user (*UC*, e.g. time, accidents, vehicle operating costs), and externality (*EC*, e.g. related to CO_2e emissions, etc.) costs.

As for agency costs (AC), the assessment of the increase of asset productivity due to the improved asset management system (sensors and monitoring) is quite complex (AASHTO Pavement Management Guide 2012; Cirilovic et al. 2014). The economic benefits which derive from an improved asset management system can be estimated based on the decrease of the costs related to variance effects (effect of spatial variability of performance) and timing effects, i.e. effect of treatment timing on repair costs (see, e.g. Peterson 1987; Smadi 2004).

Note that according to Archondo-Callao (2010), road user cost (UC) components include vehicle operating costs (fuel lubricant oil, tyre wear, crew time, maintenance labour, maintenance parts, depreciation, interest, overheads), time costs (passenger time, cargo holding time), accidents costs and emissions.

As for *user costs*, note that (i) in the case of an extended life, a reduction is expected (see Praticò et al. 2011b) and (ii) surface monitoring and sensors imply an optimization of timing and a reduction of user costs.

In more detail, note that the minimization of traffic disruption in the proposed solution can depend on (i) extended life span, (ii) timeliness in rehabilitation and maintenance (for both crack-sensor and surface texture-surface monitoring) and (iii) synergetic approach (data integration).

Importantly, fewer, more sustainable and better planned interventions increase the safety for the workers. To this end, the quantification of the impact can be derived based on probability analysis and based on the same algorithms used to quantify agency and user costs for a period of analysis (even infinite; see Weed 2001; Praticò et al. 2012).

Note that the relative importance of user and agency costs in pavement LCCA was analysed by Delwar and Papagiannakis (2001). In general, the review of the literature suggests that user costs, both roughness and delay related, can be very significant even if compared with agency costs. Overall the magnitude of *UC*s and *AC*s which result from the literature is often comparable under the considered hypotheses, due to the relevance of vehicle operating costs (VOC, as a part of the *UC*s). In more detail, Delwar and Papagiannakis (2001) evaluated a two-lane undivided asphalt cement concrete pavement. They evaluated cost streams for two different traffic levels (e.g. level of service, *C* and *E*, with AADT of 8000 and 19,000, respectively). In each case, the analysis period was 40 years. Two types of user cost relationships were used (Papagiannakis and Delwar 1999; Haugodegard et al. 1994). All alternatives were analysed over a 40-year analysis period using a discount annual rate of 4 %.

Walls and Smith (1998) found similar orders of magnitude, when comparing user and agency costs.

As for the *externality costs (EC)*, a valuable decrease of *environmental costs (EC)* is expected, because of the recycling of the existing superstructure, for all transport modes/types (pedestrian paths, bicycle paths, low-volume roads, high-volume roads). This decrease can be quantified in terms of CO_2e savings, landfill-related costs and quarry-related costs. To this end, note that: (i) CO_2e emissions related to hot mix asphalt (approximately 49 kg/t) can be roughly quantified in terms of production (45 % of the total emissions), aggregate extraction/production (12 %), bitumen processing/production (30 %), transports (12 %) and placement (1 %); (ii) recycling mainly affects emissions related to (virgin) aggregates and bitumen (30 % + 12 % = 42 %); and (iii) consequently, for recycling percentages around 60 %, CO_2e emissions around 37 kg/t are expected (approximate decrease of 12Kg/t, i.e. in the range 10–30 %).

Note that other environmental savings can derive from having fewer traffic interruptions and fewer landfill-related and quarry-related emissions (transports, technologies, management).

2.3.2 Conceptual Framework

Model algorithms are presented in Eqs. 2.1, 2.2, 2.3, 2.4, 2.5, 2.6, 2.7, 2.8, 2.9, 2.10, 2.11, 2.12, 2.13, 2.14, 2.15, 2.16 and 2.17, while the nomenclature is presented in Table 2.1.

Table 2.1 Factorial plan of experiments and LCC analysis options

Pavement type		Flexible, see table below	
Design alternatives		All options, recycling, sensors, zero option, porous asphalt, reinforcement	
Main indicators		PV_{AC}, PV_{UC}, PV'_{EX}, G_{AC}, G_{UC}, G_{EX}	
Analysis period		Infinite	
Discount rate		4%	
Nomenclature			
AC	Agency cost, $	NLO	Number of lanes open
C, C_{REH}	Initial construction cost and costs incurred in successive rehabilitations, $	O	Expected life of successive rehabilitations, years
CA_{ki}	Truck capacity, m^3	PV_{AC}	Present value of agency cost, $
C_{max}	Maximum initial construction cost, $	PV_D	Present value due to delay, $
D_{kj}	Average distance to cover, km	PV_{UC}	Present value of user cost, $
E	Expected life, year	PV_{VOC}	Present value of vehicle operating costs, $
EM_{kj}	Emission surcharge factor, $/l	R	R ratio which depends on *i* and *r* (inflation and interest rate)
EX	Externalities/externality cost, $	R_k	Rehabilitation cost, $
EX_0	Externality cost during construction, $	R_{kmax}	Maximum rehabilitation cost, $
EX_{AD}	Additional externality cost, $	S_a	Salvage value, $
EX_{EM}	Externality cost due to emission, $	SC	Stopping cost, $
EX_k	Externality cost during the kth rehabilitation, $	T	Period of analysis, year
EX_{LA}	Externality cost related to landfilling, $	TD	Traffic data
EX_{QU}	Externality cost related to quarrying, $	TOD	Traffic hourly distribution
FC_{kj}	Fuel consumption, l/km	UC	User cost, $
G	Gain, $	VN_{kj}	Volume of material needed, m^3
G_{AC}	$(PV_{AC})_{NS}$-$(PV_{AC})_S$, $, gain in agency cost for choosing S instead of NS	VUC_k	Value of user costs, $/hr
G_e	Pavement geometry	WZC	Work zone capacity, vpd
I	Interest rate, %	WZD_k	Work zone duration, hr
M	Mechanical properties	WZL_k	Work zone length, m
n_k	Expected life of kth rehabilitation, years	WZSL	Work zone speed limit, kmph
$n_1 \ldots n_N$	Expected lives, years	ΔT	Added time, hr

The conceptual framework of the model was formulated and applied based on the following steps:

(i) Estimation of the expected life of the flexible pavement for a given pavement design (see AASHTO Guide for Design of Pavement Structures 1993; MEPDG 2007)
(ii) Estimation of the present value of agency costs for each solution (see Eqs. 2.1, 2.2, 2.3 and 2.4)
(iii) Estimation of the corresponding present value of user costs (see Eqs. 2.5 and 2.6)

(iv) Estimation of the present value of externalities (Eqs. 2.7, 2.8, 2.9, 2.10, 2.11, 2.12, 2.13, 2.14 and 2.15)
(v) Derivation of the overall present value (Eq. 2.16)
(vi) Estimation of the gains (which sum up the consequences with respect to the control case; see Eq. 2.17)

Note that in the model, the analysis period (i.e. the time period over which alternate design strategies are analysed) is infinite (Weed 2001; Burati et al. 2003). Equations 2.1, 2.2, 2.3, 2.4, 2.5, 2.6, 2.7, 2.8, 2.9, and 2.10 derive from those reported in Praticò et al. (2011a, b), Praticò (2011) and Praticò et al. (2012). Equations 2.11, 2.12, 2.13 and 2.14 have been set out herein.

In pursuit of the estimation of agency costs for the given alternative (Praticò 2011; Praticò et al. 2012), it is important to observe that, for the different options considered, the initial construction cost (*C*) can be different, as well as the construction period and costs incurred in successive rehabilitation works (C_{REH}). In more detail, the initial construction costs can be higher for those solutions which present higher technological level (e.g. with sensors, porous asphalt, reinforcement). This fact can affect the present value (*PV*) of agency costs (i.e. the future amount of expenses, discounted to reflect the current value).

In Eq. 2.1, PV_{AC} is the present value of agency costs (*AC*), C is the construction cost, C_{REH} is the rehabilitation cost (*C* and C_{REH} may be the same) and *R* is the ratio between $(1+i)$ and $(1+r)$, where i is the inflation rate (typically 4 % or 0.04) and *r* is the interest rate (typically 8 % or 0.08). *D* (e.g. 20 years) is the expected life of the as-designed pavement and *E* (e.g. 15 years) is the expected life of the as-constructed pavement:

$$PV_{AC} = C + C_{REH} \cdot R^E \cdot \left(1 - R^D\right)^{-1}. \tag{2.1}$$

The same conceptual framework can be applied to estimate routine maintenance costs:

$$PV_{AC\_RES} = C_{RES} \cdot R^O \cdot \left[F + R^E \cdot \left(1 - R^D\right)^{-1}\right] \tag{2.2}$$

Overall it results:

$$PV_{AC} = C + C_{REH} \cdot R^E \cdot \left(1 - R^D\right)^{-1} + C_{RES} \cdot R^O \cdot \left[F + R^E \cdot \left(1 - R^D\right)^{-1}\right] \tag{2.3}$$

where O ($\leq E$) is the time between two successive rehabilitations or resurfacings (typically 10 years), C_{RES} is the corresponding cost and *F* is a step function (Heaviside like) of *E*:

$$F(E) = \begin{cases} 0, E \leq O \\ 1, E > O \end{cases} \tag{2.4}$$

As for user costs (UCs) (Praticò 2011; Walls and Smith 1998) (see Eq. 2.5), note that they can be related to delays (PV_D), originated by work zones (*WZ*) of given length and duration. Additionally, *UC*s can be related to vehicle operating costs (PV_{VOC}). Their effects mainly depend on the work zone duration (WZD_k) of successive maintenance and rehabilitation operations, added time (ΔT) and stopping costs (*SC*):

$$PV_{UC} = PV_D + PV_{VOC} \tag{2.5}$$

$$PV_D = f(WZL_k, WZC_k, WZD_k, WZSL_k NLO_k, TD_k, VUC_k, TOD_k, \Delta T_k, SC_k) \tag{2.6}$$

In the above equations, PV_{UC} stands for present value of user costs, PV_D stands for present value related to delays, PV_{VOC} stands for present value of vehicle operating costs, WZL_k stands for the length of lanes in the work zone area for the kth rehabilitation process, WZC_k stands for work zone capacity, WZD_k refers to the duration, NLO_k is the number of lanes open, $WZSL_k$ is the work zone speed limit, TD_k stands for traffic data, VUC_k is the value of user costs, TOD_k is the traffic hourly distribution, ΔT_k is the added time and SC_k refers to the stopping costs (Walls and Smith 1998).

As for 'external' costs (*EX*), note that the most relevant impacts of the transportation system on the environment are related to climate change, air quality, noise, water quality, soil quality, biodiversity, land take, quarries, landfills and visual impacts (Yin and Siriphong 2006; Olof 1997; Ian et al. 2009; Praticò et al. 2010). Furthermore, externalities, sustainability and smart infrastructures seem to be dynamically linked. In the future, smart technologies and systems might reduce journey times and develop material designs that can withstand natural and man-made climate changes (CO_2-related). The sustainable design of pavements can drastically reduce the externality costs, and many researchers showed the successful reuse of recycled materials in the pavement construction can reduce these costs (Nunes et al. 1996; Chesner et al. 2001; Ellis 2003; Saride et al. 2010; Chen and Wong 2013; Voiene et al. 2014; Praticò et al. 2013a). The author is aware that the quantification of externality costs is difficult; however, if some hypotheses are assumed, this objective can be pursued.

As for production-related issues (cement and lime), note that carbon dioxide is one of several heat-trapping greenhouse gases (GHGs) emitted. These gas emissions are said to be the primary cause of global warming (European Commission 2010). To this end, a carbon dioxide equivalency is usually used. It is a quantity that describes, for a given mixture and amount of greenhouse gas, the amount of CO_2 that would have the same global warming potential (GWP), when measured over a specified timescale (generally, 100 years). It can be expressed, for example, in g of carbon dioxide equivalents/km or simply gCDE/km.

It should be noted that for the production of one ton of lime, depending on many factors (type of kiln, specifications of the finished product, characteristics of the raw material and geographical location of the lime plant), the total CO_2 emissions vary from 1.0 to 1.7 t of CO_2. The ECO_2 figure of cementitious material can be considered around 950 kg CO_2 per tonne (Iwatani et al. 2010; Hanson et al. 2012; Deng et al. 2014). Based on this discussion, the following four main classes of externalities, *EX*, are considered, both in the construction (EX_0) and in the successive *k*th rehabilitation/routine maintenance (EX_k) works. These are emission-related (EX_{EM}), quarry/production-related (EX_{QU}), landfill-related (EX_{LA}) and additional (EX_{AD}) costs (Praticò et al. 2011a):

$$EX_k = EX_{kEM} + EX_{kQU} + EX_{kLA} + EX_{kAD} \text{ for } k = 0,\ 1,\ \ldots\ N. \tag{2.7}$$

If congestion, road accidents, local air pollution, and global warming are taken into account, based on current literature, fuel consumption (*FC*) (l/Km) can also become a key factor. The following tentative algorithm can be derived for this condition:

$$EX_{kEM} = \sum_j \frac{VN_{kj}}{CA_{kj}} \cdot D_{kj} \cdot EM_{kj} \cdot FC_{kj} \tag{2.8}$$

where the subscript '*k*' refers to the *k*th rehabilitation process, while the subscript '*j*' refers to the given process material. The VN_{kj} (m^3) is the volume of material needed (e.g. slab/soil/stabilizer), CA_{ki} is the average truck capacity (m^3), D_{kj} is the average distance to cover (round trip, km), EM_{kj} is an emission surcharge factor ($/l) and FC_{kj} stands for fuel consumption (l/km). For quarrying/production (for a given *j*-type material/process) externalities, the following algorithm is proposed:

$$EX_{kQU} = \sum_j VN_{kj} \cdot QU_{kj} \tag{2.9}$$

where VN_{kj} (m^3) is the volume of material needed (slab/soil/stabilizer), while QU_{kj} takes into account for externality quantification ($/$m^3$) the effect of production processes on CO_2 equivalent production and compaction levels (Giustozzi et al. 2012; Amadore et al. 2013).

Finally, landfill-related externalities are considered through the following expression:

$$EX_{kLA} = \sum_j VN_{kj} \cdot LA_{kj} \tag{2.10}$$

where VN_{kj} (m^3) is the volume of material needed (slab/soil/stabilizer), while LA_{kj} takes into account externality quantification ($/$m^3$). Finally, it results (where E_k takes into account the difference between *EX*s which refer to rehabilitation/maintenance and *EX*s which refer to operations) to

$$PV_{EX} = EX_0 + \sum_k EX_k \cdot R^{E_k} \cdot \left(1 - R^D\right)^{-1} \tag{2.11}$$

As for the externality costs, note that the quantification of the CO_2 emissions is a well-established practice. An example is offered by the EU ETS (European Union emission trading system; see (Sijm et al. 2006)). On the other hand, the fluctuation of the carbon price is a matter of fact. The ETS was launched in 2005 and prices ranged from almost zero € per tonne of carbon dioxide (in 2007) up to 32 (in April 2006). This fact is probably due to many factors, among which the fact that demand and offer are variable and 'new'. This increases the variability of the results in terms of *EX* and consequently in terms of their sum. Furthermore, it makes the attempt to 'compare apples and oranges' more difficult and less objective.

The following procedure is herein proposed in the case of $PV_{EXi} < PV_{ACi} + PV_{UCi}$ (for ech i-th project or solution among the k ones), which is the most recurrent case.

The first step is to derive the indicator which refers to the most critical relationship between internal and 'internalized' costs:

$$\nu = \min_{i=1,\ 2,\ k} \frac{PV_{ACi} + PV_{UCi}}{PV_{EXi}}. \tag{2.12}$$

The second step is to operate in this vector space of the internalized factors to get a linear magnification:

$$PV_{EXi}' = \nu \cdot PV_{EXi} \tag{2.13}$$

Consequently, in just one case, let us say j, it will result to

$$PV_{ACj} + PV_{UCj} = \nu \cdot PV_{EXj} = P_{VEXj}' \tag{2.14}$$

And

$$PV_j = PV_{ACj} + PV_{UCj} + PV_{EXj}' = 2 \cdot PV_{EXj}' \tag{2.15}$$

It is noted that the basics of the above application rely (i) on thermodynamics and (ii) on practical considerations about the variability of demand and offer equilibrium. To this end, it seems noteworthy to observe that (Feynman 2008) 'the energy available for human utility is not conserved so easily'. Given the extreme complexity of the topic, a tentative *argumentum ad absurdum* can be proposed. If the present value of the internalities (*AC* and *UC*) was greatly higher than the P_{VEX}, then the weight of *EX* in terms of decision-making would be greatly hindered.

If the opposite was true, given the well-known difficulties in *EX* estimation, decisions could be leaded by a sort of unknown parameter.

More in general, given the abovementioned variability of externality costs (CO_2 cost per tonne), short-term variations of *EX* might imply decision-making processes strongly dependent on the given historical period. The above sentences contribute to support the algorithms presented in *n*5.

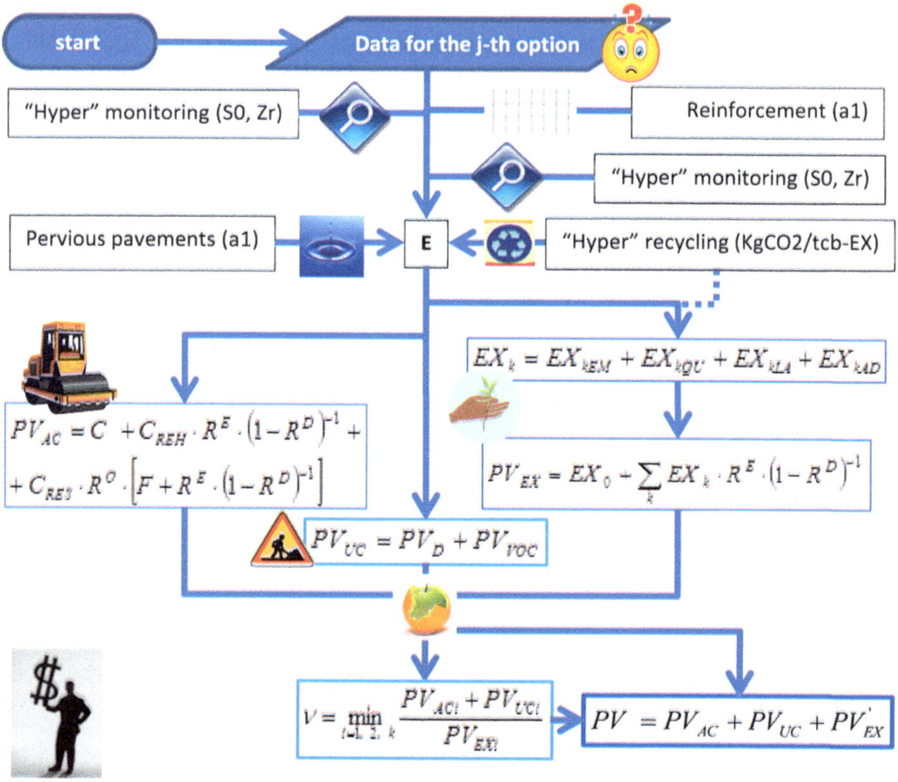

Fig. 2.1 Graphical abstract (Symbols: see Table 2.1)

Overall the approach presents several aspects which call for further research and validation, and the author is aware about that.

In pursuit of the estimation of the overall gain for each solution, note that agency costs (*AC*), user costs (*UC*) and externality costs (*EX*) contribute to the overall present value, *PV* given as

$$PV = PV_{AC} + PV_{UC} + PV_{EX} \tag{2.16}$$

Fig. 2.1 provides a visual summary of the methodology.

Based on the abovementioned three different costs (*AC*, *UC*, *EX*), it is possible to estimate the gain (*G*) originated by one solution with respect to the control case (*NS*). Note that the gain can be evaluated with consideration of only agency costs (e.g. G_{AC}) or only externalities (e.g. G_{EX}), or any other combinations as appropriate:

$$\begin{aligned} G &= (PV_{AC} + PV_{UC} + PV_{EX})_{NS} - (PV_{AC} + PV_{UC} + PV_{EX})_i \\ &= [(PV_{AC})_{NS} - (PV_{AC})_i] + [(PV_{UC})_{NS} - (PV_{UC})_i] + [(PV_{EX})_{NS} - (PV_{EX})_i] \\ &= G_{AC} + G_{UC} + G_{EX} \end{aligned} \tag{2.17}$$

It should be noted that care must be taken while deriving the externality costs for a given scenario. Reasonable assumptions are needed especially when the relevant laboratory and field data are unavailable for this purpose.

2.4 Experiments and Results

In order to apply and partly validate the LCCA-based algorithms, the following options were considered:

Zero option: the superstructure summarized in the table below. Material specifications are as *per* (ANAS 2010).
All options: zero option with sensors, reinforcement, porous asphalt (friction course), recycling.
Sensors: zero option with continuous monitoring of pavement structural condition (using self-powered sensors) with periodic high-speed inspections of the surface
Reinforcement: use of grids as a reinforcement of HMA layers
Porous asphalt: porous friction course (air void percentage: 20 %)
Recycling: high-percentage recycling of the porous asphalt concrete (friction course)

A two-lane undivided asphalt concrete pavement was considered.

Pavement structure is summarized in Table 2.2, while results are reported in Table 2.3.

Note that while agency costs are part of pavement costs, user costs can be considered as a part of 'pavement' strategy only in differential terms, i.e. in terms of delays (with respect to a given zero condition), or in terms of vehicle operating costs differential (e.g. because of the different International Roughness Index (IRI) which is related to a different Present Serviceability Index (PSI)). In the case under study, due to the fact that PSI overall variation was hypothesized to be constant (transition from $PSI_{initial}$ to PSI_{final} in E years), the corresponding IRIs were confined into a given min-max dominion. Consequently, its effect was quite negligible.

Note that the contribution of accident ratio was considered, based on traffic and under the hypotheses of prevailing wet surfaces (Hall et al. 2009). Furthermore, the contribution of the silentness (porous asphalt concrete) was considered, under given

Table 2.2 Superstructure considered

Layer	Thickness (cm)
Friction course	5.0 (porous asphalt and 'all options') or 3.0
Binder course	7.0
Base course	20.0
Cement treated base course	30.0
Unbound base course	30.0

Table 2.3 Results obtained

Solution:	All options	Recycling	Sensors	Porous asphalt	Reinforcement	Zero option
PVAC (€)	3.641	**3.920**	**3.471**	**4.385**	**3.912**	**3.920**
PVUC (€)	2.126	**2.519**	**2.285**	**2.514**	**2.509**	**2.519**
PVEX′ (€)	3.387	**4.049**	**5.756**	**5.442**	**5.979**	**6.104**
PVAC + PVUC + PVEX′(€)	9.154	**10.488**	**11.511**	**12.340**	**12.399**	**12.543**
PVAC (%)	39.8 %	**37.4 %**	**30.2 %**	**35.5 %**	**31.5 %**	**31.3 %**
PVUC (%)	23.2 %	**24.0 %**	**19.8 %**	**20.4 %**	**20.2 %**	**20.1 %**
PVEX (%)	37.0 %	**38.6 %**	**50.0 %**	**44.1 %**	**48.2 %**	**48.7 %**
Overall ranking	Best					

PVAC, PVUC, PVEX′: present value of agency, user externality costs, respectively (millions of €)
Zero option: the superstructure summarized in the above table. Material specifications are as per (ANAS 2010)
All options: zero option with sensors, reinforcement, porous asphalt (as friction course), recycling
Sensors: zero option with continuous monitoring of pavement structural condition (using self-powered sensors) with periodic high-speed inspections of the surface
Reinforcement: use of grids as a reinforcement of HMA layers
Porous asphalt: porous friction course (air void percentage: 20 %)
Recycling: high-percentage recycling of the porous asphalt concrete (friction course)

hypotheses pertaining to amenity and health (acute myocardial infarction) consequences in terms of cost (€ per household per dB change; see also Althaus et al. 2009; DEFRA 2013).

In Table 2.3, results are ranked based on the overall present value. Even if the max/min ratio is just 1.24, several observations can be expressed as follows.

By referring to the results, it is possible to observe what follows:

(a) With respect to the zero option, the use of the *porous asphalt* yielded a higher value of the present value of the agency costs (due to the slight decrease of the expected life).
(b) With respect to the zero option, the use of the system for the *continuous analysis of pavement status* and performance ('sensors') yielded a decrease of the present value of agency, user and externality costs.
(c) As *for recycling*, there was a considerable decrease in terms of externality costs, being the user costs essentially the same. Importantly, in the case only agency and user costs are considered; recycling still remains among the best solutions, while porous asphalt is still among the worst solutions.
(d) As far as the *overall balance* is considered, porous asphalt and reinforcement result the worst solutions. As for porous asphalt concretes, this result originates from the fact that positive outcomes, such as improved wet friction and reduced negative externalities (reduced noise), didn't balance the reduction in terms of expected life.
(e) With respect to the zero option (reference pavement), the *reinforcement* yielded a slight decrease of the present value of the agency costs (due to the increase of

the expected life) and a slight decrease of the user costs (due to the lower number of rehabilitation processes). Note that for the *reinforcement case study*, the present value of agency costs resulted quite satisfactory (due to the increased expected life). It is noted that from this standpoint (which, indeed, is the agency perspective), the reinforcement appears to be among the best solutions. Note that in this case, the present value of the user costs resulted quite convenient (due to the increased expected life). Furthermore, the present value of the externalities resulted quite inadequate (even if better than the 'zero option'). The main reason was the absence of positive mitigations of the externalities. The overall value of the present value resulted quite high. To this end, it seems noteworthy to highlight that from a traditional standpoint, this solution still presents well-known and sound advantages from an agency perspective (increase of the expected life). Furthermore, the prediction of the increase of the expected life is supported by many scientific evidences, while, in contrast, in the case of the hyper-monitoring approach, the increase of the expected life, even if predicted based on a rational approach, isn't supported by experimental investigations.

2.5 Limitations of the Methodology

The following limitations apply:

1. Unit costs are subject to fluctuations over time. Similarly, interest and inflation rates can change over time.
2. Many assumptions, even if based on the international literature (e.g. noise impact, accident crash statistics and their average comprehensive claim, transportation distances, production processes and their impact on CO_2e), refer to a particular case study. Consequently, their transportability to other scenarios could result problematic and should be demonstrated on a case-by-case basis.
3. The method set up presents a clear shift if compared with the previous study. This point raises the necessity for more applications and tests, in order to validate the suitability with reference to a wider spectrum of cases.

2.6 Conclusions

Based on the results obtained, the following conclusions may be drawn:

(a) Ranking appears to be strongly dependent on the criterion adopted to carry out the comparison (e.g. agency, user or externality costs). This point refers to the same definition of sustainability.
(b) Agency-based ranking yields that the continuous monitoring has a great potential.

(c) In contrast, from a user perspective, the best solution appears to be the sensor-based solution.
(d) Finally, when externality or overall costs are considered, merging synergistically all the solutions is the option which yields the best results (minimization of the costs).
(e) It seems noteworthy to highlight that from an overall perspective (total *PV*), the cost of a porous asphalt solution is well comparable to the cost of the zero solution. This fact can be considered as an indirect proof of the consistency of the method. At the same time, this particular would point out that the permeable solution comes out from an implicit consideration of a wider spectrum of instances (*AC*, *UC*, *EX*).
(f) *In terms of innovation and competitiveness, note that synergies between ICT and civil engineering seem to be able to promote and* improve the innovation capacity of the system and to strengthen the competitiveness and growth of construction industry, by developing ICT-related innovations.
(g) Overall, the quantification of the expected impacts of innovative pavements (recycling, monitoring, reinforcing and porous) on the productivity of the asset yields *a reduction of (the present value of the) assets costs, which implies an increase of asset productivity. It is not*ed that in the 'all solutions' case, transdisciplinary efforts are merged: information communication technology (as a tool to gather data and to improve asset management), material science (surface and bulk properties of recycled mixes) and construction management (reduction of traffic impacts due to maintenance activities, optimisation of the overall cost of the transport).
(h) *As for technology readiness, note that* the green properties (silent, permeable, contributing to a better water management) of the infrastructure system almost pertain to the highest level of readiness level (system prototype demonstration in operational environment). As for the long-life properties (in intermodal and multimodal applications), they relate to the mechanical properties of the infrastructure solution, and the readiness level can be considered between the level of 'technology concept formulated' and the level 'technology validated in laboratory'.

References

AASHTO (2012) Pavement management guide. AASHTO, Washington, DC

Al-Qadi IL, Elseifi M, Loulizi A (2000) Geocomposite membrane effectiveness in flexible pavements, Final report project TRA-00-002. The Roadway Infrastructure Group, Virginia Tech Transportation Institute, Blacksburg

Althaus HJ, de Haan P, Scholz RW (2009) Traffic noise in LCA. Int J Life Cycle Assess 14 (6):560–570

Amadore A, Bosurgi G, Pellegrino O (2013) Identification of the most important factors in the compaction process. J Civ Eng Manag 19(suppl1):S116–S124

American Association of State Highway and Transportation Officials (1993) AASHTO guide for design of pavement structures. AASHTO, Washington, DC

ANAS (2010) Gestione Delle Pavimentazioni Stradali – Linee Guida di Progetto e Norme Tecniche Prestazionali [Pavement management – design basics and performance-based standards]. Retrieved September, 2014, from: http://www.geologiveneto.it/wp-content/uploads/Prosperi_NUOVO-CAPITOLATO-ANAS.pdf

Appendix EE: reliability and variance estimates are proposed in the AASHTO guide for design of pavement structures NCHRP project 20-7/24 – May1985

Archondo-Callao R (2010) World bank, HDM-4 road user costs, Model Version 2.00. Retrieved September, 2014, from: http://web.worldbank.org/WBSITE/EXTERNAL/TOPICS/EXTTRANSPORT/EXTROADSHIGHWAYS/0,,contentMDK:20483189~menuPK:1097394~pagePK:148956~piPK:216618~theSitePK:338661,00.html#RUC2

Asphalt recycling. United States Patent 4540287

Burati JL, Weed RM, Hughes CS, Hill HS (2003) Optimal procedures for quality assurance specifications, Report No. FHWARD-02-095. Office of Research, Development, and Technology, FHWA, McLean

Button JW, Lytton RL (2007) Guidelines for using geosynthetics with hot mix asphalt overlays to reduce reflective cracking. Presented at the 86th annual meeting of the Transportation Research Board (CD-ROM), Washington, DC

Canadian Patent 2,131,429 dated November 11, 2003

Ceylan et al (2013) Highway infrastructure health monitoring using Microelectromechanical Sensors and Systems (Mems). J Civ Eng Manag 19(Supplement 1):S188–S201

Chen MJ, Wong YD (2013) Porous asphalt mixture with 100% recycled concrete aggregate. Road Mater Pavement Des 14(4):921–932

Chena Y, Tebaldi G, Roque R, Lopp G, Exline M (2013) A mechanistic test to evaluate effects of interface condition characteristics on hot-mix asphalt overlay reflective cracking performance. Road Mater Pavement Des 14(Supplement 1, Special Issue: EATA):262–273

Chesner WH, Simon MJ, Eighmy TT (2001) Recent federal initiatives for recycled materials use in highway construction in the United States. In: Beneficial use of recycled materials in transportation applications. pp 3–10

Cirilovic et al (2014) Developing cost estimation models for road rehabilitation and reconstruction: case study of projects in Europe and Central Asia. J Constr Eng Manag 140(3):04013065

DEFRA, Department for Environment, Food & Rural Affairs (2013) Noise pollution: economic analysis

Delwar M, Papagiannakis AT (2001) Relative importance of user and agency costs in pavement LCCA. In: Fifth international conference on managing pavements, Seattle, Washington, 11–14 August 2001, Conference Proceedings

Deng Y, Gao HO, Qian S (2014) Environmental impact of concrete pavement construction based on life cycle assessment. In: TRB 93rd annual meeting compendium of papers, Transportation Research Board

Dermot et al (2012) Stress intensity factors at the tip of a surface initiated crack caused by different contact pressure distributions. Procedia Soc Behav Sci 48:733–742

Douglas DG, Molenaar RK (2004) Life-cycle cost award algorithms for design/build highway pavement projects. J Infrastruct Syst 10(4):167–175

Ellis SJ (2003) Recycling in transportation. In: Frost mW, Jefferson I, Faragher E, Rofft EJ, Fleming PR (eds) Transportation geotechnics symposium.Thomas Telford, London, pp 177–188

European Commission (2010) Reference document on best available techniques in the cement, lime and magnesium oxide manufacturing industries

Feynman RP (2008) Six easy pieces. Penguin Books, Camberwell

FHWA (2002) Life-cycle cost analysis primer 2002. Office of the Asset Management, United States Department of Transportation, Washington, DC, p 25

FHWA-SA-98-042, Pavement recycling guidelines for state and local governments, participant's reference book. December 1997, Publication No. FHWA-SA-98-042

Giustozzi F, Toraldo E, Crispino M (2012) Recycled airport pavements for achieving environmental sustainability: an Italian case study. Resour Conserv Recycl 68:67–75

Hall JW, Smith KL, Titus-Glover L, Wambold JC, Yager TJ, Rado Z (2009) Guide for pavement friction. NCHRP Project 01–43

Hanson CS, Noland RB, Cavale KR (2012) Life-cycle greenhouse gas emissions of materials used in road construction. Transportation Research Record: J Transp Res Board 2287:174–181. doi:10.3141/2287-21, Transportation Research Board of the National Academies, Washington, D.C

Haugodegard T, Johansen J, Bertelsen D, Gabestad K (1994) Norwegian public roads administration: a complete pavement management system in operation. In: Proceedingsof the third international conference on managing pavements, vol 2. San Antonio, Texas

Hosokawa H, Gomi A, Tamai A, Kasahara A (2014) Hot in-place recycling of porous asphalt concrete. Retrieved August, 2014, from: http://www.green-arm.com/area/japan/050817Mairepav4/Paper%20134%20Hot%20inplace%20recycling%20of%20porous%20asphalt%20concrete.pdf

Hudson WR, Haas RCG (1994) What are the true costs and benefits of pavement management? In: 3rd international conference on managing pavements

Ian WH, Parry IWH, Timilsina GH (2009) Pricing externalities from passenger transportation in Mexico City. Policy Research Working Paper 5071. The World Bank, Development Research Group – Environment and Energy Team

Iwatani Y, Kawai K, Aoki Y, Fujiki A (2010) Optimum road pavement from the viewpoint of CO2 emission reduction. In: Second international conference on sustainable construction materials and technologies, Università Politecnica delle Marche, Ancona, Italy, June 28–June 30, 2010

Japanese Patent 1524756 dated October 12, 1989

Kringos JN, Scarpas A (2008) Physical and mechanical moisture susceptibility of asphaltic mixtures. Int J Solids Struct 45(9):2671–2685

Lajnef N et al (2013) Smart pavement monitoring system, Report no. FHWA-HRT-12-072. Turner-Fairbank Highway Research Center, McLean

Malal K et al (2013) Long-term skid resistance of asphalt surfacings: correlation between Wehner–Schulze friction values and the mineralogical composition of the aggregates. Wear 303 (1–2):235–243

MEPDG (2007) Guide for mechanistic-empirical design of new and rehabilitated pavement structures. National Cooperative Highway Research Program, Transportation Research Board, Washington, DC. www.trb.org/mepdg

Netherlands Pavements Consultants (1990) The influence of HaTelit reinforced asphalt in comparison with unreinforced asphalt. Hoevelaken, The Netherlands

Nunes M, Bridges M, Dawson A (1996) Assessment of secondary materials for pavement construction: technical and environmental aspects. Waste Manag 16:87–96

Olof J (1997) Optimal road-pricing: simultaneous treatment of time losses, increased fuel consumption, and emissions. Transp Res Part D Transp Environ 2(2):77–87

Papagiannakis AT, Delwar M (1999) Methodology to improve pavement investment decisions. Final Report to National Cooperative Highway Research Program for Study. Transportation Research Board, Washington, DC, pp 1–33

Pearson M, Eaton M, Pullin R, Featherston C, Holford K (2012) Energy harvesting for aerospace structural health monitoring systems, modern practice in stress and vibration analysis. J Phys Conf Ser 382:012025

Perkins SW (2001) Mechanistic-empirical modeling and design model development of geosynthetic reinforced flexible pavements. Final report. Montana State University, Bozeman

Peterson DE (1987) Good roads cost less- Report No. UDOT-MR-77-8, Utah DOT, Salt Lake City, UT

Praticò FG (2011) Pay adjustment in construction engineering. In: 2011 international conference on business intelligence and financial engineering, Hong Kong, 12–13 December 2011

Praticò FG (2014) On the dependence of acoustic performance on pavement characteristics. Transp Res D 29:79–87

Praticò FG, Ammendola R, Moro A (2010) Factors affecting the environmental impact of pavement wear. Transp Res D Transp Environ 15(3):127–133. ISSN 1361–9209. doi: 10.1016/j.trd.2009.12.002. Elsevier Science Ltd.

Praticò FG, Casciano A, Tramontana D (2011a) Pavement life cycle cost and asphalt binder quality: a theoretical and experimental investigation. J Constr Eng Manag 137(2):99–107. ISSN: 0733–9364

Praticò FG, Sireesh S, Puppala AJ (2011b) Comprehensive life cycle cost analysis for the selection of stabilization alternatives for better performance of low volume roads. U.S. Transportation Research Board, TRB, Transportation Research record No 2204, vol 2, pp 120-129.

Praticò FG, Casciano A, Tramontana D (2012) The influence of dispersion and location on pay adjustment in construction engineering. J Constr Eng Manage 138(10):1943–7862 (online). doi::http://dx.doi.org/10.1061/(ASCE)CO.1943-7862.0000540. ISSN: 0733–9364 (print)

Praticò FG, Vaiana R, Giunta M, Iuele T, Moro A (2013a) Recycling PEMs back to TLPAs: is that possible notwithstanding RAP variability? Appl Mech Mater 253–255:376–384© (2013)

Praticò F, Vaiana R, Giunta M (2013b) Pavement sustainability: permeable wearing courses by recycling porous European mixes. J Archit Eng 19(3):186–192

PRIN (2008) Research Project "Drenante da drenante"

Romeo E, Freddia F, Montepara A (2014) Mechanical behaviour of surface layer fibreglass-reinforced flexible pavements. Int J Pavement Eng 15(2):95–109

Saride S, Puppala AJ, Williammee R (2010) Assessing recycled/secondary materials as pavement bases, special issue on sustainability in ground improvement projects. Proceedings of ICE, Ground Improvement GI 163(GI1):3–12

Shahin MY, Welborn W, Vansteenburg G, Rutland C (2013) Pavement management using the PAVER system. In: 17th IRF world meeting & exhibition, Riyadh, Saudi Arabia, November 2013

Sijm J, Neuhoff K, Chen Y (2006) CO2 cost pass through and windfall profits in the power sector. Clim Pol 6(1):49–72

Smadi O (2004) Quantifying the benefits of pavement m0anagement. In: 6th international conference on managing pavements

Vaiana R, Iuele T, Gallelli V, Tighe SL (2014) Warm mix esphalt by water contatining methodology: a laboratory study on workability properties versus micro-foaming time. Can J Civ Eng 41(3):183–190

Walls J, Smith MR (1998) Life-cycle cost analysis in pavement design, FHWA report FHWA-SA-98-079. Federal Highway Administration, Washington, DC

Weed RM (2001) Derivation of equation for cost of premature pavement failure. Transp Res Rec 1761(1):93–96

Wilde WJ, Waalkes S, Harrison R (1999) Life cycle cost analysis of Portland cement concrete pavements, Report No. FHWA/TX-00/0-1739-1. U.S. Department of Transportation, Austin

Wolters A, Zimmerman K, Schattler K, Rietgraf A (2011) Implementing pavement management systems for local agencies implementation guide – a synthesis of ICT-R27-87, Implementing Pavement Management Systems for Local Agencies Illinois Center for Transportation, August 2011

Yin Y, Siriphong L (2006) Internalizing emission externality on road networks. Transp Res D Transp Environ 11(4):292–301

Yun et al (2014) Smart wireless sensing and assessment for civil infrastructure. Struct Infrastruct Eng 10(4):534–550

Chapter 3
Bitumen-Stabilised Materials for Sustainable Road Infrastructure

M.K. Nivedya and A. Veeraragavan

Abstract Cold in-place recycling (CIPR) consists of milling existing distressed pavement-layer materials to a predetermined depth and mixing it with foamed bitumen (FB) or emulsion along with virgin aggregates. The mixture of reclaimed asphalt pavement material, foamed binder, active filler and fresh aggregates is called as bitumen-stabilised material (BSM). The mix design, material characterisation and design of pavement structures with BSM are currently in the development stages world over.

One of the key questions to be answered while designing and constructing pavements with BSM mix is the characterisation of the BSM under all the ranges of temperature, confinement conditions and life. It is expected that BSM mix can show behaviour similar to that of a granular material or a bituminous material at different stages of compaction, active filler content and confinement conditions during the design life.

In this chapter, the factors affecting the foaming process are discussed. The characterisation of BSM mixes carried out by the researchers is explained. The evaluation of sections which were constructed using CIPR technology in India is also discussed, and the needed research is highlighted.

Keywords Cold in-place recycling • Foamed bitumen • Bitumen-stabilised material • Portable seismic pavement analyser • Field studies

3.1 Introduction

Pavements deteriorate with time with increased traffic and also due to fatigue of the materials. Conventionally, an overlay is provided over distressed pavements to increase the structural condition of the pavements. This process requires virgin materials, viz. aggregates and binder, and considering the large network of roads in India, the use of virgin materials for the maintenance of the pavements is to be

M.K. Nivedya • A. Veeraragavan (✉)
Department of Civil Engineering, Indian Institute of Technology Madras, Chennai, India
e-mail: nivedyamk@gmail.com; av@iitm.ac.in

G.L. Sivakumar Babu et al. (eds.), *Sustainability Issues in Civil Engineering*,
Springer Transactions in Civil and Environmental Engineering,
DOI 10.1007/978-981-10-1930-2_3

relooked. For the construction of 40 mm-thick overlay of bituminous concrete (BC) per km of a two-lane road, approximately 240 cum of aggregate and 27 tonnes of bitumen are required. Similarly for construction of 75 mm-thick dense bituminous macadam (DBM) and 40 mm-thick bituminous concrete overlay for a two-lane road, approximately 700 cum of aggregates and 80 tonnes of bitumen are required per km. Recycling of existing pavements for rehabilitation of roads has gained considerable importance due to depletion of aggregates and high cost of transportation of road construction materials. Considering the material and construction costs alone, it is estimated that by using recycled materials, savings ranging from 14 to 34 % can be achieved during the rehabilitation of a pavement.

Large-scale rehabilitation demands have seen the adoption of CIPR as the preferred technique. Around the world, the experience and choice of technology for cold in-place recycling varies largely due to different pavement compositions, traffic conditions, availability of virgin aggregates, type of bitumen and emulsion. With the development of specialised equipments over the last 2 decades and from the experience gained from field trials, CIPR with foamed bitumen has gained popularity over other methods of cold recycling due to less curing time, speedy construction, stockpiling of the mix, energy savings and better performance.

3.1.1 Foamed Bitumen and Bitumen-Stabilised Material

The pulverised bituminous material from the existing distressed pavement is known as reclaimed asphalt pavement (RAP). Cold in-place recycling can be carried out by the addition of foamed bitumen to the RAP materials. Foamed bitumen is produced when a small quantity of cold water is injected into the hot bitumen under pressure. The bitumen foams and expands up to 18–20 times its original volume. The foamed bitumen bubble collapses in less than a minute (Asphalt Academy 2009). Figure 3.1 shows a schematic of the foaming process. Wirtgen (2012) explains the mechanism associated with foamed binder mixing in the following manner: "During the mixing with aggregates, the bubbles burst and break into tiny bitumen splinters that disperse throughout the aggregate by adhering to the finer particles. Due to large surface area of the bubbles and breaking of bubbles into tiny bitumen splinters, the foamed bitumen completely coats the finer particles and these finer particles spot welds the larger size aggregate which provides better cohesion in the mix".

RAP materials treated with foamed bitumen are called as bitumen-stabilised material (BSM) (Asphalt Academy 2009). Here, the existing bituminous layers are pulverised, and depending on the gradation available and the targeted gradation, additional virgin materials are added. In CIPR technology, water is added to the pulverised material to ensure that the foamed binder can be dispersed easily. Active fillers (cement, lime, cement kiln dust, fly ash C) are also added to provide the initial strength. This mixture thus consists of pulverised bitumen-coated aggregate

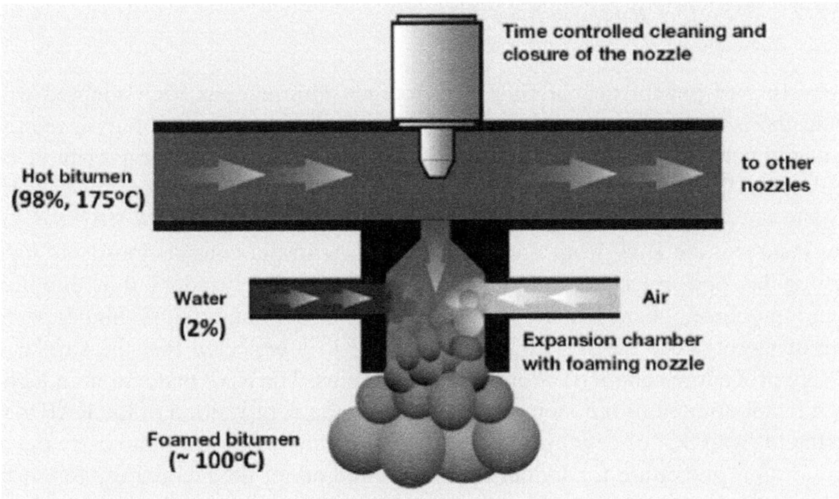

Fig. 3.1 Schematic of foaming process (Wirtgen 2012)

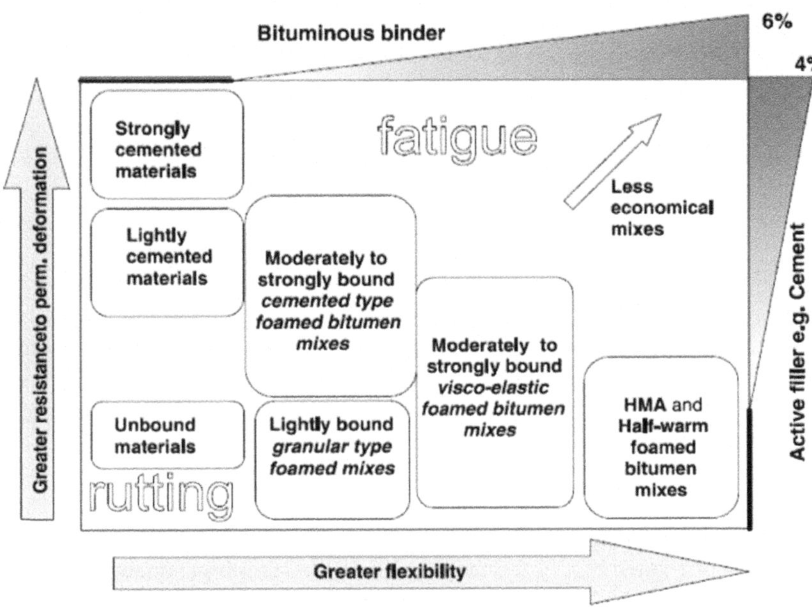

Fig. 3.2 Various manifestations of mixes (Jenkins et al. 2007)

materials (with aged binders) mixed with fresh granular materials (mostly fines), water, foamed binders and active fillers. The behaviour of BSM mixes is shown in Fig. 3.2, and it can be seen that the BSM mixes can exhibit behaviour in between that of a bituminous mix and moderately bound cemented material.

3.1.2 Needed Research in India

Recycling of pavement materials with foamed bitumen has been adopted worldwide and is in use for over a decade. However in India, the technology is relatively new and very few field trials have been carried out. The major component of BSM mix is RAP, which is a site-specific material. Different countries have evolved widely varying mix design procedures depending on the quality of RAP available. For instance, the RAP from South Africa and Australia consists mostly of natural aggregates and cracked cement-stabilised layers as they adopt thin bituminous wearing course, whereas in the United States, the RAP material has higher percentage of aggregates coated with binder. Hence, it is expected that the similar mix design procedure cannot be adopted for all regions. The RAP material in India has a significant amount of bitumen present in it, but the composition of the RAP is very different from that of developed countries. It would be interesting to carry out cold mix design procedure for Indian conditions and adopt the technology for sustainable road infrastructure in the years to come.

3.1.3 Characterisation of Foamed Bitumen

Foamed bitumen is characterised by two parameters and they are expansion ratio (ER) and half-life (HL). Expansion ratio of foamed bitumen is the ratio between the maximum volume achieved in the foamed state and the final volume of the binder after the foam has decayed (Jenkins 2000). Half-life is the time measured in seconds for the foamed bitumen to subside from the maximum volume to half of the maximum volume (Jenkins 2000). The measurement of ER and HL is highly dependent upon the individual's estimation and judgement because of rapid foaming and settling of bitumen, and the timings are recorded manually using a stopwatch. Hence, the measurements made for determining ER and HL are empirical. Apart from ER and HL, foam index (FI) and modelling of foam decay are also carried out to characterise the foam (Jenkins 2000). The area under the decay curve for particular foamed bitumen is called foam index. The decay curve is a graph between expansion ratio and the rate at which the foam decays. Higher FI value indicates better foaming quality.

Saleh (2006) stated that FI was deduced from two empirical quantities which are ER and HL, and hence FI values are also empirical. He introduced a new approach for characterisation of foamed bitumen, in which the average viscosity over the first 60 s was measured. Saleh's method seems to be more logical but has its own limitation, as it was observed that for the same bitumen, the viscosity will vary with different shear rates. Saleh (2006) has not reported whether the experiments were carried out at a constant shear rate or shear rate sweep. To characterise the foamed bitumen using the method suggested by Saleh (2006), all the bitumen samples should be subjected to the same shear rate. However, almost all guidelines and reports generally characterise foamed bitumen by ER and HL due to the ease of carrying out experiments involving simple measurements. There is a need to investigate the foaming characteristics of binders more scientifically in order to understand the effect of foaming temperature and foaming water content.

3.1.4 Influence of Foaming Water and Temperature on the Foaming Process

The foaming water content (FWC) and the foaming temperatures are the most important factors which influence the ER and HL of foamed bitumen. With the increase in temperature and water application rate, the ER increases and the HL decreases (Sunarjono 2007; Asphalt Academy 2009). The optimum FWC and the foaming temperature are determined to achieve the maximum ER and HL and also ensure that these values are well above the minimum acceptable limits. Leek and Jameson (2011) reported that the bitumen should be in expanded state for coating the fine particles; the more it expands and the more it stays in expanded form during mixing, higher coating of fine particle is achieved which imparts more cohesion in the mix.

Figure 3.3 shows the Wirtgen's approach for determining the optimum FWC at a particular temperature. The minimum acceptable ER and HL are considered as 8 times and 6 s, respectively. The average foamed water contents at which the minimum ER and HL occur are considered as the optimum foamed water content (Wirtgen 2012).

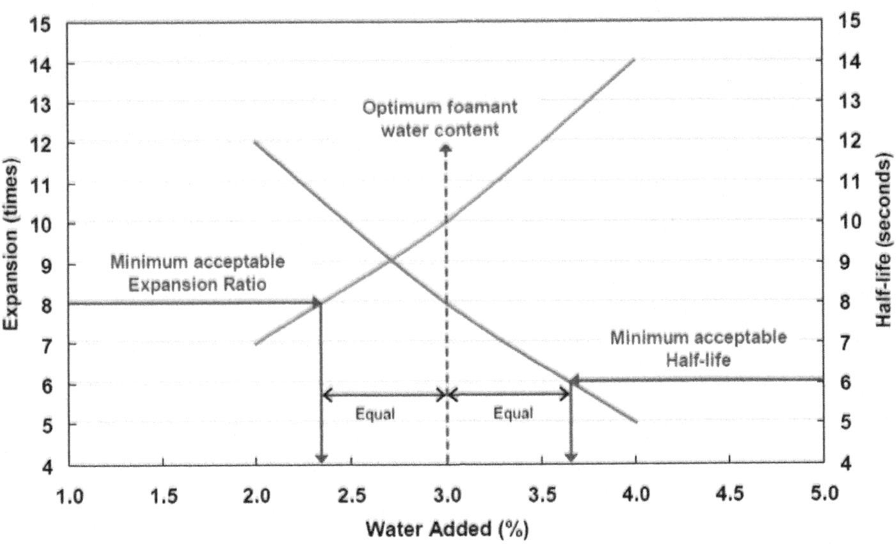

Fig. 3.3 Determination of optimum foamed water content (Wirtgen 2012)

Characterisation of foamed binder is considerably challenging due to the existence of several constituents at the same time – binder, water and air. Furthermore, the process is significantly dependent on temperature – a factor that is sometimes relatively difficult to maintain in the field. Therefore, proper characterisation of the foaming process is needed that will allow the designers to accurately predict the effect of the different significant factors affecting the foaming process and hence conduct a robust mix design and construction procedure. From the perspective of mixing and compaction, it is necessary that one characterises the response of the material, and it can then result in relooking at the empirical parameters such as expansion ratio and half-life (Nivedya et al. 2013).

The decay of foam immediately after foaming is a complex process. The interest in foam decay is not only related to how "stable" the foam is before mixing with aggregate particles but how such decay can influence the workability or otherwise of foamed bituminous mixtures. It is understood here that the apparent viscosity of the binder is expected to strongly influence the decay properties of the foam. It is postulated that such manifestation of foam decay could be captured by means of rheological investigation conducted at a range of temperature that is well below the temperature at which the material was foamed. Such information will help in making a rational choice for the foamed binder production temperature.

3.1.5 Influence of Foaming Water Content on Stiffness of BSM Mix

The determination of foaming characteristics of the foamed bitumen independently cannot predict the performance of foamed bituminous mixes in terms of fatigue resistance, permanent deformation and shear parameters. Saleh (2006) determined the optimum FWC as 2.5 % which produced the lowest rotational viscosity of the foamed bitumen. Sunarjono (2007) studied the influence of FWC on the indirect tensile strength modulus (ITSM) of foamed bituminous (FB) mixes. In Fig. 3.4, it can be observed that with increase in FWC, the ER and FI increase and HL decreases and becomes constant after a particular value of FWC. Sunarjono (2007) found that at FWC of 5 %, maximum stiffness of foamed bituminous was achieved. The bitumen was foamed at 180°C at different FWCs, and stiffness of the mix was represented by indirect tensile strength modulus (ITSM). The specimens were compacted with 200 gyrations and cured for 3 days at 40°C temperature. Figure 3.4 shows the graph between FWC and ITSM at test temperature of 5, 20 and 40°C. It can be observed in Fig. 3.4 that maximum ITSM values are achieved at 5 % FWC.

It is more important to study the combined effect of foaming temperature and FWC with the BSM mixes on its mechanical properties such as indirect tensile strength (ITS), ITSM, shear parameters and dynamic modulus rather than

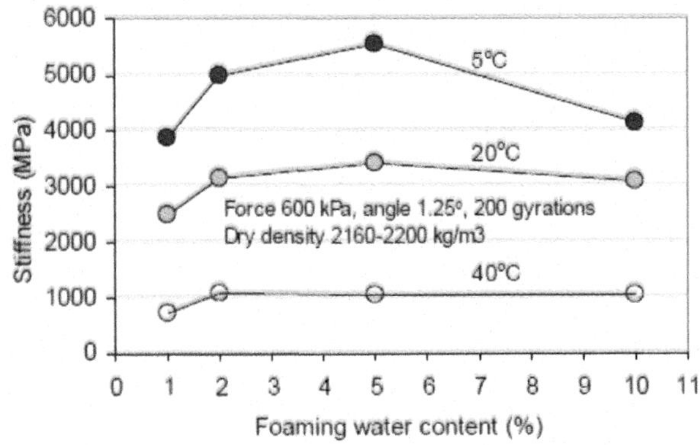

Fig. 3.4 Influence of FWC on stiffness of BSM mixes (Sunarjono 2007)

determining the foaming characteristics of foamed bitumen alone, as BSM mix is the final product of this process.

3.1.6 Mix Design Considerations of BSM Mixes

A wide range of mix design procedures exist for BSM mixes. Each of them differs in terms of gradation, method for determination of percentage of foamed binder content, type and content of active filler, moisture content and the methodologies associated with size of sample, mixing, curing and compaction. The mix design procedures are also strongly dependent on the type and quality of reclaimed asphalt pavement (RAP) available. Since different countries have different pavement cross sections with the mix constitution varying considerably, it is obvious that the mix design procedure also varies. However, it should be emphasised that the overall framework towards the design of BSM mix remains more or less the same. The overall objective of the BSM mix design procedure is to determine the appropriate quantities of foamed bitumen, RAP material, virgin aggregates, active filler, moisture content and foamed bitumen at optimum temperature and foaming water content, such that the mix when laid and compacted in the field should withstand repeated traffic load repetitions and perform satisfactorily over its design life.

Asphalt binder can be made workable at moderate temperatures either by "fluxed bitumen", bitumen emulsions or foamed bitumen (Lesueur et al. 2004). Foamed bitumen is produced by injecting small quantity of water and compressed air into hot bitumen in an expansion chamber. The foaming can be carried out at 140–180°C at water contents varying from 1 to 6 %. The bitumen will expand 10–15 times, and this foamed bitumen will be injected into the pug mill mixer, which contains the RAP material. Moisture at the optimum moisture content (OMC) will be mixed with RAP using a twin-shaft pug mill mixer. During mix production, RAP materials will be mixed dry with the active filler, followed by the addition of water and the injection of foamed bitumen into the pug mill mixer. When the foamed bitumen is mixed with the aggregates, it is expected that the finer particles are coated first which then "spot-welds" the coarse aggregates (Jenkins 2000). The addition of filler will depend on the materials passing 0.075 mm sieve size (Mohammad et al. 2003).

Xu and Hao (2012) investigated the influence of aggregate gradation on the mix design of foamed bituminous mixtures and showed that the optimum foamed bitumen content was dependent on fines' content (passing 0.075 mm sieve) and independent of the coarse aggregate fractions (percent passing 19 mm and retained on 4.75 mm). RAP materials with fines' (passing 0.075 mm) contents between 5 and 20 % are considered ideal for foamed bitumen treatment (Ramanujam and Jones 2007). He and Wong (2007) carried out repeated loading dynamic creep test at 35 ° C and found that the increase in RAP content reduces the permanent deformation of foamed bituminous mixes. They also found that the ITS decreased with increase of RAP content under both dry and soaked conditions. Lee (1981) found that percent passing 75-micron sieve plays a key role in judging the suitability of stabilisation with foamed bitumen, and he also suggested an upper limit of 35–40 % of fine content.

The selection of gradation of the BSM mix depends upon the gradation of the RAP material. It is postulated that if the RAP is obtained from milling the bituminous concrete (BC) layer, then the desired gradation as per Wirtgen (2012) can be utilised.

The mixing and compaction mechanics of hot mix asphalt involve achieving adequate coating and workability by reducing the viscosity of bitumen. In the case of BSM mixes, coating/binding is achieved through the foamed bitumen and workability is attained by the addition of water to the RAP materials. The role of water is to separate the fine particles to "allow channels through which the binder can be dispersed" (Jenkins 2000). The water also aids in providing workability of the foamed bituminous mixtures even at ambient temperature. There is a need to investigate the effect of water content on workability of the mix and quantify the optimum water content that has to be added to the mix to produce sufficient workability. Saleh (2006) stated that the mixing has to be carried out within half-life and the effect of foam stability in mixing needs investigation. A study conducted by Pengcheng et al. (2011) has shown that the production temperature (145–158 °C) and foaming water (3–4 %) content for the same asphalt binder did not affect the mix strength.

The optimum binder content is evaluated on the basis of the engineering parameters of the mix. The ITS test is assumed to quantify the "strength" parameter for low to medium traffic, i.e. less than 6 msa. For high traffic, i.e. greater than 6 msa, the resilient modulus is suggested to be used (Asphalt Academy 2009). The fact that resilient modulus test has been suggested to quantify the influence of binder content on the mechanical properties of BSM underlines the inherent assumption that the BSM is a stabilised form of granular materials. Clearly, the viscoelastic nature of such materials has been ignored and this is an aspect that should be explored. Ruckel et al. (1983) have proposed a design chart to estimate the optimum asphalt contents of foamed asphalt mixes based on the fines' contents. Muthen (1999) has selected the optimum foamed asphalt content based on the relationship between indirect tensile strength and foamed asphalt content. Many researchers recommended that the optimum foamed asphalt content should correspond to the highest resilient modulus value (Maccarrone et al. 1994; Lancaster et al. 1994; and Saleh 2003). Mohammad et al. (2003) determined the optimum foamed asphalt content based on the maximum retained ITS. Lee and Kim (2003) evaluated both Marshall stability and ITS test procedures. The BSM mixes were treated as an unbound granular material, and various properties such as unconfined compressive strength, resilient modulus and moisture susceptibility have been studied (Ramanujam and Jones 2007; Huan et al. 2010). He and Wong (2006) found out that the extent of ageing of RAP material significantly affects moisture susceptibility of mixes in permanent deformation and lower-grade bitumen and less aged RAP material help mixes to improve their moisture susceptibility in permanent deformation. The extent of ageing was not clearly given in the study by He and Wong (2006), and it was merely based on the softening point of the extracted binder. Ebels and Jenkins (2007) assumed that in a cold mixing process, no blending of the aged and fresh bitumen takes place and the RAP aggregate should be regarded as "black rock". However, the use of foamed bitumen at an elevated temperature could hypothetically result in blending with RAP asphalt, and there is a need for further investigations to study the effect of blending of aged and fresh bitumen on the strength parameters.

Since the gradation and mechanical properties of RAP available are key to designing and optimising BSM mixes, it is necessary to explore in detail the various mix design methods available. Tacit in such attempts is the quality and gradation of the RAP as is available in India.

3.1.7 Characterisation of the Mechanical Response of BSM Mixes

To accurately predict the expected performance in the field, the major task that needs to be accomplished is the characterisation of the mechanical response of the material. The primary failure mode that has been observed for in-place foamed bituminous mixtures is permanent deformation and not fatigue cracking (Gonzalez et al. 2009; Asphalt Academy 2009). The distress prediction of foamed bituminous mixtures needs further attention. It is not clear whether the foamed bituminous mixtures will fail by fatigue cracking or any other mode of failure (Jenkins 2012). Since considerable field-related distress data of these layers are not available, the distress transfer functions within the context of mechanistic-empirical pavement design (M-E PD) method are not available with the expected rigour. At this point, recourse to empirical laboratory performance data is preferably the only way possible. Gonzalez et al. (2009) conducted an accelerated full-scale experiment to study the effect of foamed bitumen and cement on performance and concluded that that shrinkage cracks were developed in a 10-year-old pavement when the cement content was more than 1 %.

The density and mix volumetric properties have been used to determine the optimum foamed bitumen content (Bowering 1970). The BSM mixes depict both stress-dependent characteristics of granular materials and temperature and loading-time dependency of bituminous materials (Jenkins 2012). The classical fatigue behaviour under repeated loading is not applicable due to the non-continuous bond nature of these mixtures. Kim and Lee (2009) conducted dynamic modulus tests on BSM mixes and constructed master curves which depicted steeper slope for mixtures with higher foamed asphalt content. They also inferred that BSM mixes should be classified as viscoelastic materials in the M-E PD framework.

3.1.8 Experimental Investigations by the Authors

3.1.8.1 Mix Design

The wet sieve analysis was done for the RAP material collected from a typical National Highways project road in Tamil Nadu, India, and it was found that the gradation lacks the minimum requirement of 4% material passing 0.075 mm sieve. Fresh aggregates in the form of material passing 4.75 mm was added to achieve the required gradation both for laboratory testing and job execution in the field. Wet sieve analysis of crusher dust and dry sieve analysis of cement were carried out. A blend of 80 % RAP material, 19 % crusher dust and 1 % filler (OPC 53 Grade) met the grading requirements for bitumen-stabilised material.

The maximum dry density (MDD) and the optimum moisture content (OMC) of the untreated blended material (RAP + crusher dust) were determined by establishing the moisture-density relationship of the material when prepared and compacted with modified AASHTO compaction procedure at different moisture contents. The OMC and MDD were found to be 6.53 and 2092 kg/m^3. Hygroscopic moisture content of the blended aggregate was determined as per ASTM D2216. The hygroscopic moisture content was found to be in the range of 0.18–0.2 %. This test is carried out to assess the existing moisture content present in the material and subtract it from OMC at the time of addition of water to the mix.

The foaming temperature was chosen as 180°C, and FWC was chosen as 6 % as the expansion ratio of 15 and half-life of 15 s have been obtained against a minimum requirement of 8 and 6 s. The samples were prepared with different foaming binder contents for fabrication of 100 mm-diameter specimen which were cured and tested for indirect tensile strength (ITS) for dry and wet condition, and the optimum foaming bitumen content was found to be 2 %.

3.1.8.2 Testing

The repeated triaxial test was carried out as per AASHTO T307-99 (2007) protocol on the samples that were prepared at 15 % air void. The resilient modulus of the BSM samples was found to be in the range of 500 MPa, whereas the recommended value in IRC-37 ranges from 600 to 1200 MPa. The dynamic modulus values of the cores obtained from the field were found to be in the range of 1800 MPa at 35 °C (Dhiraj et al. 2015). Thus, it can be inferred that the variation in gradation, cement content, compaction level, etc. in the field can result in widely varying strength values for the BSM mix. The studies are under progress to assess the strength values from field cores periodically, so that the change in the modulus with time and traffic can be assessed.

3.1.8.3 Cold In-Place Recycling: Case Studies in India

1. **Chennai-Tada road section**

Rehabilitation of NH-5 from Chennai to Tada was carried out using CIPR. The scope of the work was to widen the existing four-lane highway to six lanes. The mix design and field evaluation were carried out at Indian Institute of Technology Madras. South African mix design procedure (Asphalt Academy 2009) was adopted.

2. **Baroda-Halol road section**

Rehabilitation of 11 km of Baroda-Halol Highway was carried out using CIPR with foamed bitumen. The photos showing the construction procedure are shown in Fig. 3.5.

Fig. 3.5 Various stages of cold in-place recycling at Baroda-Halol section (Wirtgen 2012)

3.1.8.4 Field Investigation of Mechanistic Properties of BSM Layer with Portable Seismic Pavement Analyser (PSPA)

The study carried out by Dhiraj et al. (2015) used non-destructive testing tools and techniques to evaluate the performance related to "stiffness" and density of BSM layer. They also compared data obtained from experiments conducted on field cores and samples fabricated in the laboratory.

Portable seismic pavement analyser (PSPA), a non-destructive testing tool, was used for the field investigations. PSPA provides a means for estimating the in situ seismic modulus of Portland cement concrete and asphalt concrete pavement (Bell 2006). The seismic modulus is measured using ultrasonic surface waves (USW). The heterogeneity in the material and high air void content affected its performance on BSM layers. The modulus of hot mix asphalt mix was adjusted to a reference temperature of 25 °C using a linear relationship suggested by Li and Nazarian (1994). The data were collected when the pavement temperatures varied between 30 and 48 °C. Celaya et al. (2006) found that, in the case of bituminous pavements, the variability of PSPA was less than 3% when repeated measurements were made without moving the device. When the device was moved within a small area, the variability was found to be 7 % (Celaya et al. 2006). The data collected during the study conducted by Dhiraj et al. (2014) indicated that a higher variability was expected in a BSM layer with heterogeneity in the material and high air void content which was also reported by Xu and Hao (2012). PSPA was used along the project stretch at periods between 2 and 6 weeks after construction. The data collected by Dhiraj et al. (2014) were across each cross section at offsets of 0.4 m from the median, and the data were reduced to a common temperature of 25 °C and plotted. An evaluation of moduli was conducted at different cross sections to assess the quality and is shown in Fig. 3.6.

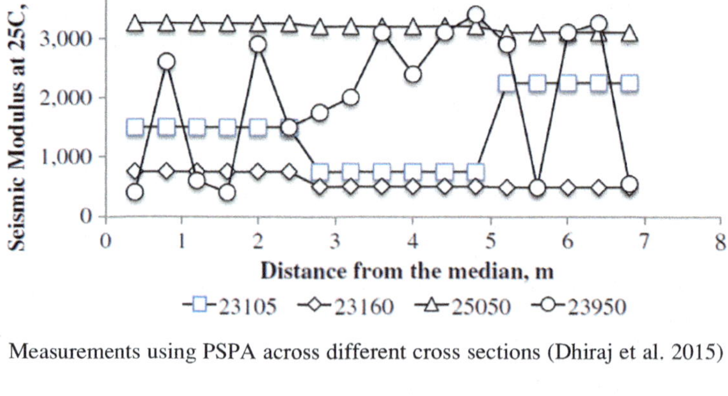

Fig. 3.6 Measurements using PSPA across different cross sections (Dhiraj et al. 2015)

3.2 Conclusions

The main observation and conclusions made from this study are as follows:

1. It is practicable to construct BSM layers with modulus in the range of 600–1200 MPa considering the RAP materials and the quality of construction prevalent in India. Differences in materials used at site and quality control issues related to the field can be expected to cause significant difference between field and those properties obtained from lab.
2. The quality of the recycled layer during construction can be evaluated using a surrogate parameter such as density. It can be measured with a non-destructive density gauge.
3. PSPA can be used in the evaluation of BSM layers in the field, as more number of measurements can be taken. The sections where there was inadequacy of material and process control can be detected.
4. Seismic measurements and dynamic modulus measurements can be compared for BSM mixes.
5. Good quality control practices can be developed with a significant amount of research for cold in-place recycling projects with foamed bitumen.
6. Strict quality control of CIR projects to assess the density and stiffness properties can be carried out by non-destructive testing.

3.3 Lessons Learnt from Field Studies in India and Needed Research

7. The study reiterates the importance of selection of candidate projects for CIP recycling. In such projects, gradation control in the field can only be exercised by monitoring the quality of fines/crusher dust used. The control on the gradation of blend can be exercised only by the material control of the crusher dust. This should be done by gradation and tray tests on the spreading of crusher dust and cement.
8. The use of bulk bitumen tanker with heating arrangement is an important requirement. Inadequate bitumen temperature can lead to 40–45 % loss in strength as observed in laboratory simulation. The laboratory ITS_{dry} and ITS_{wet} values dropped down to 59 and 52 % of the design blend, respectively, when the sample was prepared by foaming at 150 °C as against mix design recommendation of 180 °C.
9. Laboratory simulation indicates that deficiency of fines passing 0.075 mm sieve below specified gradation limits manifests as loss in ITS of the mix. However, this was not observed in any of the ITS tests done by the field laboratory. The likely cause for this appears to be high strength gain due to excess cement. This raises questions about the efficacy of ITS as a quality control (QC) test. The use of only ITS as an independent index test to maintain

QC is thus not recommended. It needs to be supplemented with other QC methods.

10. PSPA exhibited potential in evaluation of BSM, especially in terms of the number of measurements possible. It demonstrated its ability to detect sections where material and process control had failed.
11. Studies are required to establish a methodology to determine the time to adequate initial strength gain, in order to permit trafficking. This especially assumes importance in Indian conditions in view of the high traffic levels. This would require the initial strength to be defined based on the anticipated traffic and ideally a field test to confirm that the same has been attained.
12. Taking a cue from the initial trails using Plaster of Paris, a rapid methodology to reduce the effect of surface irregularities in the field to enable PSPA measurements at all desired locations can be attempted.
13. Material-specific temperature correction equations for typical Indian mixes are to be developed to reduce the PSPA measurements to a common reference temperature.
14. Studies are recommended to be carried out using the PSPA and FWD in conjunction, in order to compare the measurements of the two instruments in BSM.
15. The effect of gradation, bitumen content, active filler content and the amount of water content added to the mix, on mix performance, should be studied. The mix design procedure which suits the Indian construction practices has to be identified.
16. Guidelines/code of practice for cold recycling of bituminous pavement layers is the need of the hour in India.

Acknowledgments The authors acknowledge M/s Wirtgen, Germany, for providing us with the laboratory foaming equipment for the mix design and L&T IDPL, the concessionaire, for having permitted the research team to carry out the tests in the field.

References

Asphalt Academy (2009) Technical guidelines: bitumen stabilized materials, guidelines for the design and construction of bitumen emulsion and foamed bitumen stabilized materials. Asphalt Academy, Pretoria

Bell HP (2006) Operating the portable seismic properties analyzer, geotechnical and structures laboratory. U.S. Army Engineer Research and Development Center

Bowering RH (1970) Upgrading marginal road building materials with foamed asphalt. Highway Engineering in Australia, Mobil Oil of Australia, Melbourne South

Celaya M, Nazarian S, Zea M, Tandon V (2006) Use of NDT equipment for construction quality control of hot mix asphalt pavements. Research Report. 574, Arizona Department of Transportation

Dhiraj M,Veeraragavan A, Mallick RB (2015) Laboratory and field investigation of mechanistic properties of foamed asphalt recycled base course materials for high volume pavements in

India. J Perform Constr Facil. Accepted for publication in the ASCE Journal of Performance of Constructed Facilities, 29(6)

Ebels LJ, Jenkins K (2007) Mix design of bitumen stabilized materials: best practice and considerations for classification. In: Proceedings of the Conference on Asphalt Pavements for Southern Africa (CAPSA), Gaborone, Botswana, pp 213–232

Gonzalez A, Cubrinovski M, Pidwerbesky BD, Alabaster D (2009) Full scale experiment on foam bitumen pavements in CAPTIF accelerated testing facility, Transportation Research Record 2094. TRB, Washington, DC, pp 21–29

He G, Wong W (2006) Decay properties of the foamed bitumens. Constr Build Mater 20 (10):866–877

He G, Wong W (2007) Laboratory study on permanent deformation of foamed asphalt mix incorporating reclaimed asphalt pavement materials. Constr Build Mater 21(8):1809–1819

Huan Y, Siripun K, Jitsangiam P, Nikraz H (2010) A preliminary study on foamed bitumen stabilization for Western Australian pavements. Sci Res Essays 5(23):3687–3700

Jenkins KJ (2000) Mix design considerations for cold and half-warm bituminous mixes with emphasis on foamed bitumen. PhD thesis, University of Stellenbosch, Stellenbosch

Jenkins K (2012) Cracking behavior of Bitumen Stabilized Materials (BSMs): is there such a thing?. In: Proceedings of the 7th RILEM international conference on cracking in pavements, vol 4, 1007–1015

Jenkins KJ, Long FM, Ebels LJ (2007) Foamed bitumen mixes = shear performances? Int J Pavement Eng 8(2):85–98

Kim Y, Lee H (2009) Dynamic modulus and repeated load tests of cold-in-place recycling mixtures using foamed asphalt. J Mater Civ Eng 21(6):279–285

Lancaster J, McArthur S, Warwick R (1994) VICROADS experience with foamed bitumen stabilization. In: Proceedings of the 17th ARRB conference, vol 17. Australian Road Research Board, Gold Coast, pp 193–211

Lee DY (1981) Treating marginal aggregates and soil with foamed asphalt. In: Proceedings of the Association of Asphalt Paving Technologists, vol 50. pp 211–250

Lee H, Kim Y (2003) Development of a mix design process for cold in-place rehabilitation using foamed asphalt. J Mater Civ Eng 8(1):54–63

Leek C, Jameson G (2011) Review of foamed bitumen stabilization mix design methods. Technical Report, Austroads AP-T178/11, Sydney, Australia

Lesueur D, Clech H, Aline B, Christian S, Cazacliu B, Koenders B, Cerino PJ, Bonvallet J (2004) Foamability and foam stability. Road Mater Pavement Des 5(3):277–302

Li Y, Nazarian S (1994) Evaluation of aging of hot-mix asphalt 399 using wave propagation techniques, engineering properties of asphalt 400 mixtures and the relationship to their performance. ASTM STP 1265, 401 American Society for Testing and Materials, Philadelphia, pp 166–179

Maccarrone S, Holleran G, Leonard DJ, Hey S (1994) Pavement recycling using foamed asphalt. In: Proceedings of the 17th ARRB conference, vol 17. Australian Road Research Board, Gold Coast, pp 349–365.

Mohammad LN, Abu MY, Wu Z, Abadie C (2003) Louisiana experience with foamed recycled asphalt pavement base materials. In: 82nd TRB annual meeting (CD-ROM), TRB, Washington, DC

Muthen KM (1999) Foamed asphalt mixes—mix design procedure. CSIR Transportek, Report No. CR-98/077, South Africa

Nivedya MK, Roja KL, Veeraragavan A, Krishnan JM (2013) Rheological investigations on foamed bitumen. In: Airfield and Highway Pavement: Sustainable and Efficient Pavements, Los Angeles, California, 9–12 June 2013, pp 967–976

Pengcheng F, David J, John TH (2011) The effects of asphalt binder and granular material characteristics on foamed asphalt mix strength. Constr Build Mater 25(2):1093–1101

Ramanujam JM, Jones JD (2007) Characterization of foamed-bitumen stabilization. Int J Pavement Eng 8(2):111–122

Ruckel PJ, Acott SM, Bowering RH (1983) Foamed-asphalt paving mixtures: preparation of design mixes and treatment of test specimens. Transp Res Rec 911:88–95

Saleh MF (2003) New Zealand experience with foam bitumen stabilization. In: 82nd TRB annual meeting (CD-ROM), TRB, Washington, DC

Saleh MF (2006) Characterisation of foam bitumen quality and the mechanical properties of foam stabilised mixes. In: Proceedings of the 10th International Conference on Asphalt Pavements (ICAP), Quebec City, Canada

Sunarjono S (2007) Tensile strength and stiffness modulus of foamed asphalt applied to a grading representative of Indonesian road recycled pavement materials. Dinamika Teknik Sipil 7 (1):1–10

Wirtgen (2012) Wirtgen cold recycling manual, 1st edn. Wirtgen GmbH, Windhagen

Xu J, Hao P (2012) Study of aggregate gradations in foamed asphalt mixes. Road Mater Pavement Des 13(4):660–677

Chapter 4
Sustainable Transportation for Indian Cities: Role of Intelligent Transportation Systems

Partha Chakroborty

Abstract This chapter looks at how intelligent transportation systems can help create a sustainable transportation plan for a city through improvements in the efficiency of system. The chapter attempts to argue that a proper measure of efficiency of transportation systems will lead to a more sustainable system. The chapter also highlights the issues that need attention in order to create a sustainable urban mobility plan.

Keywords Sustainable transportation • Urban mobility • Intelligent transportation system • Efficient transportation • Public transportation

4.1 Introduction

Sustainability, as an important issue in development, came into public focus with Carson's book titled *Silent Spring* (1962). Schumacher's thoughts in *Small Is Beautiful* (1973) brought the issues of sustainability and the then development models into sharper focus. Since then, policymakers, by and large, have been mindful of sustainability while formulating policies.

Oxford English Dictionary (2010) defines "sustainable" as an adjective meaning "capable of being maintained at a certain rate or level". *Merriam-Webster Dictionary* (2011) defines it as "of, relating to, or being a method of harvesting or using a resource so that the resource is not depleted or permanently damaged". One of the most oft-quoted definitions of sustainable development appeared in the Brundtland Report (1987); according to this definition, sustainable development is that which "meets the needs of the present without compromising the ability of future generations to meet their own needs".

This manuscript was first published in Current Science (2011), Vol. 100 (9), pp. 1386–1390. It is reproduced here with permission.

P. Chakroborty (✉)
Department of Civil Engineering, Indian Institute of Technology,
Kanpur, Kanpur 208016, India
e-mail: partha@iitk.ac.in

G.L. Sivakumar Babu et al. (eds.), *Sustainability Issues in Civil Engineering*,
Springer Transactions in Civil and Environmental Engineering,
DOI 10.1007/978-981-10-1930-2_4

From the above definitions, few points about sustainability emerge. These and other thoughts on sustainability and sustainable transportation are the subject matter of the next section. The third section provides a brief description of the issues related to urban mobility that are foremost in India. The fourth section suggests how intelligent transportation system (ITS) can help India move towards a sustainable urban mobility plan. The fifth section summarizes the discussions in this paper.

4.2 Sustainability and Sustainable Transportation

The definitions of the word "sustainable" indicate that the concept of sustainability includes the following features: (i) processes need to be maintained (or carried on with) over a period of time, and (ii) harvesting of resources is inevitable for processes to run. It is the contention of the author that systems which remain efficient over a period of time and over space are the ones which can be maintained and hence are the only sustainable systems. Of course, the word efficient is used in a broader sense than it is generally used while describing efficiency of engineering systems. It must be accepted that engineering interventions (like infrastructure) which affect the society at large and use significant resources cannot be viewed and evaluated in isolation and must be looked at as a part of the habitat; that is, the efficiency of such systems must be defined in a more inclusive manner.

In the particular case of urban transportation, one needs to define what this habitat includes, what are the resources that one is dealing with and how one should measure efficiency (in the broad sense that is envisaged here). In the rest of this section, these points are expanded.

Urban transportation facilities or the processes of achieving mobility in an urban setting are a part of the urban habitat. The question is: what does this habitat include other than the roads, the intersections, the bus stops, the rail lines and so on? The urban habitat includes the people – rich, middle class and poor – the work places, the services (like the hospitals, the fire services, etc.), the residential areas, the recreational facilities, the educational institutions, the commercial establishments and the like. The way this habitat is organized creates the transportation demand and supply patterns.

The resources that urban transportation systems deal with are, broadly speaking, as follows:

1. Material resources such as fuel, aggregates, bitumen, etc.
2. Space on land, water and air
3. Time
4. People (and sometimes certain types of animals)
5. Environment
6. Opportunity

Of the six resource classes listed, opportunity needs a little exposition. It is felt that generally human endeavour with the use of other resources creates opportunities which if utilized improve the quality of life. However, if unutilized, opportunities often perish. This in some sense causes inefficient use of other resources on which opportunities are built. A definite mandate of any transportation system should be to allow the users of the system to efficiently harvest the opportunities.

What types of demand patterns are created and how they are met (the supply patterns) through the use of the resources have a large bearing on whether the transportation system is sustainable, i.e. whether the transport system will remain efficient for over a period of time and space (i.e. the system must be efficient not only for a restricted area but also regionally). However, as mentioned earlier, how the efficiency is measured is also important.

Efficiency has been a driving force in engineering design. If a system is seen to be inefficient, then effort is expended to improve the efficiency. The problem arises in the way this efficiency is often measured. A couple of examples on how a traditional view of efficiency can lead to non-sustainable development and how a more inclusive definition of efficiency could have led to sustainable development will highlight this issue better.

Few decades ago, it was felt that good roads should be provided to achieve fast and safe transportation of people and goods. The efficiency of the road system would be measured in how well it met the stated goals. Hence, when roads got congested, one built even more roads, roads without at-grade intersections, limited access roads and so on. Two-lane roads became four-lane highways; four-lane highways became six-lane expressways; and this would have continued but for the realization that there is no end to this. If on the other hand the efficiency definition was more inclusive and had features like (i) the amount of exhaust that would be created if more people drove (which would happen, and did happen, as the roads provided quicker mobility), (ii) the amount of fossil fuel that will be consumed, etc., then obviously roads which encouraged more automobile traffic would no longer be thought of as efficient. Planners and engineers would have had to look for other solutions. There are many other such examples.

In the Indian context, many medium-sized cities, like Kanpur, having allowed an uncontrolled growth of jitney services[1] today, face the problem of a largely inefficient urban transportation system. As a city grew, so did the need for public modes of transport. The few jitneys that were on the roads (when the city was small) were not enough, and waiting times for passengers were high; so more of the same were allowed to ply on the roads. For some time the problems seemed to be solved, but soon there was a need for more. As with the previous example, this constant striving to get efficiency in the short term eventually created an inefficient transportation system. Had the definition of efficiency included issues like congestion

[1]Motorized cabins with capacity of six to eight persons which typically ply on a fixed route but on a flexible schedule

(in the long run), environmental degradation and the like, in addition to issues of waiting time, network coverage, etc., then at some point planners and engineers would have surely moved away from the small cabin-based jitney services to larger cabin-based bus services, light rail transit systems and so on.

The preceding text tries to introduce three aspects that are important to the creation of a sustainable urban transportation system. As mentioned before, these are (i) the habitat of which the transportation system is a part, (ii) the resources that such a system will need to harvest and (iii) the measure of efficiency that should be employed to evaluate such a system. In this context the definition of sustainable transportation as put forward by the European Union Council of Ministers of Transport is particularly important to note. Hence, this definition, as quoted in "Sustainable Transportation and TDM" (Victoria Transport Policy 2010), is reproduced here.

> A sustainable transportation system is one that:
>
> 1. Allows the basic access and development needs of individuals, companies and society to be met safely and in a manner consistent with human and ecosystem health, and promotes equity within and between successive generations.
> 2. Is affordable, operates fairly and efficiently, offers a choice of transport mode, and supports a competitive economy, as well as balanced regional development.
> 3. Limits emissions and waste within the planet's ability to absorb them, uses renewable resources at or below their rates of generation, and uses non-renewable resources at or below the rates of development of renewable substitutes while minimizing the impact on the use of land and the generation of noise.

Another way of looking at the above definition is that it tries to enunciate, although not exhaustively, the basic principles which will lead to an efficient transportation system over space and time.

Before discussing how ITS technologies can help in achieving sustainable transportation for urban India, the issues related to mobility in urban India are briefly discussed.

4.3 Urban Mobility Issues in India

This section highlights some of the issues that need to be looked at in order to provide an efficient (in the long run), and hence sustainable, urban transportation system for Indian cities. These issues, which are enumerated in no particular order, are definitely not exhaustive, and there are other issues (often specific to a city) which are also important:

(i) India is a populous country. As per the 2001 census (Office of the Registrar General & Census Commissioner, India 2010), close to 300 million live in urban settlements. If one counted India's urban population alone, even then "urban India" would easily become the fourth largest country[2] (behind China, "rural India" and the USA) (World Population 2010). Not surprisingly, among other things, Indian cities are teeming with pedestrians. Any transportation system which does not properly account for pedestrian movement is bound to be inefficient in an Indian context.

(ii) Further, a large section of the urban population, especially in medium-sized cities, uses bicycles as the mode of transport. This natural tendency of the population to use non-polluting modes of transport should be encouraged, and the urban transportation plan should integrate bicycles into the system.

(iii) Also, a large section of the population in the cities uses public transportation. In fact, by some estimate, the larger is the city, the greater is the share of public transportation (Pucher et al. 2004). This implies that the success of urban transportation lies in providing an efficient public transportation network. However, a large percentage of the public transportation users are captive users with meagre income levels; this implies that the cost of public transportation has to be kept low – often artificially so.

(iv) In the last two decades, Indian economy has grown at unprecedented levels. This has meant that affordability of the urban population has gone up manifolds. Concurrently, India has produced more affordable vehicles. The net effect has been a brisk increase in the number of vehicles in urban India. Today, urban India also demands good roads and efficient intersections for mobility and adequate parking facilities at the destinations. Although one can argue that improving these facilities will encourage automobile use and hence may hurt the cause of sustainable development, the reality is not that straightforward. The road network currently available in most Indian cities is far from where it should be and does not even come close to a state where questions can be raised as to whether more should be done to improve the network.

(v) The larger number of vehicles on the roads has increased the possibility of accidents; further with improvements in speeds, chances of more severe accidents have also increased. Adding to the heightened safety concerns is the fact that Indian cities have large pedestrian population; this increases the chances of human-vehicle conflict. In short, safety-related issues are important more than before in achieving an efficient and maintainable transportation system.

(vi) Due to cities becoming rich employment sources and because of increased cost of living in the cities, movements of people into and out of the city from and to the suburbs have also increased. Integrating the urban transport system with the regional transport network is now important; consequently, multimodal transportation has also become important.

[2]The word country is used here to indicate that even if we looked at Indian cities as a unit, it will be larger, in population, than most other countries. Of course, urban and rural India together is one nation.

4.4 Role of ITS in Achieving Long-Term Efficiency in Urban Transportation

Solutions to the above problems, of course, lie in prudent policies and efficient implementation of these policies. The scope of this paper, however, is limited to discussing how elements of intelligent transportation systems (ITS) can help in achieving a sustainable urban transportation system. It may be noted here that ITS can help in implementing some policy measures which can lead to sustainability. ITS per se cannot create sustainability; it is the policies (which may use ITS) that can ultimately lead to a sustainable transportation system.

As argued earlier, sustainable development can be achieved through systems which remain efficient over a period of time and over space (or regionally). Hence, in this section the discussion will focus on how ITS can aid in achieving (i) efficient road transportation and (ii) efficient public transportation. It may be noted that the word "efficient" is to be understood in the broad sense envisaged here. Further, it must be understood that sustainability cannot be achieved without an efficient road and public transportation system; however, the converse is not true. Unless one develops a balance between road and public transportation system of a city, neither of them will remain efficient in the long run and hence will not create a sustainable system.

For the purposes of this paper, the term ITS will be used to refer to the area of transportation which deals with the application of (modern) information and communication technologies in transportation. It may be noted that information technology is assumed to include sensor technologies also.

4.4.1 Efficient Road Transportation Through ITS

Without going into the details, one can say that efficient road transportation can be achieved through (i) improved travel times, (ii) improved safety and (iii) improved terminals (parking). Figure 4.1 provides a general outline of how these can be achieved.

In the next few paragraphs, how ITS can aid in achieving some of the measures enunciated in the lower-level boxes of Fig. 4.1 is described.

ITS can help *modify demand* characteristics and *improve capacity* of roads in many ways. Some of these are mentioned here:

(i) Telecommuting (where much of the work can be done from remote locations) can reduce the need to travel, thereby reducing demand.
(ii) Ease of toll collection through ITS makes congestion pricing easy to implement; this allows the engineer the use of a very effective tool to modulate demand.

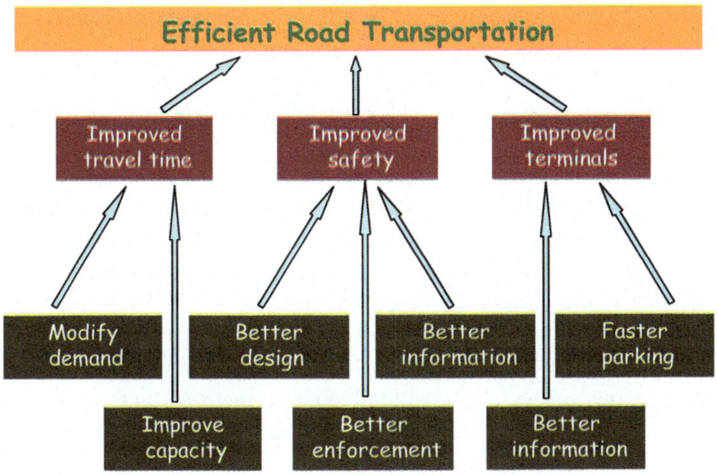

Fig. 4.1 Measures which can improve the efficiency of road transportation

(iii) Enforcement has improved through the use of ITS; hence, high-occupancy vehicle (HOV) lanes can be better enforced which will then create a greater incentive to pool cars, thereby reducing demand.

(iv) Collecting information on traffic status on large number of road sections and disseminating this information as well as disseminating predictions of travel times on these sections to the population are now feasible; it is envisaged that such information will allow travellers to choose paths and departure times more judiciously. This in turn will lead to modulation of demand through spatial and temporal distribution of the demand. It is also not too far-fetched to imagine that such information will also lead to people making better choice of modes which will again lead to a modulation of demand.

(v) A more direct way in which ITS can help in temporal modulation of demand is through ramp metering. The process of monitoring the traffic flow on freeways (expressways) and accordingly modulating the inflow at different intake points is a strategy which has been around for some time now and has proven to help the system by achieving a smoother flow on its primary roadways. As can be imagined, ITS does play an important role in improving the effectiveness of ramp-metering strategies.

(vi) ITS can also help in improving capacity through better implementation and enforcement of reversible lanes as well as the use of shoulders from time to time. Reversible lanes and the use of shoulders help augment capacity in the peak direction of travel.

(vii) Incidences (accidents) often reduce capacity of roads significantly. Better identification of incidences and quicker response to the incidences can reduce the impact of such events on the capacity of a road. ITS can help in improving both the aspects related to incidence management enunciated earlier. It may be noted here that efficiency of incidence management depends largely on how quickly relevant information is obtained from the incidence site and how quickly appropriate instructions reach the right agencies (like towing companies, ambulance service, hospitals, fire services, etc.). Not surprisingly then, ITS can play (and does play) an important role in improving incidence management and through it improving capacity.

(viii) Signal efficiency can be improved through, among other things, adaptive phases, reduced start-up loss times (through information to waiting drivers on the time left for the next green, etc.) and better pedestrian control. As these features rely on collecting and disseminating accurate information on the state of the system, ITS features can play an important.

(ix) Often, especially in India, the full capacity of the road is not realized due to non-compliance with driving rules and lack of driving decorum. The flow at intersections (especially unsignalized intersections) and rail-highway grade crossings are some examples of how improper driving behaviour limits efficient use of roads. It is felt that ITS can improve this through better driver education and improved enforcement of rules.

Application of modern information and communication technologies can improve safety of the road network through *better design* (primarily of vehicles), *better enforcement* and *better information* dissemination to the drivers. In the following, as before, some of the ways in which ITS can achieve these are mentioned:

(i) High-end vehicles come equipped with collision avoidance systems, lane-change assistance systems and intelligent anti-skid devices which all assist in improving safety. Most of these systems are a part of ITS, and with further improvements in ITS technologies, many of these features will become routine features of all cars and will make a considerable impact on the improved safety of roads.

(ii) Overspeeding and non-compliance with signal indications are major causes for accidents the world over; India is no different. ITS technologies, like cameras which record defaulters, can and does play an important and impartial role in enforcing speed limits and signals. Such enforcements improve the safety of roads.

(iii) Information on surface conditions and incidences help improve safety. ITS plays an important role in providing this type of information to the drivers. Real-time information on driving errors, like (i) problems in longitudinal control where a vehicle closes in on another in a dangerous manner, (ii) problems in lateral control where a vehicle fails to stay within a lane or (iii) problems in speed control where a driver overspeeds, to the driver often helps the driver to correct these problems. These corrections in turn help avoid possible accidents and improve the safety of the road.

ITS can be used to improve the *efficiency of terminal (parking) facilities* in an urban area. Information on nearby parking lots, their current occupancy status and information within the parking lot about the general location where space is available improve the efficiency of parking facilities. This helps in reducing the time drivers spend on the road searching for an appropriate parking facility. This further helps in reducing the demand (especially of the unproductive kind, where the only purpose is to find a place to park) as well as safety hazards as drivers' exasperation levels are low.

4.4.2 *Efficient Public Transportation Through ITS*

The long-term efficiency of public transport systems depends on improvements in right-of-way categories, technology issues and certain operational issues. Public transport systems, like the bus system, which utilize the road network can improve their efficiency manifold and become even more attractive if such systems get an improved right-of-way category. One way where ITS technologies can help is by allowing buses to get preferential treatment at signals.

However, the biggest impact that ITS can have on improving the efficiency of the public transportation system is through improvements in operational issues. Some of ways in which operations can be improved are:

(i) Improvements in routes and schedules to increase coverage and reduce in-vehicle travel times and waiting times.
(ii) ITS can help improve fleet management, thereby making operations more economical; this helps in keeping the fares down without relying much on government subsidy.
(iii) ITS can also help make transit systems more user-friendly by (a) providing personalized route plans though handheld devices like cell (mobile) phones, (b) providing current information about the operation (routes and schedules) both inside the vehicle and outside (e.g. at terminals), (c) improving transfers through schedule coordination, (d) making fare payment easier and faster, (e) making ticketing for multimodal systems seamless and the like.

4.5 Summary

This short paper or note tries to highlight the relationship between sustainability (or maintainability) and efficiency of transport systems. It also outlines some of the problems and issues that exist in providing mobility to urban Indians. Finally it enumerates some of the ways in which application of (modern) information and communication technologies (ITS) can help improve the efficiency of the transportation system and ultimately help achieve a sustainable urban transportation system.

References

Carson R (1962) Silent spring. Fawcett Publications, INC., Greenwich

Merriam-Webster Dictionary (2011) http://www.merriam-webster.com. 3 Jan 2011

Office of the Registrar General & Census Commissioner, India (2010) http://censusindia.gov.in/Census_Data_2001/India_at_glance/rural.aspx. 3 Jan 2011

Oxford English Dictionary (2010) http://www.oed.com. 20 Dec 2010

Pucher J, Korattyswaroopam N, Ittyerah N (2004) The crisis of public transport in India: overwhelming needs but limited resources. J Public Transp 7(4):1–20

Report of the World Commission on Environment and Development: Our Common Future (1987). http://www.un-documents.net/wced-ocf.htm. 23 Dec 2010

Schumacher EF (1973) Small is beautiful. Blond and Briggs Ltd., London

Victoria Transport Policy Institute (2010) Sustainable Transportation and TDM. http://www.vtpi.org/tdm/tdm67.htm. 26 Dec 2010

World Population (2010) http://en.wikipedia.org/wiki/World_population. 5 Jan 2011

Chapter 5
Economic Sustainability Considerations in Asphalt Pavement Design

Animesh Das

Abstract The present article briefly highlights the thought process involved in economizing an asphalt pavement structure, while its thicknesses are designed from structural considerations. The need for building a sustainable infrastructure encourages development of alternative materials and technologies. This brings in changes to the original design solution. Thus, the present article discusses how the economic sustainability considerations are involved in the asphalt pavement design process.

Keywords Asphalt pavement • Pavement design • Pavement maintenance • Optimal design • Economic sustainability

5.1 Introduction

Structural design of asphalt pavement involves provision of appropriate thicknesses to the pavement structure so that it survives up to the design period. If the thickness is less than what is required, the pavement is expected to fail prematurely. If the thickness is more than what is required, the design may be considered uneconomical. However, it may not always be a simple task to estimate the best thickness for a pavement structure. These aspects have been addressed in the present article.

The concept of sustainability specific to road infrastructure can be broadly classified as economic (e.g., expenditure involved in construction and maintenance of road, cost of operating vehicles on road, etc.), social (e.g., health and safety of people, mobility, traffic delay, etc.), and environmental (e.g., gas emission from vehicles, noise generation, energy spent during construction, etc.) aspects (Gopalakrishnan 2011; Muench and Anderson 2009; Van Dam et al. 2012, 2015). The focus of the present article is to discuss the economic sustainability considerations related to structural design of asphalt pavements.

A. Das (✉)
Department of Civil Engineering, Indian Institute of Technology Kanpur,
208 016 Kanpur, India
e-mail: adas@iitk.ac.in

G.L. Sivakumar Babu et al. (eds.), *Sustainability Issues in Civil Engineering*,
Springer Transactions in Civil and Environmental Engineering,
DOI 10.1007/978-981-10-1930-2_5

5.2 The Best Pavement Design

For every project location, the pavement design may be different depending on the design input. Even, for the same project location, alternative pavement designs are possible. These individual pavements are expected to have different cost implications. If the pavement is designed for a higher longevity, the design solution is expected to be costlier (refer Fig. 5.1).

A pavement may be considered economical if, for the same longevity, the cost of construction is cheaper between two design solutions (e.g., as per Fig. 5.1, "C" is cheaper than "D" though longevities are the same) or for the same cost of construction, the longevity is higher (e.g., as per Fig. 5.1, "A" has more longevity than "B" though the costs are the same).

The parameters which contribute to the cost of construction of a pavement structure include cost of (i) land, (ii) mobilization of equipment and manpower, (iii) earthwork, (iv) construction materials, (v) laying, compacting, finishing, and so on. Considering all other cost components to be the same between two alternative pavement designs, the cost of pavement construction (between these two alternative designs) will be different due to the differences in the material costs. The material costs for two alternative designs may be different because the design thicknesses of these alternative designs are different.

Structural design of pavement, based on mechanistic-empirical (M-E) design approach, involves estimation of stresses or strains at critical locations (AI 1999; Austroads 2004; Das 2015; IRC 2012; NCHRP 2004; Shell 1978; Theyse et al. 1996). Thickness of the individual layers affects the stress or strain values. The stress or strain values (anywhere within the pavement structure) decrease with the increase of layer thickness (refer Fig. 5.2a). Increase of stiffness of material also reduces stress or strain within the pavement structure (refer Fig. 5.2b). For a given distress type, the critical stress or strain (typically strain is used for design of asphalt pavement) is compared with the limiting value, and the pavement design is finalized iteratively. For most of the types of distresses, the limiting values of the strain are functions of the expected number of traffic load repetitions (Das 2009, 2015). Such functions are generally developed through various laboratory tests and subsequent

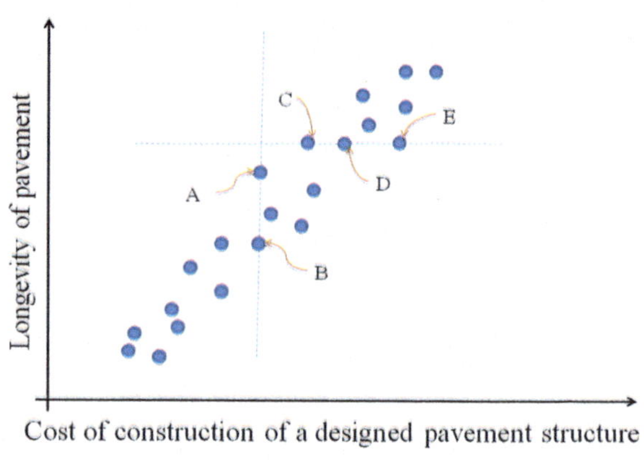

Fig. 5.1 Schematic diagram representing variation of cost of construction of a designed pavement structure with longevity

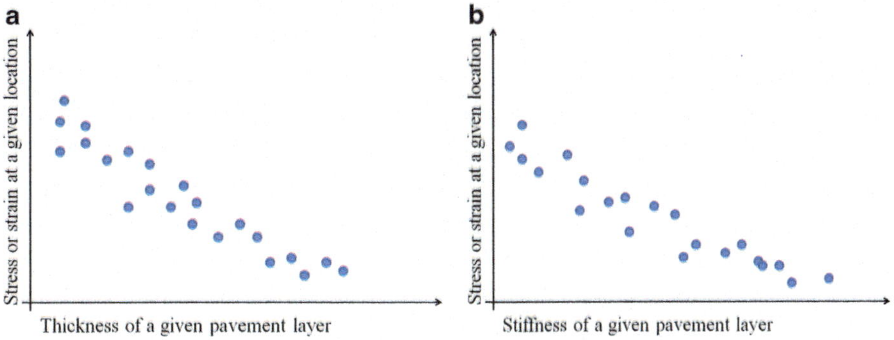

Fig. 5.2 Schematic diagram representing variation of stress or strain at a given location within the pavement structure (**a**) with thickness and (**b**) with stiffness modulus of a given pavement layer

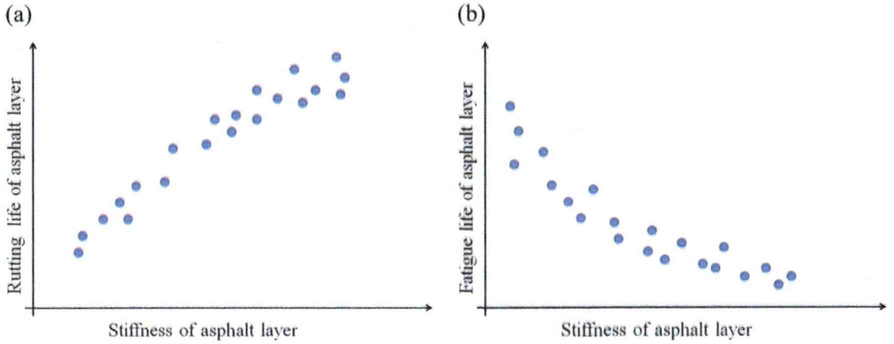

Fig. 5.3 Schematic diagram representing possible variation of (**a**) rutting life of asphalt layer and (**b**) fatigue life of asphalt layer with the stiffness of asphalt mix

calibration based on field performance (AI 1999; Austroads 2004; IRC 2012; NCHRP 2004; Shell 1978; Theyse et al. 1996).

The layer thickness or stiffness may have different effects on the different modes of structural failures against which a pavement is designed. An asphalt pavement is generally designed for its survival against load fatigue, thermal fatigue, rutting (permanent deformation along the most traversed wheel path), low-temperature shrinkage cracking, and top-down cracking (NCHRP 2004). Increase of thickness may reduce the chances of occurrence of certain distresses, but for certain other distresses, the dependency may not be clearly known (NCHRP 2004, Rajbongshi and Das 2008b). Similarly, increase of stiffness may be beneficial against certain distresses and may not be beneficial against certain other distresses. For example, higher stiffness of asphalt mix may increase the rutting life of the asphalt layer (refer Fig. 5.3a), but it may reduce its fatigue life (Baburamani 1999) (refer Fig. 5.3b).

Thus, even though higher stiffness is beneficial in reducing the stress or strain values in asphalt layer (refer Fig. 5.2b), it may not be beneficial from some other considerations (refer Fig. 5.3(b)). Therefore, interplay between selection of material type and thickness is an important consideration in asphalt pavement design (Das 2015;

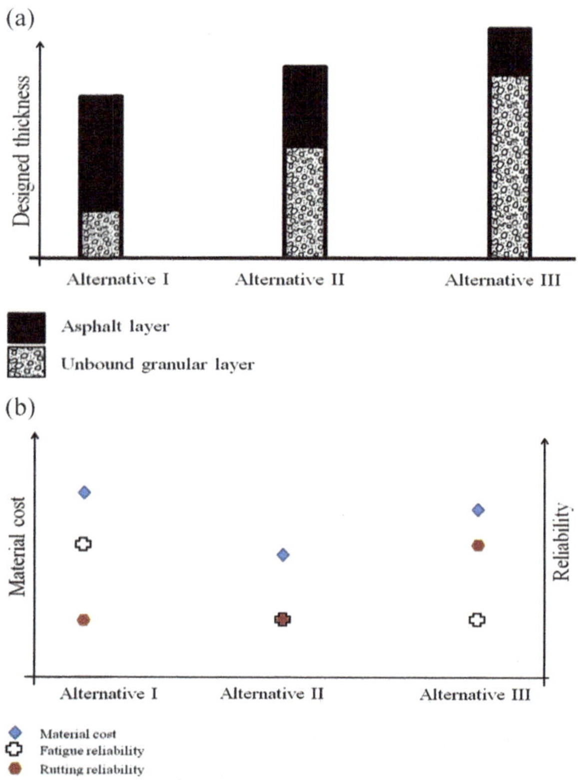

Fig. 5.4 Schematic diagram showing that alternative pavement designs (for the same design input) have (**a**) different thicknesses and (**b**) different cost and reliability values

Das and Gogoi 2012). By varying the constituents and their relative proportions, it is possible to change one or more properties of the pavement material (Bari and Witczak 2006; Harvey and Tsai 1996; Monismith et al. 2001) and hence the design thickness.

A pavement is constituted with a number of layers. Thus, structural design of an asphalt pavement requires estimation of thicknesses of the individual layers. For the same design input parameters, the individual layer thicknesses and the total pavement thickness values of the alternative design solutions may be different (Narasimham et al. 2001; Rajbongshi and Das 2008a). Figure 5.4a schematically represents such alternative designs, considering a three-layer asphalt pavement structure (i.e., asphalt layer, unbound granular layer, and subgrade).

Relative economy can be considered to select the suitable design thicknesses in such a situation. Figure 5.4b schematically represents a comparison of material costs among these alternative pavement designs. The material cost depends on the thicknesses and on the unit costs of the layers (Narasimham et al. 2001). It may be noted that the reliability values of these alternative solutions can be different (refer Fig. 5.4b) (Das 2013; Rajbongshi and Das 2008a). Reliability, in this case, is defined as the probability that the pavement survives up to a given number of traffic repetitions (say, design traffic) for a given distress (Das 2013; Maji and Das 2008).

Thus, the optimal pavement design solution will be the one which will maximize the reliability while minimizing the material cost (Rajbongshi and Das 2008a). However, this is about the design of a pavement which is built for the first time. The pavement, within the entire time period of existence (i.e., the analysis period), may undergo various forms of maintenance, for example, minor maintenance, rehabilitation, and recycling at different points of time (refer Fig. 5.5). At the end, when no further maintenance is possible (i.e., end of life), the pavement may be reconstructed.

The pavement design, within the analysis period, is optimal when all these maintenance actions and the corresponding application timings are planned a priori, and the total cost is minimized. This is done using life cycle cost analysis (Cooper III et al. 2012; Walls III, and Smith 1998; Lamptey et al. 2005; Rangaraju et al. 2008) and is routinely employed (AAPTP 2011; Caltrans 2007; NCHRP 2004; SANRA 2013; Van Dam et al. 2015) for economic analysis for comparing across (i) various maintenance options within a project and (ii) various road projects.

The cost typically involves two components, namely, agency cost (e.g., the cost of maintenance) and user cost (e.g., the vehicle operating cost). The total cost discounted to some base year is used for comparison purpose. Figure 5.6 schematically shows an idealized situation where a fixed rehabilitation (e.g. overlay) measure is undertaken at regular time interval for a given stretch of road. It is expected that the agency cost will gradually be lesser as the time interval between successive rehabilitation cycles increases (refer Fig. 5.6a); similarly, it is expected that the user cost will gradually be higher as the time interval between successive rehabilitation cycles increases (refer Fig. 5.6b). The total cost (i.e., life cycle cost) may take a shape as shown in Fig. 5.6c indicating that an optimal point may exist. This optimal point decides the extent of the rehabilitation (e.g., overlay) and its time interval. However, in realistic situation, multiple maintenance and rehabilitation options are available, and the application timings may not be fixed (refer Fig. 5.5). Various approaches have been suggested in the literature for deciding type and extent of maintenance and rehabilitation (Irfan et al. 2012; Labi and Sinha 2005; Li and Madanat 2002; Peshkin et al. 2004; Zhang, et al. 2010).

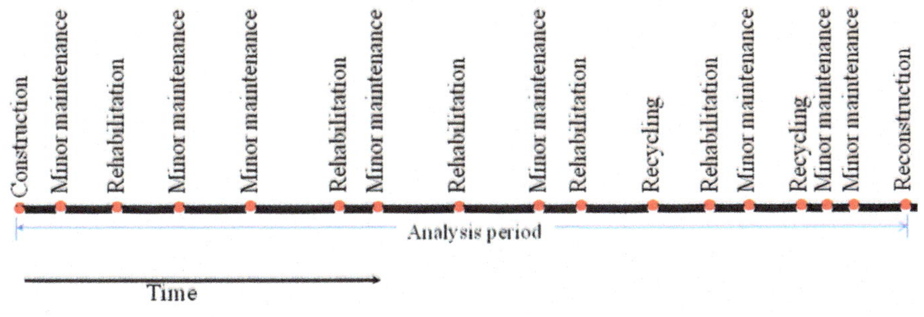

Fig. 5.5 Schematic diagram showing different types of maintenance a pavement undergoes within its analysis period

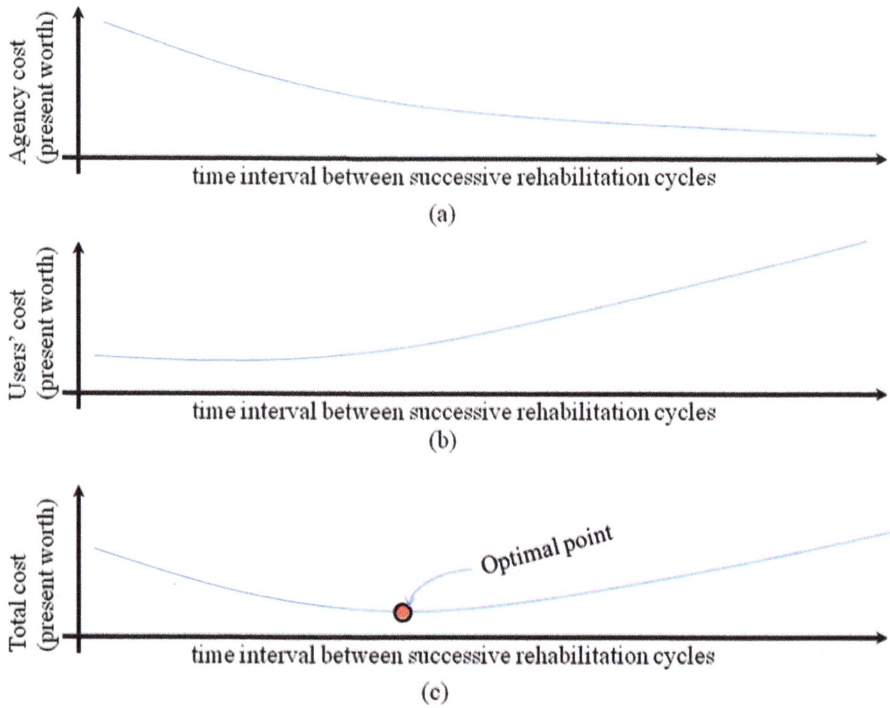

Fig. 5.6 Schematic diagram showing variation of (**a**) agency cost, (**b**) users' cost, and (c) total cost with the interval between successive rehabilitation cycles

5.3 Sustainable Pavement Design

For available construction materials and for a given technology (analysis, design, and construction), one can arrive at the optimal planned pavement design solution, as discussed above. In a life cycle assessment (LCA) process, the environmental and social impacts are further considered to assess the overall sustainability of the pavement being designed (Chen et al. 2015; Muench 2010; Muench and Anderson 2009; Santhanam and Gopalakrishnan 2013; Liu et al. 2014; Santero et al. 2011; Santos et al. 2015; Tsai and Chang 2012). Various ratings, credit- or point-based system, are used to quantify the environmental and social impact of a proposed project (Muench and Anderson 2009; Santhanam and Gopalakrishnan 2013; Van Dam et al. 2015).

Innovations in the design, construction and in material utilization can bring in further economy to the pavement design, for example:

(i) Cold mix or warm mix asphalt reduces requirement of heat energy during the production of the asphalt mix (Cooper III et al. 2012; Kucukvar et al. 2014; Technical Guidelines 2009).
(ii) Asphalt recycling (hot or cold, in situ or plant, or full depth) saves energy and resources (Al-Qadi et al. 2007; Alkins et al. 2008; Karlsson and Isacsson 2006; Stroup-Gardiner 2011).

(iii) Rich bottom asphalt pavement enhances the fatigue life of asphalt mix without compromising with the stiffness requirement (Das and Pandey 2000; Harvey and Tsai 1996; Monismith et al. 2001).
(iv) Heat recycling reduces the urban heat island effect and reduces rutting of asphalt layer (EPA 2009; Mallick et al. 2009).
(v) Cemented base/subbase with marginal aggregates reduces the haulage cost for acquiring good-quality aggregates (Austroads 2004; IRC:37 2012; Wen et al. 2014).
(vi) Stabilization of subgrade reduces the design thickness of other layers (Little and Nair 2009).
(vii) Porous pavement helps to recharge the ground water (EPA 2009; NAPA 2008).
(viii) Perpetual pavement enhances the longevity of the pavement by reducing the need of subsequent rehabilitation (Amini et al. 2012; Timm and Newcomb 2006).
(ix) Binder modification (Roque et al. 2005) and mix reinforcement (Abtahi et al. 2010) enhance the mix property (refer Fig. 5.3), and so on.

Naturally available construction materials (or the raw materials which are processed to create construction materials) are depleting rapidly. Thus, possibilities are being explored to recycle the existing deteriorated road materials (Chui et al. 2008) and utilize some of the waste materials (Das and Krishna Swamy 2014; FHWA 2015; Schimmoller et al. 2000; Stroup-Gardiner and Wattenberg-Komas 2013). These waste materials originate from industry, domestic, construction and demolition, and agricultural sources. Some of these materials, after due processing, can be utilized/recycled as bulk fill, partial/full replacement of aggregates, or admixture to asphalt binder (Das and Krishna Swamy 2014; FHWA 2015; Stroup-Gardiner and Wattenberg-Komas 2013).

When a new/recycled material is decided to be used in pavement structure (after it is adjudged suitable) and the extent of replacement of conventional materials is decided (Al-Qadi et al. 2007; Aravind and Das 2007b; Oliveira et al. 2013), the new material properties will be used as input to the pavement design (Das 2015). The optimal pavement thicknesses may then be different than the pavement design with conventional material. Even if the design thickness (with new/recycled material) is computed to be more than the pavement design with conventional material (Aravind and Das 2007a), the design solution may still be cheaper, because of utilization of existing deteriorated/waste materials (cost of acquiring deteriorated material may be cheaper than that of virgin material). But if the thickness becomes more than certain threshold level, the design solution may not remain cheaper any longer. Referring to Fig. 5.1, if the design with conventional material is presented as point "D," a revised design point as "C" is preferred than a design point as "E," when new/recycled materials are used. Other environmental and social sustainability issues are as well involved to decide the quantity of reutilization of material (Aurangzeb and Al-Qadi 2014).

5.4 Closure

Structural design of pavement involves consideration of a number of parameters and factors. For a given set of design input, a number of alternative design solutions are possible. Optimally designing a pavement and planning for all its subsequent maintenance and rehabilitation measures is a complex task (Das 2015), more so because of the difficulty in prediction of pavement performance (before and after maintenance), which is generally described empirically in the design process (NCHRP 2004; Perera et al. 1998; Roque et al. 2010; RRD-307 2006).

The present article has briefly discussed the economic sustainability issues connected to the structural design of asphalt pavements. With the advent of new construction technologies and application of innovative materials, new experience will be gained on the performance of pavement. This will in turn help us to design pavements which will be more durable, economical, and sustainable.

References

Abtahi SM, Sheikhzadeh M, Hejazi SM (2010) Fiber-reinforced asphalt-concrete – a review. Construct Build Mater 24:871–877

Alkins A, Lane B, Kazmierowski T (2008) Sustainable pavements environmental, economic, and social benefits of in situ pavement recycling, Transportation research record, no. 2084. TRB, Washington, DC, pp 100–103

Al-Qadi IL, Elseifi M, Carpenter SH (2007) Reclaimed asphalt pavement – a literature review. FHWA-ICT-07-001. Illinois Department of Transportation

Amini AA, Mashayekhi M, Ziari H, Nobakht S (2012) Life cycle cost comparison of highways with perpetual and conventional pavements. Int J Pavement Eng 13(6):553–568

Aravind K, Das A (2007a) Pavement design with central plant hot-mix recycled asphalt mixes. Construct Build Mater 21(5):928–936

Aravind K, Das A (2007b) Preliminary constituent proportioning for central plant hotmix asphalt. J Mater Civ Eng 19(9):740–745

Asphalt Institute (AI) (1999) Thickness design—asphalt pavements for highways and streets, 9th edn, Manual series no.1. Asphalt Institute (AI), Lexington

Aurangzeb Q, Al-Qadi IL (2014) Asphalt pavements with high reclaimed asphalt pavement content: economic and environmental perspectives, Transportation research record, no. 2456. TRB, Washington, DC, pp 161–169

Baburamani P (1999) Asphalt fatigue life prediction models – a literature review. ARRB Transport Research Ltd, Vermont South

Bari J, Witczak MW (2006) Development of a new revised version of the Witczak E* predictive model for hot mix asphalt mixtures. Proc Assoc Asphalt Paving Technol 75:381–423

Chen F, Zhu H, Yu B, Wang H (2015) Environmental burdens of regular and long-term pavement designs: a life cycle view. Int J Pavement Eng. doi:10.1080/10298436.2014.993189

Chui CT, Hsu TH, Yang WF (2008) Life cycle assessment on using recycled materials for rehabilitating asphalt pavements. Resour Conserv Recycl 52:545–556

Cooper SB III, Elseifi M, Mohammad LN, Marwa H (2012) Performance and cost-effectiveness of sustainable technologies in flexible pavements using the mechanistic-empirical pavement design guide. J Mater Civ Eng 24(2):239–247

Das A (2009) A reliable design of asphalt pavement from structural considerations, editor's corner. Int J Pavement Res Technol 2(1): p.IV

Das A (2013) Reliability considerations in asphalt pavement design. In: Proceedings of International Symposium on Engineering Under Uncertainty: Safety Assessment and Management (ISEUSAM- 2012), Bengal Engineering and Science University, Springer-India, 4–6 January 2012, pp 345–354

Das A, (2015) Structural design of asphalt pavements: principles and partices in various design guidelines. Transp Dev Econ 1:25–32

Das A, Gogoi R (2012) An integrated approach to mix design and structural design for bituminous pavements. In: Souvenir national get-together on road research and its utilization, Central Road Research Institute, New Delhi, 1–2 March 2012, pp 42–44

Das A, Krishna Swamy A (2014) Reclaimed waste materials in sustainable pavement construction. Chapter 15, Climate change, sustainability, energy, and pavements. In: Gopalakrishnan K, Steyn WJ, Harvey J. Springer-Verlag, Berlin, pp 419–438

Das A, Pandey BB (2000) Economical design of bituminous pavements with two grades of bitumen in the surfacing. In: Seminar on road financing, design, construction and operation of highways in 21st century, IRC, New Delhi, 24–25 September 2000, pp II-35–II-42

Federal Highway Administration (FHWA) (2015) User guidelines for waste and byproduct materials in pavement construction. http://www.fhwa.dot.gov/publications/research/infrastructure/structures/97148/cbabs1.cfm. Last accessed 16 April 2015

Gopalakrishnan K (2011) Sustainable highways, pavements and materials an introduction. Transdependenz LLC

Guidelines for the design of flexible pavements. IRC:37-2012, 3rd Revision, Indian Roads Congress, New Delhi

Hansen K (2008) Porous asphalt pavements for stormwater management. National Asphalt Pavement Association (NAPA), Lanham

Harvey JT, Tsai BW (1996) Effects of asphalt content and air void content on mix fatigue and stiffness, Transportation research record, 1543. TRB, Washington, DC, pp 38–45

Irfan M, Khurshid MB, Bai Q, Labi S, Morin TL (2012) Establishing optimal project-level strategies for pavement maintenance and rehabilitation – a framework and case study. Eng Optim 44(5):565–589

Karlsson R, Isacsson U (2006) Material-related aspects of asphalt recycling – state-of-the-art. J Mater Civ Eng 18(1):81–92

Kucukvar M, Noori M, Egilmez G, Tatari O (2014) Stochastic decision modeling for sustainable pavement designs. Int J Life Cycle Assess 19:1185–1199

Labi S, Sinha KC (2005) Life-cycle evaluation of flexible pavement preventive maintenance. J Transp Eng 131(10):744–751

Lamptey G, Ahmad MZ, Labi S, Sinha KC (2005) Life cycle cost analysis for INDOT pavement design procedures. Report # FHWA/IN/JTRP-2004/28, Joint Transportation Research Program. Indiana Department of Transportation and Purdue University, West Lafayette

Li Y, Madanat S (2002) A steady-state solution for the optimal pavement resurfacing problem. Transp Res A 36:525–535

Life cycle cost analysis for airport pavements (2011) AAPTP 06-06, Airfield Asphalt Pavement Technology Program

Life cycle cost analysis procedures manual (2007) Caltrans, State of California, Department of Transportation

Little DN, Nair S (2009) Recommended practice for stabilization of subgrade soils and base material, Web-only document 144. NCHRP, Transportation Research Board, Washington, DC

Liu X, Cui Q, Schwartz C (2014) Greenhouse gas emissions of alternative pavement designs: framework development and illustrative application. J Environ Manage 132:313–322

Maji A, Das A (2008) Reliability considerations of bituminous pavement design by mechanistic-empirical approach. Int J Pavement Eng 9(1):19–31

Mallick RB, Chena BL, Bhowmick S (2009) Harvesting energy from asphalt pavements and reducing the heat island effect. Int J Sustain Eng 2(3):214–228

Mechanistic-empirical design of new and rehabilitated pavement structures (2004) NCHRP design guide. National Cooperative Highway Research Program, Project 1-37A, Washington, D.C. http://www.trb.org/mepdg/guide.htm. Last accessed 16 Apr 2015

Monismith CL, Long F, Harvey JT (2001) California's Interstate-710 rehabilitation: mix and structural design, construction specifications. Proc Assoc Asphalt Paving Technol 70:762–799

Muench ST (2010) Roadway construction sustainability impacts – review of life-cycle assessments, Transportation research record, No.2151. TRB, Washington, DC, pp 36–45

Muench ST, Anderson JL (2009) Greenroads: a sustainability performance metric for roadway design and construction. Final technical report, TNW 2009-13, WA-RD 725.1, University of Washington

Narasimham KV, Misra R, Das A (2001) Optimization of bituminous pavement thickness in mechanistic pavement design. Int J Pavement Eng Asphalt Technol 2(2):59–72

Oliveira JRM, Silva HMRD, Jesus CMG (2013) Pushing the asphalt recycling to the limit. Int J Pavement Res Technol 6(2):109–116

Pavement design (2004). Austroads, Sydney

Perera RW, Byrum C, Kohn SD (1998) Investigation of development of pavement roughness, FHWA-RD-97-147. US Department of Transportation, Federal Highway Administration , McLean

Peshkin DG, Hoerner TE, Zimmerman KA (2004) Optimal timing of pavement preventive maintenance treatment applications, , Report 523. NCHRP, TRB, Washington, DC

Rajbongshi P, Das A (2008a) Optimal asphalt pavement design considering cost and reliability. J Transp Eng ASCE 134(6):255–261

Rajbongshi P, Das A (2008b) Thermal fatigue considerations in asphalt pavement design. Int J Pavement Res Technol 1(4):129–134

Rangaraju PR, Amirkhanian S, Guven Z (2008) Life cycle cost analysis for pavement type selection, Report # FHWA-SC-08-01. Southern California Department of Transportation, Columbia

Reducing urban heat islands: compendium of strategies – cool pavements (2009) United States Environmental Protection Agency, EPA, Washington, DC. www.epa.gov/heatisland/resources/pdf/CoolPavesCompendium.pdf. Accessed 18 Apr 2015

Roque R, Birgisson B, Drakos C (2005) Guidelines for the use of modified binders, Final report, report no 4910-4504-964-12. Florida Department of Transportation, Gainesville

Roque R, Zou J, Kim YR, Thirunavukkarasu S, Underwood BS, Guddati MN (2010) Top-down cracking of hot-mix asphalt layers: models for initiation and propagation, NCHRP web only document 162. TRB, Washington, DC

RRD-307, Independent review of the mechanistic-empirical pavement design guide and software (2006) National Cooperative Highway Research Program (NCHRP). http://onlinepubs.trb.org/onlinepubs/nchrp/nchrp_rrd_307.pdf. Last accessed 19 Apr 2015

Santero NJ, Masanet E, Horvath A (2011) Life-cycle assessment of pavements. Part 1: critical review. Resour Conserv Recycl 55(9–10):810–818

Santhanam GR, Gopalakrishnan K (2013) Pavement life-cycle sustainability assessment and interpretation using a novel qualitative decision procedure. J Comput Civ Eng 27(5):544–554

Santos J, Ferreira A, Flintsch G (2015) A life cycle assessment model for pavement management: methodology and computational framework. Int J Pavement Eng 16(3):268–286

Schimmoller VE, Holtz K, Eighmy TT, Wiles C, Smith M, Malasheskie G, Rohrbach GJ, Schaftlein S, Helms G, Campbell RD, Van Deusen CH, Ford B, Almborg JA (2000) Recycled materials in European highway environments: uses, technologies, and policies. Office of International Programs, Office of Policy, Federal Highway Administration U.S. Department of Transportation, Washington, DC

Shell pavement design manual—Asphalt pavement and overlays for road traffic (1978) Shell International Petroleum Company Limited, London

South African Pavement Engineering Manual (2013) South African National Roads Agency (SANRA) Ltd.

Stroup-Gardiner M (2011) Recycling and reclamation of asphalt pavements using in-place methods, NCHRP synthesis of highway practice 421. TRB, Washington, DC

Stroup-Gardiner M, Wattenberg-Komas T (2013) Recycled materials and byproducts in highway applications, a synthesis of highway practice, vol 1–8, NCHRP synthesis 435. Transportation Research Board, Washington, D.C. http://www.trb.org/Publications/NCHRPSyn435.aspx

Technical guidelines: bitumen stabilized materials 2: a guideline for the design and construction of bitumen emulsion and foamed bitumen stabilized materials, TG 2 (2009) 2nd edn. Asphalt Academy, CSIR Built Environment, Pretoria

Theyse HL, Beer M, Rust FC (1996) Overview of the South African mechanistic pavement design analysis method, Transportation research record, 1539. TRB, Washington, DC, pp 6–17

Timm DD, Newcomb DE (2006) Perpetual pavement design for flexible pavements in the U.S. Int J Pavement Eng 7(2):111–119

Tsai CY, Chang AS (2012) Framework for developing construction sustainability items: the example of highway design. J Clean Prod 20:127–136

Van Dam T, Taylor P, Fick G, Gress D, VanGeem M, Lorenz E (2012) Sustainable concrete pavements: a manual of practice. National Concrete Pavement Technology Center, Iowa State University, Ames

Van Dam TJ, Harvey JT, Muench ST, Smith KD, Snyder MB, Al-Qadi IL, Ozer H, Meijer J, Ram PV, Roesler JR, Kendall A (2015) Towards sustainable pavement systems: a reference document, FHWA-HIF-15-002. Federal Highway Administration, Washington, DC

Walls III J, Smith MR (1998) Life-cycle cost analysis in pavement design – interim technical bulletin. FHWA-SA-98-079. Federal Highway Administration (FHWA)

Wen H, Muhunthan B, Wang J, Li X, Edil T, Tinjum JM (2014) Characterization of cementitiously stabilized layers for use in pavement design and analysis, NCHRP report 789. TRB, Washington, DC

Zhang H, Keoleian GA, Lepech MD, Kendall A (2010) Life-cycle optimization of pavement overlay systems. J Infrastruct Syst 16(4):310–322

Chapter 6
Sustainable Design of Indian Rural Roads with Reclaimed Asphalt Materials

Sireesh Saride, Anu M. George, Deepti Avirneni, and B. Munwar Basha

Abstract Designing low-volume roads (LVRs) or rural roads without having premature failures, in terms of heavy rutting and fatigue cracking, has always been a challenging task for the design engineers. This is especially due to the design procedures followed to obtain the pavement thickness which is completely based on the subgrade soil properties and the expected traffic volume. The current design procedure in India neglects the properties of the pavement material in base/subbase layers which will, in principle, provide structural support to the traffic loading. Hence, there is a need to revisit the design methodology by adopting the damage analysis of these pavement layers. Besides, the longevity of the pavement system can be improved by adopting sustainable pavement materials such as a combination of reclaimed asphalt pavement (RAP) material and virgin aggregates (VA). This study focuses on developing a design chart for selecting the thickness of base layer for low-volume roads when 80 % RAP material and 20 % VA mixture treated with different dosages of fly ash are used based on damage analysis. For the damage analysis, resilient modulus and unconfined compressive strength of the pavement layers are necessary to adopt in pavement analyses software such as KENLAYER and IITPAVE. The resilient modulus data was obtained from repeated loading triaxial tests in this study. Damage analysis has been performed in KENLAYER to obtain the desired pavement thickness calculated corresponding to a design life of 10 years. In addition, the fatigue and rutting strains developed in the pavement layers due to a standard wheel load obtained from KENLAYER are compared with the data obtained from IITPAVE software. It is noticed that based on the damage analysis, the base layer thickness has been reduced by about 50 % when 80R:20A mix stabilized with 40 % fly ash against conventional virgin aggregates.

Keywords Reclaimed asphalt pavement • Rural roads • Pavement design • Damage analysis • Sustainability

---

S. Saride (✉) • A.M. George • D. Avirneni • B. Munwar Basha
Department of Civil Engineering, Indian Institute of Technology Hyderabad, Kandi, Sangareddy, Telangana 502 285, India
e-mail: sireesh@iith.ac.in; ce13m1011@iith.ac.in; ce11p1002@iith.ac.in; basha@iith.ac.in

G.L. Sivakumar Babu et al. (eds.), *Sustainability Issues in Civil Engineering*,
Springer Transactions in Civil and Environmental Engineering,
DOI 10.1007/978-981-10-1930-2_6

6.1 Introduction

Rural roads are the key component for social and economic growth of a developing country like India. As per Pradhan Mantri Gram Sadak Yojana (PMGSY), 80 % of Indian roads are rural roads. The Government of India envisaged to invest about Rs. 29,000 crores during the 11th five-year plan period to upgrade the existing rural roads of about 1.16 lakh km. In India, the design of low-volume roads (LVRs) or rural roads (RR) is performed based on Indian Roads Congress (IRC) guidelines specified in IRC-SP-72 2007. According to this guideline, the design thickness of a pavement is solely based on the subgrade's California bearing ratio (CBR) and cumulative traffic in terms of equivalent single-axle load (ESAL) application. It does not consider the strength and stiffness properties of the base materials in the design of pavements, even though the performance of the pavement is highly dependent of the quality and behavior of the pavement materials. This will result in, often, an overestimation of the pavement layer thickness. Several other established pavement design methods such as the American Association of State Highway and Transportation Officials (AASHTO) and Asphalt Institute (AI) methods consider the base material properties including resilient modulus (M_r) in determining the pavement layer thickness. The resilient modulus is defined as a ratio of cyclic deviatoric stress to the elastic component of total settlement observed under the corresponding cyclic deviatoric stress. The major failure criteria adopted for the design of pavements are fatigue cracking and rutting apart from the CBR-based penetration resistance. The fatigue strains are horizontal tensile strains developed at the bottom of the bituminous layer (just above the base or interlayer), and the rutting strains are vertical compressive strains developed on top of the natural soil subgrade/roadbed. The fatigue and rutting strains at these critical locations are calculated by performing damage analysis using KENLAYER, a pavement analysis software developed based on AASHTO pavement design method (AASHTO 1993). KENLAYER considers the material properties of individual pavement layers and also the cumulative traffic ESALs which will result in an optimum pavement design thickness. Adopting such failure criteria for the design of LVRs would reduce the premature failures.

A major sustainable alternative to reduce the cost of rural road construction and maintenance is by replacing the conventional virgin aggregates (VA) with recycled materials like reclaimed asphalt pavement (RAP) along with appropriate stabilizer. The RAP materials are degenerated bituminous pavement that has been recovered, usually by milling, and is to be used in part or as a whole in a new pavement by mixing it with other virgin aggregates, asphalt, cement, lime, and/or other stabilizing materials. The major challenge in using RAP as a base material is that the lack of proper aggregate packing. The aggregate packing can be increased by treating the RAP-VA mix with some cementitious material like lime, cement, or fly ash which will act as a binder material between the aggregates, thereby improving the aggregate packing. In this study, fly ash, a coal combustion product obtained from thermal power plants, is used as the stabilizing material. In current practice of pavement construction, only a small percentage of RAP material along with VA is used as base layer. Hence, the reduction in the usage of the VA is very small. For the sustainable and economic pavement construction, it is necessary to increase the percentage of RAP in the base layers. A 100 % RAP as a base course is not possible to use due to lack of aggregate interaction; hence, in this study a combination of 80 % RAP and 20 % VA (denoted as 80R:20A) treated with different percentages of fly ash (10, 20, 30, and 40 %) has been used as a base material. The materials used were graded according to the Ministry of Rural Development (MoRD), Government of India, specifications. A series of repeated load triaxial tests were performed to find the resilient modulus (M_r) of the mixes. The results obtained were incorporated in KENLAYER software to perform damage analysis to find the design thickness of the pavement for a design period of 10 years. The strains obtained from KENLAYER using AASHTO method are compared with that obtained using IITPAVE software developed based on IRC:37 (2012).

6.2 Literature Review

The design and laboratory performance of asphalt mixtures with a high percentage of RAP were studied by Aurangzeb et al. (2012), and the study concluded that mixtures including RAP could provide equal or better performance in resistance against moisture susceptibility, rutting, and fatigue failure. Taha et al. (2002) worked on cement stabilization of RAP material and concluded that cement-stabilized VA-RAP mixtures seem to be a viable alternative to dense-graded aggregate used in road base construction. Later, the studies of Senol et al. (2006) showed that Class C fly ashes have been effective in stabilizing the soft and fine-grained subgrade soils. Wen et al. (2010) conducted a series of laboratory tests to evaluate the performance of pavement materials including crushed aggregate, untreated recycled pavement material (RPM), and treated RPM with high carbon fly ash and observed that fly ash-stabilized RPM has more desirable characteristics than untreated RPM and grade 2 gravel, in terms of CBR, M_r, and permanent deformations. Similar works were performed by Li et al. (2008) and Berthelot and Gerbrandt (2002) on fly ash for treating base course and were found to be effective.

Past research experience indicates that the usage of RAP in base courses is limited to 30 % as a replacement of VA. However, a combination of 80 % RAP and 20 % VA is proposed in this study to promote higher quantities of RAP usage in Indian low-volume roads.

Past literature discussed about the performance improvement of either VA-RAP mixes or VA-RAP mixes treated with fly ash in terms of CBR, UCS, and M_r. However, the design of base layer mixes incorporating recycled materials was not performed based on damage analysis, which uses the stress-strain behavior of the mixes.

6.3 Objective

The main objective of the study is to incorporate the damage analysis in the design of low-volume rural roads and to incorporate the base material properties in designing the thickness of the pavement; and to develop a design chart for selecting the thickness of pavement when 80R:20A mixture treated with different dosages of fly ash (10 %, 20 %, 30 %, and 40 %) is used as a base material.

6.4 Materials Used

RAP material used in this study was collected from an ongoing milling project on NH-5 at Nellore in Andhra Pradesh. The virgin aggregate was obtained from a local quarry. Fly ash was obtained from a thermal power plant, Neyveli Lignite Corporation (NLC), Tamil Nadu.

6.5 Experimental Study

The important tests to verify the performance of the pavements include CBR, unconfined compressive strength (UCS) at 28-day curing period if stabilizing compounds are used, and resilient modulus (M_r). These tests provide very important data if the design of pavements is performed according to the mechanistic-empirical design guidelines such as the one followed by AASHTO (1993).

RAP material along with small percentage of VA treated with different proportions of Class C fly ash is used to conduct the performance tests. The RAP, fly ash, and VA materials were graded according to the Ministry of Rural Development (MoRD) as well as the Ministry of Road Transport and Highways (MoRTH) specifications. Figure 6.1 shows the gradation curves for both fly ash and RAP used in the study. The gradation of RAP was modified to meet the average gradation of wet mix macadam (WMM) base as per the MoRTH specification. The average asphalt content of RAP material is about 3.8 % by weight of the mix. The 100 % RAP material possesses a CBR value of 26 %.

The compaction tests were performed on different mixtures of 80R:20A treated with different percentages of fly ash (10 %, 20 %, 30 %, and 40 %). Figure 6.2 presents the compaction characteristics of 100 % RAP and 80R:20A mixes. With an increase in fly ash content increment from 10 to 40 % with 10 % interval, the optimum moisture content (OMC) has increased from 6 to 8% and the maximum

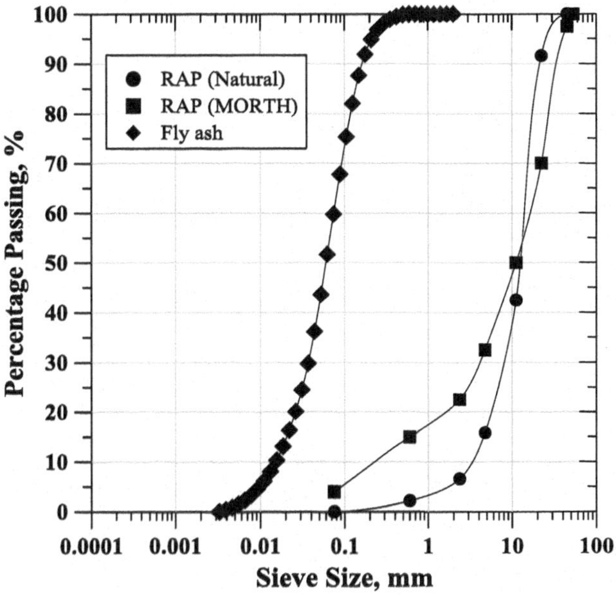

Fig. 6.1 The gradation curves of fly ash and natural/modified RAP material

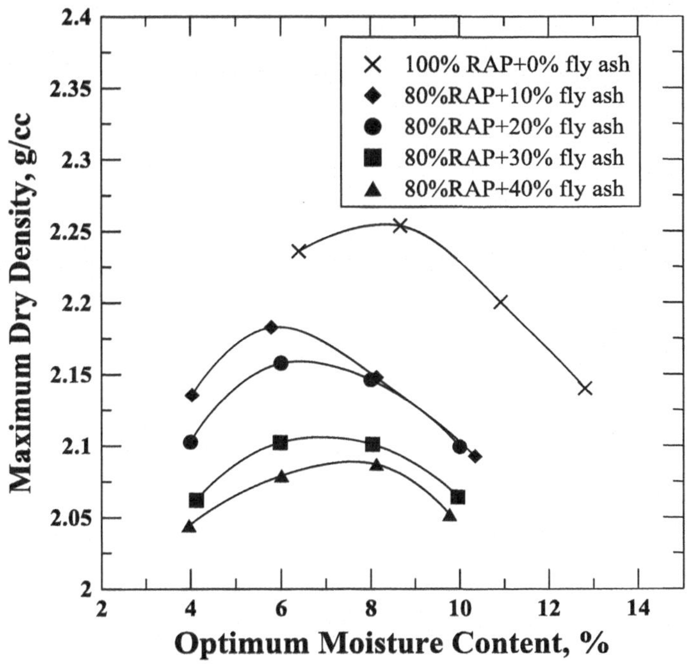

Fig. 6.2 Compaction characteristics of RAP-VA mixes with fly ash

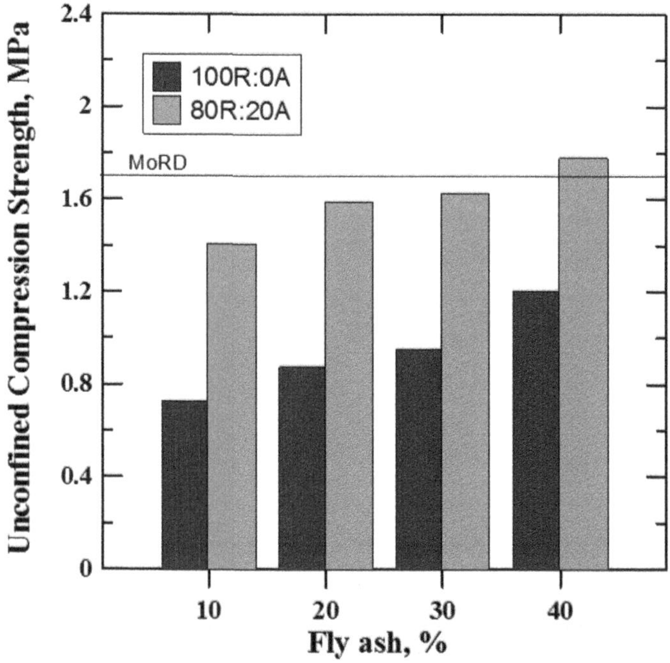

Fig. 6.3 Variation of UCS with fly ash and RAP content at 28-day curing period

dry density (MDD) has decreased from 2.18 g/cc to 2.08 g/cc. Once the compaction characteristics are obtained, the UCS and resilient modulus properties of each mix were determined. The samples were prepared to determine the UCS and resilient modulus in specially designed split cylindrical molds of size 100 mm diameter and 200 mm height in five layers, each layer giving 25 blows and then subjected to 28-day curing period. The samples were stored in a stability chamber where the relative humidity and temperature are maintained at 70 % and 20 °C, respectively.

Figure 6.3 shows the variation of UCS with fly ash dosage at 28-day curing period. The compressive strength of 100 % is also presented for comparison studies. The UC strength of 80R:20A mixes is about 40 % higher than 100 % RAP mixed with different percentages of fly ash. The UC strength of 80R:20A mix stabilized with 40 % fly ash only meets the strength requirement (1.7 MPa) proposed by MoRD and IRC-SP-72 2007 for low-volume roads. However, to perform damage analysis, the resilient modulus of these layers is a crucial information.

6.5.1 Resilient Modulus Test Procedure

The resilient modulus test provides a basic relationship between the cyclic stress and the deformation of pavement materials for the structural analysis of a layered

Table 6.1 Resilient modulus corresponding to various 80R:20A mixes

Fly ash content (%)	M_r (MPa)
VA with 0 % fly ash	130
10 % fly ash	322
20 % fly ash	363
30 % fly ash	402
40 % fly ash	526

pavement system. The laboratory test to find the resilient modulus was based on AASHTO T-307-99 (2003). A repeated axial cyclic stress of fixed magnitude, load duration (0.1 s), and cyclic duration (1.0–3.1 s) is applied to a cylindrical test specimen. During testing, the specimen is subjected to a dynamic cyclic stress and static confining stress provided by means of a triaxial pressure chamber. The total resilient axial deformation response of the specimen is measured and used to calculate the resilient modulus (AASHTO T-307-99 2003). The resilient moduli obtained from the repeated triaxial tests for different mix combinations are presented in Table 6.1. It can be seen that the M_r value increases with an increase in the fly ash content.

6.5.2 *Resilient Modulus* (M_r) *Data Interpretation*

The M_r of a material depends on the state of stress during the loading. AASHTO T-294 recommends using the bulk stress model, also known as K-θ model, to express the results of resilient modulus. The K-θ model in its general form can be expressed as

$$M_r = K_1 \theta^{K_2} \tag{6.1}$$

where $\theta = \sigma_1 + \sigma_2 + \sigma_3$ (in kPa).

In a triaxial test, $\sigma_2 = \sigma_3 =$ confining pressure in kPa.

K_1 and K_2 are the parameters obtained by fitting the test data to the model.

The parameter K_1 represents the intercept on the Y-axis, and K_2 represents the slope of the line when Eq. 6.1 is represented on log-log scale as shown in Fig. 6.4. The K_1 and K_2 parameters obtained from K-θ model for the design mix tests are given in Table 6.2.

6.6 Pavement Analysis

Pavement design is performed using the resilient modulus values obtained from the experimental results by adopting both AASHTO method and IRC method of pavement design. The AASHTO method is based on the modulus of resilience of the pavement materials, which is obtained by conducting repeated loading triaxial tests.

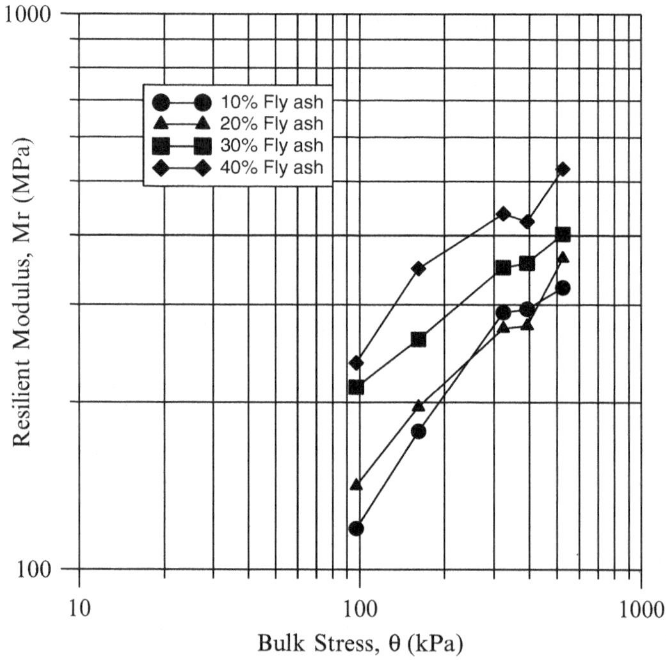

Fig. 6.4 The variation of resilient modulus with bulk stress to find K_1 and K_2 parameters

Table 6.2 K_1 and K_2 for various mixes of 80R:20 V

Fly ash content (%)	K_1 (MPa)	K_2
VA*	44.85	0.55
10 % fly ash	95.288	0.4841
20 % fly ash	104.85	0.4825
30 % fly ash	184.46	0.4392
40 % fly ash	213.62	0.6018

**Note: K_1 and K_2 values of virgin aggregates are taken from Kalcheff and Hicks* (1973)

The design life of the pavement is then calculated using the KENLAYER software which is based on AASHTO method. Several studies done by Suh et al. (2011) and Gedafa (2006) show that KENLAYER can be successfully used to predict the stresses, strains, and permanent deformations developed in the pavement layers. KENLAYER performs a nonlinear analysis of the pavement layers by incorporating the variation of M_r value with respect to the depth of the pavement.

IRC method of LVR design is based on the cumulative traffic ESALs and subgrade CBR value obtained from the laboratory or plate bearing tests. However, the design can be verified through IITPAVE software which will provide the stresses and strains within the pavement layers. Hence, the fatigue strains and rutting strains obtained from the AASHTO method are then compared with the corresponding strains obtained from the IRC method by using IITPAVE software.

6.7 Failure Criteria for Pavements

Strains due to fatigue cracking and rutting have been considered most critical for the design of bituminous pavements. A horizontal tensile strain ε_t developed at the bottom of the bituminous layer causes fatigue cracking, and vertical compressive strain ε_c on the surface of subgrade causes a permanent deformation or rutting. The life of a new pavement can be predicted using distress models by assuming pavement configuration.

6.7.1 Fatigue Cracking Criteria

Cumulative damage concept developed by Miner (1945) has been widely used to predict the fatigue cracking. The amount of damage is expressed as a damage ratio, which is expressed as the ratio between the predicted and allowable number of load repetitions. Damage occurs when the sum of damage ratio reaches 1.

The allowable number of load repetitions (N_f), to prevent fatigue cracking, can be computed by

$$Nf = f1(\varepsilon t)^{-f2}(E1)^{-f3} \tag{6.2}$$

where v_t is the tensile strain at the bottom of the bituminous layer,
E is the modulus of elasticity of the bituminous layer, and
f, f_2, and f_3 are fatigue coefficients given in Table 6.3.

6.7.2 Rutting Criteria

Rutting models are used to limit the vertical compressive strain on the top of the subgrade. The allowable number of load repetitions (N_d) to limit rutting is related to the vertical compressive strain (ε_c) on top of the subgrade by

$$N_d = f_4(\varepsilon_c)^{-f_5} \tag{6.3}$$

where f_4 and f_5 are the rutting coefficients given in Table 6.3.

Table 6.3 Coefficients for rutting and cracking models

Coefficients	Asphalt institute	IRC-37-2012
f_1	0.0796	$2.2*10^{-4}$
f_2	3.291	3.89
f_3	0.854	0.854
f_4	$1.365*10^{-9}$	$0.4166*10^{-5}$
f_5	4.477	4.534

6.8 Damage Analysis Using KENLAYER

KENLAYER is a computer program developed at the University of Kentucky and is used for the solution of an elastic multilayered system under a circular loaded area as shown in the schematic (Fig. 6.5). Its calculation principle is based on the Burmister's multilayered elastic theory similar to other programs based on the analytical method. Solutions are superimposed for multiple wheels like dual or dual-tandem wheels. The superiority of KENLAYER over the other elastic layer programs is its capability of solving systems either linear elastic, nonlinear elastic, or viscoelastic (2004). Damage analysis is performed using KENLAYER to evaluate the design life of a pavement. KENLAYER uses distress models to find the damage ratio of the pavements. Fatigue cracking and rutting models, as discussed earlier, are the major two distress models used in the KENLAYER.

6.9 Pavement Design

6.9.1 Design Methodology

In KENLAYER software, a nonlinear resilient model is defined in the input phase, to calculate the stresses and strains at specified points. The calculation procedure is as follows:

- Dividing the granular layer into several sub-layers (Fig. 6.5)
- Calculating the stresses at the midheight of each sub-layer
- Application of nonlinear model defined in input phase

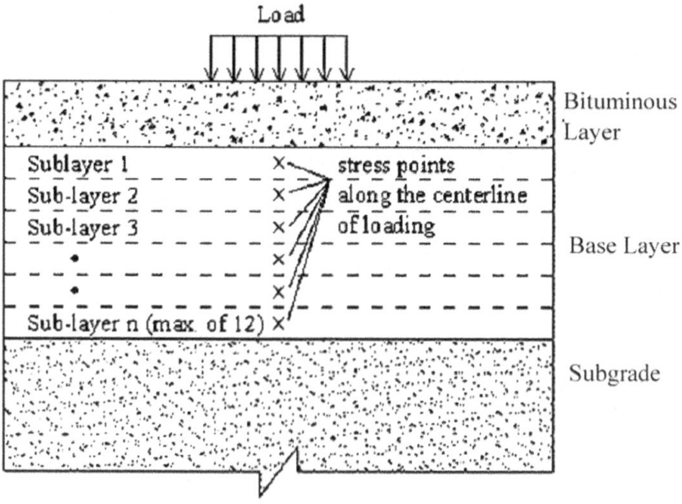

Fig. 6.5 Schematic showing principle of simplified iterative approach used in KENLAYER

Table 6.4 Properties of bituminous layer and subgrade of the pavement

Properties	Bituminous layer	Subgrade
Unit weight	22	18
Poisson's ratio	0.35	0.4
Modulus of resilience (MPa)	3000	77

Table 6.5 Parameters used for KENLAYER analysis

Parameters	
Type of base material	Nonlinear
Total number of layers	7
Total number of nonlinear layers	5
Contact pressure	550 kPa
Contact radius	15 cm
Cumulative traffic ESAL	200,000

- Calculation of new modulus values for each sub-layer
- Iterative process until the values of elastic modulus of each layer converge

6.9.2 Design Parameters

Pavement design analysis using fly ash-stabilized RAP-VA blend was considered. A typical rural pavement consists of a bituminous surface, a base layer, and natural subgrade. The surface and the subgrade properties were maintained constant in the analysis and are given in Table 6.4. Only the base material properties were varied. The addition of fly ash to RAP has a more pronounced effect on the resilient modulus than with the addition of only VA. The parameters used for KENLAYER analysis are given in Table 6.5.

6.10 Pavement Analysis Using IITPAVE

IITPAVE is a pavement design software developed under the Research Scheme R-56 for layered system analysis. From IRC-SP-72 (2007), the design thickness of the pavement is found corresponding to a subgrade CBR of 6 % and cumulative traffic ESAL of 2 lakhs. The obtained pavement layer thicknesses are given in Table 6.6. The layer thickness along with their material properties and wheel loads was incorporated in the IITPAVE to obtain the stresses, strains, and deformations in the pavement layers.

Table 6.6 Pavement layer thickness obtained from IRC-SP-72 2007 for a subgrade CBR = 6 and cumulative ESAL = 2 lakhs

Layers	Thickness (mm)
Bituminous	75
Base/subbase	250
Subgrade	–

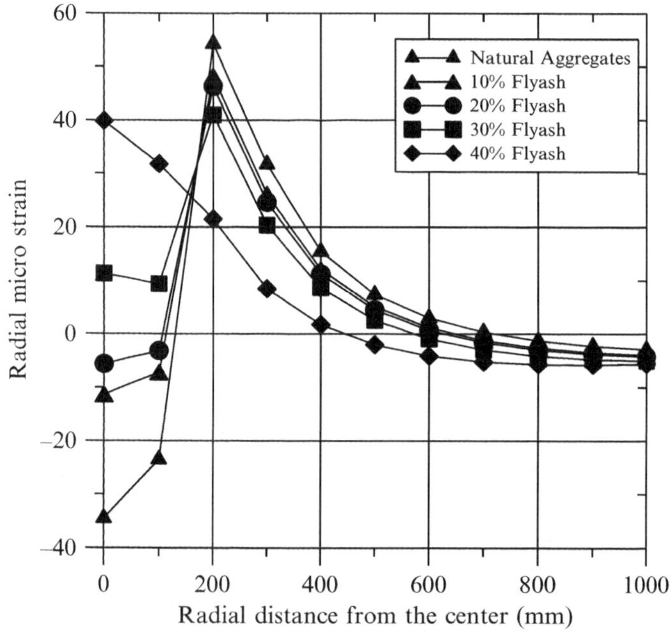

Fig. 6.6 Variation of radial microstrain along radial direction at the bottom of bituminous layer

6.11 Results and Discussions

A parametric study is performed to find the optimum thickness of the pavement. The critical responses of the pavement, i.e., fatigue strain, rutting strain, and permanent deformations, were calculated using KENLAYER software and are then compared with the data obtained from IITPAVE software. The sign convention adopted is positive in compression and negative in tension.

The variation of radial micro strain along the radial direction at the bottom of the bituminous layer is shown in Fig. 6.6. The radial microstrain developed is higher when natural aggregate is used as a base material and the value got reduced when the base material is replaced with fly ash-treated RAP material. At 40 % fly ash

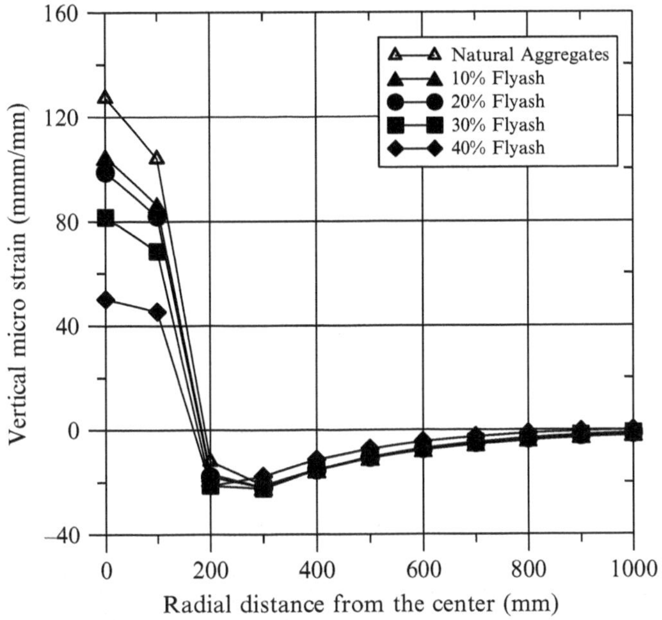

Fig. 6.7 Variation of vertical microstrain along radial direction at the top of subgrade

content, the radial microstrain decreases from 39.81 μmm/mm to −5.61 μmm/mm along the radial direction under the loading area.

Figure 6.7 shows vertical microstrain variation along the radial direction at the top of the subgrade. A considerable reduction in the vertical microstrain can be seen near the center of the loading area by the usage of fly ash-treated RAP as base material, and it goes on reducing with an increase in the radial distance from centre of the loading. The vertical strains are observed to be minimum for 40 % fly ash-treated 80R:20A mixture. The reduction in the base layer thickness due to the replacement of natural aggregate with fly ash-treated RAP material is shown in Fig. 6.8. A reduction in the base thickness of about 50 %, i.e., from 31.5 cm to 15.5 cm, can be observed. As the natural aggregate is replaced by a mixture of 80R:20A with 40 % fly ash.

Figure 6.9 shows the comparison plot of the thickness obtained from IRC (2012) and AASHTO (1993) methods. It can be seen that the thickness obtained from IRC method is constant irrespective of the fly ash content or base material of the pavement as the design is completely based on the subgrade CBR and the cumulative traffic ESAL, whereas the thickness of the base layer reduced with an increase in the percentage of fly ash in AASHTO method as this method considers the material properties like M_r of the base layer.

As the design thickness of the pavement layers obtained using IRC method is constant irrespective of the base material used and the design thickness obtained using AASHTO method is varying with respect to the base material, it will be

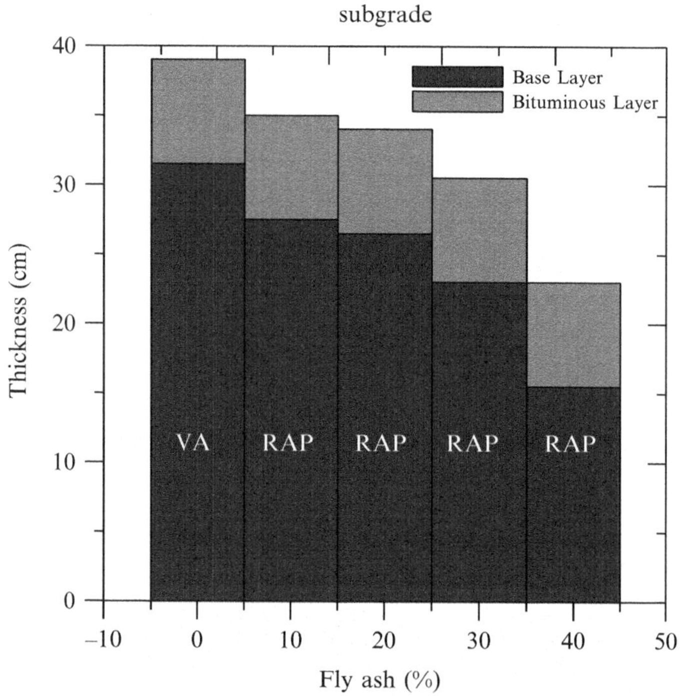

Fig. 6.8 Variation of pavement thickness with increase in fly ash content

difficult to compare the strains obtained from both the analyses. Hence, in this study, the fatigue and rutting strains obtained from these analyses are compared by considering two cases:

(a) The thickness of the pavement layers is found using IRC method (shown as IITPAVE (a) in Figs. 6.10 and 6.11) and AASHTO method (shown as KENLAYER in Figs. 6.10 and 6.11), and they are incorporated in the IITPAVE and KENLAYER software, respectively, to obtain the fatigue and rutting strains at the corresponding critical locations.
(b) The thickness of the pavement layer is kept constant for each fly ash combination, and then the strains developed at the critical locations which are calculated using IITPAVE (shown as IITPAVE (b) in Figs. 6.10 and 6.11) and KENLAYER (shown as KENLAYER in Figs. 6.10 and 6.11) software are compared.

The comparison plot for fatigue strain obtained from the analysis is shown in Fig. 6.10. For case (a), it can be observed that the fatigue strain obtained from the IITPAVE is considerably larger as compared to that obtained from KENLAYER. That is, IITPAVE is overestimating the fatigue strain which is measured at the bottom of bituminous layer. In KENLAYER, the fatigue strain is reduced from

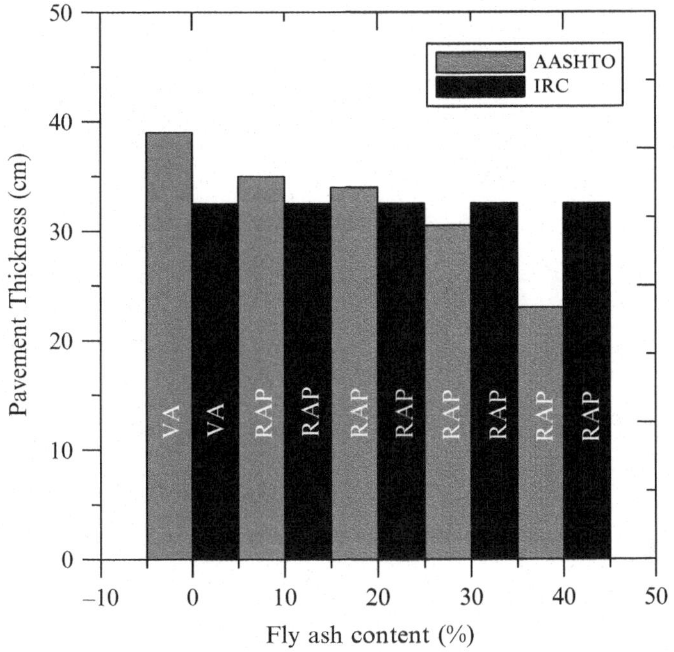

Fig. 6.9 Comparison of pavement design thickness obtained from IRC and AASHTO methods

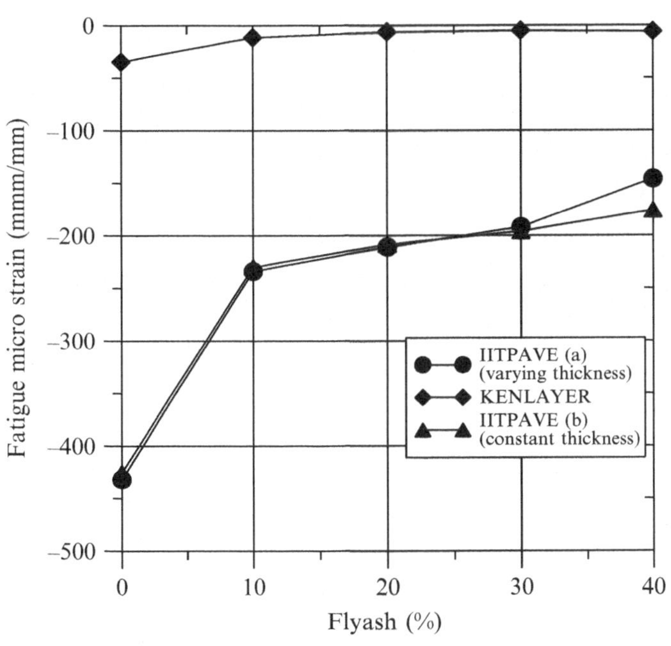

Fig. 6.10 Comparison of fatigue strain obtained from IITPAVE and KENLAYER

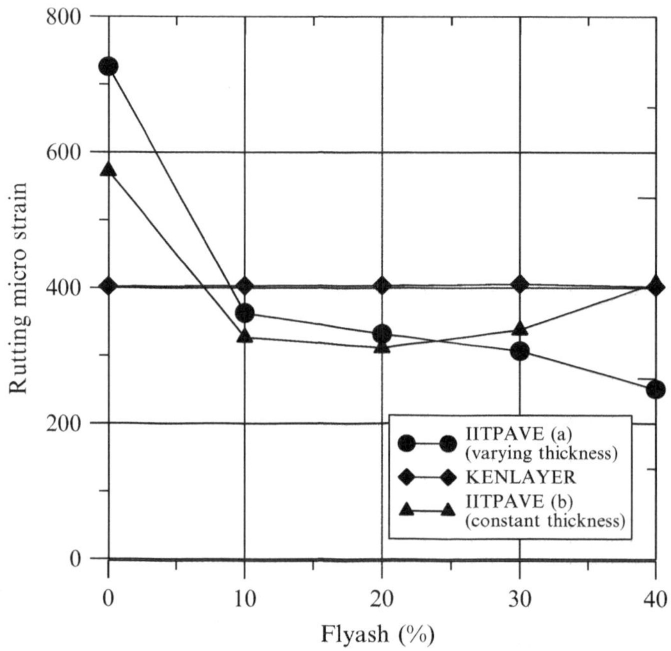

Fig. 6.11 Comparison of rutting strain obtained from IITPAVE and KENLAYER

−34.54 μmm/mm for natural aggregates to −5.78 μmm/mm for 40 % fly ash base material. For case (b), a similar plot as that of case (a) is obtained with slight increase in fatigue strain at a combination of 40 % fly ash. Hence, it can be inferred that the fatigue strain is independent of the thickness of base layer as it is measured beneath the bituminous layer.

Similarly, the comparison plot for rutting strain is given in Fig. 6.11. For case (a), it can be observed that the rutting strain obtained from IITPAVE software shows considerable variation from 726 μmm/mm to 251 μmm/mm when the natural aggregate is replaced with 40 % fly ash containing base material. On the other hand, KENLAYER is giving a constant average value of 403 μmm/mm as the rutting microstrain irrespective of the base material. The constant strain is because of the reduction in base layer thickness due to the replacement of natural aggregate with fly ash-treated RAP material. Case (b) also follows the trend similar to that of case (a).

6.12 Conclusions

Based on the damage analysis performed on natural aggregates and fly ash-treated RAP aggregates based on AASHTO and IRC methods of pavement designs, the following conclusions are drawn:

- Experimental results demonstrated that 80R:20A mix stabilized with 40 % fly ash has shown desired unconfined compressive strength and resilient modulus properties.
- As IRC method ignores the base material properties, the thickness of the pavement layers obtained from the IITPAVE analysis will remain the same irrespective of the base material type. Though the thickness obtained will be on conservative side, the cost of construction due to overconsumption of pavement materials is quite high as the design is overestimating the pavement thickness.
- In AASHTO method of pavement design used in KENLAYER software, the base material properties are also taken into account. A considerable reduction in the thickness of the base layer, i.e., about 50 % can be observed, when a mixture of 40 % fly ash along with 80 % RAP and 20 % VA is used.
- A higher percentage of RAP material with a small percentage of VA can be used as a base material for low-volume roads if the material is stabilized with fly ash.
- The combination of 80R:20A mix with 40 % fly ash is found to be optimum as it results in the reduction of the base layer thickness by 50 %.
- The radial and vertical microstrains at the critical locations got reduced considerably when fly ash content in the RAP-VA mix is increased from 10 to 40 %.
- Overall, reclaimed asphalt pavement materials stabilized with fly ash can provide sustainable solution to the Indian rural roads when the design analysis is carried based on the resilient properties of the pavement layers.

References

AASHTO (1993) Guide for design of pavements. AASHTO, Washington, DC

AASHTO T-307-99 (2003) Standard method of test for determining the resilient modulus of the soils and aggregate materials. AASHTO, Washington, DC

Aurangzeb Q, Al-Qadi IL, Abuawad IM, Pine WJ, Trepanier JS (2012) Achieving desired volumetrics and performance for mixtures with high percentage of reclaimed asphalt pavement. J Transp Res Board 2294:34–42. Transportation Research Board of the National Academies, Washington, DC

Berthelot C, Gerbrandt R (2002) Cold in-place recycling and full-depth strengthening of clay-till subgrade soils: results with cementitious waste products in northern climates. J Transp Res Board 1787:3–12. TRB, National Research Council, Washington, DC

Gedafa DS (2006) Comparison of flexible pavement performance using KENLAYER and HDM-4. In: Fall student conference midwest transportation consortium, Kansas State University Manhattan

Huang YH (2004) Pavement analysis and design, 2nd edn. Prentice Hall, Upper Saddle River

IRC SP-72 (2007) Guidelines for the design of flexible pavements for low volume roads. Indian Roads Congress, New Delhi

IRC: 37 (2012) Guidelines for the design of flexible pavements second revision. Indian Roads Congress, New Delhi

Kalcheff IV, Hicks RG (1973) a test procedure for determining the resilient properties of granular materials. In: 76th annual meeting of the american society for testing and materials, Philadelphia, Pennsylvania, 25–29 June 1973

Lin L, Benson CH, Edil TB, Hatipoglu B (2008) Sustainable construction case history: fly ash stabilization of recycled asphalt pavement material. Geotech Geol Eng 26(2):177–188

Senol AA, Edil TB, Bin-Shafique MS, Acosta HA, Benson CH (2006) Soft subgrades stabilization by using various fly ashes. Resour Conserv Recycl 46(4):365–376

Suh YC, Cho NH, Mun S (2011) Development of mechanistic–empirical design method for an asphalt pavement rutting model using APT. Constr Build Mater 25(4):1685–1690

Taha R, Al-Harthy A, Al-Shamsi K, Muamer A-ZM (2002) Cement stabilization of reclaimed asphalt pavement aggregate for road bases and sub-bases. J Mater Civ Eng 14(3):239–245

Wen H, Warner J, Edil T, Wang G (2010) Laboratory comparison of crushed aggregate and recycled pavement material with and without high carbon fly ash. Geotech Geol Eng 28:405–411

Chapter 7
Sustainability of Railway Tracks

Sarvesh Chandra and Devanshee Shukla

Abstract The infrastructural growth of the railways has brought the railway track sustainability into a fresh perspective. Safe, stable, durable, and sustainable tracks are of paramount importance for the efficient function of the railways. In this paper, the development of the railway track from its inception to the present-day scenario has been presented. Numerous studies, conducted by researchers all over the world, that are aimed at studying the track response and improving its performance have been discussed. The analytical and numerical models developed to study the stresses and deformations in the track structure have been examined. It has been established that reinforcement of the track structure with geosynthetics reduces its settlement, increases the bearing capacity, and improves the drainage performance.

Keywords Sustainability • Railway track • Ballast-less track • Geosynthetics • High-speed trains

7.1 Introduction

Railways, since their inception, have inextricably acted as arteries establishing faster and more efficient connections between regions, paving the way for infrastructure growth and economic development. In today's fast-paced world, new and fast connections are required within any country to bring economic opportunities, growth, and prosperity to the country and its people.

The increase in passenger and freight traffic has ushered in an era of high-speed trains. The objective is for these trains to serve as an alternative to air travel. The axle loads and the speed of trains are increasing with time. To accomplish this, there is a need to build tracks that are safe, stable, durable, and sustainable. It is necessary to enhance the track structure sustainability and develop and employ technology that brings about reduced rates of deterioration and maintenance along with mitigating the environmental impact of vibrations, noise, and the materials used.

S. Chandra (✉) • D. Shukla
Department of Civil Engineering, IIT Kanpur, Kanpur 208016, India
e-mail: sarv@iitk.ac.in

G.L. Sivakumar Babu et al. (eds.), *Sustainability Issues in Civil Engineering*,
Springer Transactions in Civil and Environmental Engineering,
DOI 10.1007/978-981-10-1930-2_7

7.2 Development of the Railway Track

Since the beginning of the railways, the ballasted track structure consisting of a framework of rails and sleepers supported on ballast has been most widely in use. The ballast layer rests on a sub-ballast layer, which transmits the load over the subgrade. The principle of this track structure has evolved over time and still continues to do so in order to incorporate elements that work to increase its durability and stability and reduce the maintenance requirements and cost.

Sleepers made of timber were used for almost two centuries after the birth of the railway track (refer Fig. 7.1). The properties of wood were suited to provide a resilient track with good dynamic vibration absorption capacity. The wooden sleepers were light and easy to transport and required little or no specialist equipment for fastening and subsequent maintenance. Gauge-width control was easily possible with timber sleepers, and the damage in case of derailments was also less.

However, these sleepers were replaced over time with alternative materials due to their short service life, susceptibility to fire, decay, wear and tear, and most importantly the scarcity of wood.

Cast-iron sleepers followed timber sleepers. These were extensively used in the Indian Railways for almost five decades and provided better longitudinal and lateral stability to the track and had a longer life span of 35–50 years.

The use of cast-iron sleepers was however discontinued due to its poor damping characteristics of high frequency vibrations, unsuited to modern maintenance methods and excessive damage caused during derailment.

The nonavailability of suitable species of timber and the inherent disadvantages associated with cast-iron sleepers brought in steel as a suitable contender in the construction of sleepers. Around 30 % of the track in India is laid out on steel sleepers. Steel sleepers came with a long service life, had great dimensional accuracy, were easy to manufacture, and had low maintenance requirements.

Fig. 7.1 Ballasted track with timber sleepers

The high cost of steel was the main disadvantage of this variant of sleeper material.

The use of concrete as a sleeper material gained significance in post-Second World War era due to improvements in concrete technology, better understanding, and use of prestressing techniques. The scarcity of timber and relatively high cost of steel were also instrumental in the turn toward RCC and prestressed concrete sleepers. Today, these sleepers are most widely used throughout the world. These are relatively easy to manufacture, have long service life, and provide better freedom to design and construct.

The two main types of concrete sleepers in use are twin-block sleeper and mono-block sleeper. Twin-block sleepers consist of two blocks of reinforced concrete connected by a coupling rod or a synthetic pipe filled with reinforced concrete. These sleepers have been used on many TGV tracks in France. Mono-block sleepers are based on the shape of a beam. They are of lower cost, have little susceptibility to cracking, and can be prestressed.

The latest development in the field of railway track is use of composite sleepers. These sleepers are mainly HDPE based, and many also contain other variants of waste plastic such as PVC, polyethylene, etc. As the sleepers are made from nonbiodegradable material, they have a long service life of 40–50 years. These sleepers are considered to be better replacement of traditional sleepers as these have flexibility equivalent to that of timber sleepers and cost significantly less. The manufacture of composite sleepers ensures a proper use of waste plastic, which is otherwise creating environmental hazards. The performance of the sleepers in static load test, impact load test, and dynamic load test has been evaluated and found to be satisfactory. Thus, the Indian Railways has used these sleepers on some of its tracks.

Traditional ballasted tracks undergo deterioration due to the pressure exerted by the passing load. The amount of track deterioration is directly proportional to the amount of tonnage passing and its speed over the track. This makes regular maintenance necessary. To address the drawbacks associated with the traditional ballasted track structure, new developments are being introduced to improve the efficiency and reduce the maintenance requirements. Some such modified designs are wide sleepers and frame sleepers.

Wide sleeper track superstructures were first attempted in China, former Czechoslovakia, and former East Germany (refer Fig. 7.2). It was Germany, however, that perfected the design and construction of the wide sleeper track.

The wide sleeper is 2.40 m long and 57 cm wide with sleeper weight of 560 kg, which is double the weight of the standard sleeper. With an axle load of 22.5 tonnes, the average surface pressure works out to be 2 kg/cm^2, which is nearly half of the average surface pressure in case of the standard sleeper. This results in reduced pressure on the track and homogeneous distribution of pressure onto the ballast. Due to the high supporting area, there is less strain on the subsoil and substructure along with an increase in the sideways' stability of the track. This results in lesser maintenance frequency and costs. The construction costs for such tracks are

Fig. 7.2 Ballasted track with wide sleepers

10–20 % higher than for normal tracks, but these additional costs can be compensated in the medium term through lower maintenance cost requirements.

Frame-sleeper structures have been developed and implemented mainly in Austria. The frame sleeper consists of cross sleepers placed at regular distances, combining a continuous longitudinal beam with cross members. These sleepers replace the load-transmitting structure of a traditional ballast track by a girder grid system. The wheel load is transmitted onto the ballast bed in a continuous manner, thereby reducing the pressure under the sleepers significantly. These sleepers offer very high lateral resistance and frame stiffness and yield reduced settlements, thus making the track more durable.

The traditional ballasted track systems, when used in high-speed operations, have been found to be maintenance intensive. Since maintenance operations are becoming more and more difficult to carry out due to increased freight and passenger traffic, the conventional tracks pose a serious drawback. Ballast-less tracks or slab tracks have much lower maintenance requirements and provide more stability, passenger comfort, and durability and are the preferred option for light rail as well as high-speed lines.

Slab tracks have numerous advantages over ballasted tracks. The need for maintenance in these tracks is minimal. They have a long service life of 40–50 years. They provide high lateral and longitudinal stability, thereby reducing the deviations in track alignments. Slab tracks are highly suited for locations such as tunnels and viaducts where the acceptable settlement criteria are very stringent. In India, slab tracks are used by DMRC and in recently completed Pir Panjal Railway Tunnel in J&K line (Fig. 7.3).

The construction costs of slab tracks are considerably higher than that of ballasted tracks. However, this factor does not pose a major drawback as the investment mindset is now tilting from focusing solely on initial investment costs to adopting the robust principle of life cycle costing. This principle makes comparative cost assessments over a period of time, taking into account the initial capital costs as well as future operational and asset replacement costs.

With India taking significant steps in the direction of establishing high-speed railway (HSR) lines, the transition from ballasted to ballast-less tracks is inevitable.

Fig. 7.3 View of a slab track in Germany

However, the railway track network in India is very extensive, and it is not feasible to convert the entire existing ballasted track network to a slab track network. Therefore, it is imperative that technology capable of making the existing tracks sustainable, durable, and better suited to faster trains is explored and focused upon.

7.3 Development of Numerical Models to Study Track Sustainability

The structure of the railway track can be divided in two parts – the superstructure and the substructure. The track superstructure consists of the rail, sleeper, and fastenings, while the ballast, sub-ballast, and the subgrade constitute the substructure. It was realized that the sustainability of the track structure depends mainly on the performance of the substructural components. Thus, over time many researchers have attempted to develop reliable numerical models of the track structure to study the stresses and deformations that the track structure is subjected to due to applied load. Some of these models have been discussed in the subsequent sections.

Mathews (1958) analyzed the dynamical problem of vibrations of a beam resting on an elastic foundation. The beam was assumed to be uniform and of infinite length. It was subjected to an alternating load whose point of application was considered to move along the beam at a constant velocity. The author investigated the nature of vibrations in the system when damping was absent and defined the critical parameter characteristic of the system. The general mathematical solution to the problem of a beam subjected to a force whose point of application moved along the beam was obtained through this study.

Fryba (1972) studied the response of an infinite elastic medium subjected to a moving load by considering all possible values of the velocities of the moving load.

Viscous damping was considered to be present in the system. The solution was obtained using the technique of triple Fourier integral transformation. The study identified a critical speed for the moving load by determining the equivalent stiffness of the supporting structures. It was shown that the deflections of the beam increased asymptotically at the critical speed.

Duffy (1990) studied the response of an infinite railroad track when subjected to a moving and vibrating mass. The railroad track was modeled as an elastic beam lying on a Winkler foundation, and the vibrations caused by the passing of a moving, vibrating load were analyzed. Duffy (1990) obtained solutions for moving as well as stationary vibrating loads as a function of three parameters, namely, the mass of the load, its driving frequency, and the physical properties of the track.

It was concluded that for a stationary vibrating load, resonance occurred at lower frequencies when the mass of the load was increased, whereas for a moving vibrating load, resonance occurred at lower frequencies when the mass and velocity of the load were increased.

A number of multilayer track models were developed since 1970s to analyze and determine the stresses and deformations in the track structure.

Chang et al. (1980) developed one such model named GEOTRACK, which was an elastic, multilayer, three-dimensional model capable of analyzing stresses and strains in all the major track components, such as rails, sleepers, ballast, sub-ballast, and subgrade. The impact of factors like applied loads; thickness and properties of the ballast, sub-ballast, and subgrade; and track geometry, on the response of the track structure, was evaluated through this model.

Selig and Waters (1994) used the GEOTRACK model and performed an extensive parametric study of deviator stress present in the subgrade. The study established that the deviator stress in the subgrade was the most influenced parameter affected by the granular layer thickness and resilient moduli of the granular layers and the subgrade. It was shown that the deviator stress decreased with an increase in thickness and resilient modulus of the granular layer (Fig. 7.4).

Shahu et al. (1999) developed a three-dimensional linear elastic finite element model to investigate the effect of various track parameters, such as major principal

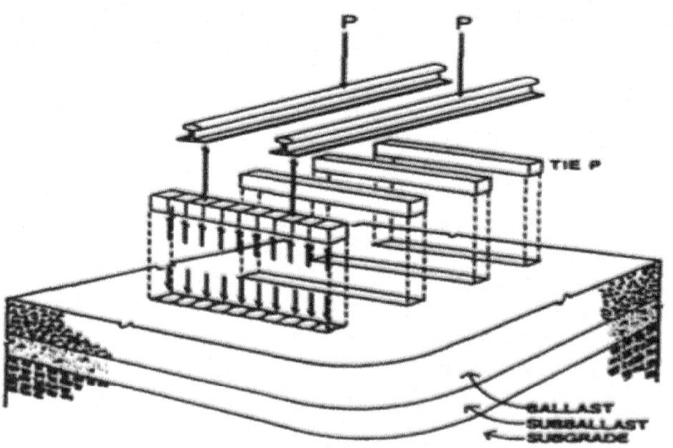

Fig. 7.4 Schematic representation of track forces and elements in GEOTRACK model (Chang et al. 1980)

stresses, deviatoric stresses at subgrade level, displacement of the sleeper, and track modulus, on the overall track response.

The model adopted in the study was a 3D20N element that used 20-noded isoparametric brick elements for various track components, one-dimensional beam elements for the rail elements, and 16-noded zero-thickness surface elements to simulate the interface between different layers. The authors, using a practical range of industry applicable track variables, also performed a detailed parametric study of the track response.

The study concluded that the most influential track response parameter was the subgrade modulus. The sub-ballast depth, rail moment of inertia, and tie spacing were found to be next most important factors that influenced the track responses (Fig. 7.5).

Shahu et al. (2000) developed a rational design method to determine the thickness of the formation of a railroad track (refer Fig. 7.6). The design approach adopted was based on keeping the maximum deviator stress induced in the subgrade layer below the threshold value of stress of the subgrade soil by providing a suitable thickness of the track formation.

The method developed by the authors used recent developments in the evaluation and prediction of threshold stresses and three-dimensional finite element modeling of the railway track to predict the stresses induced in the system. This

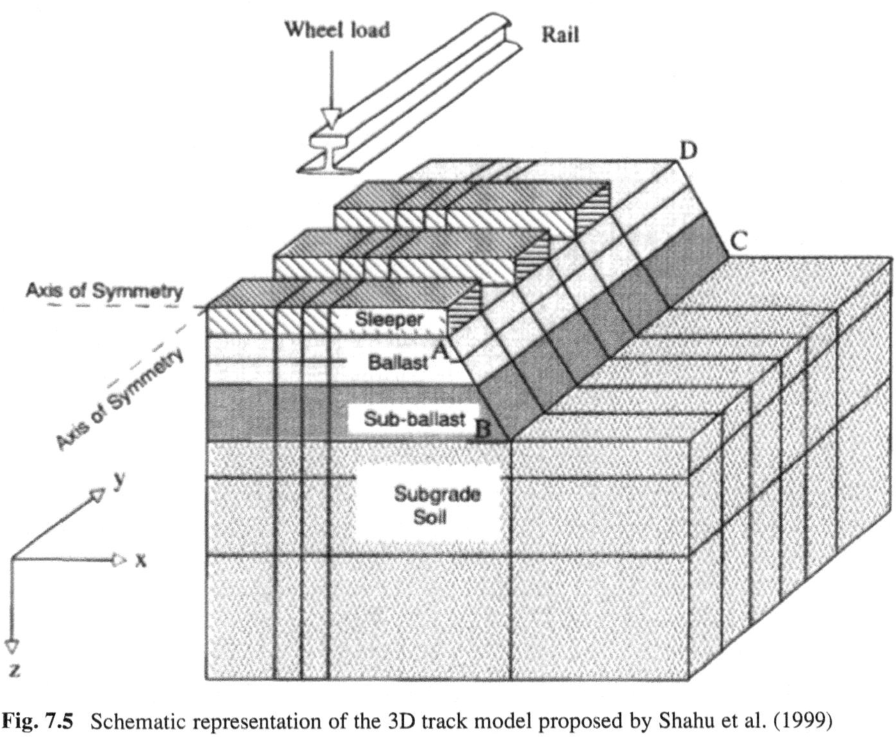

Fig. 7.5 Schematic representation of the 3D track model proposed by Shahu et al. (1999)

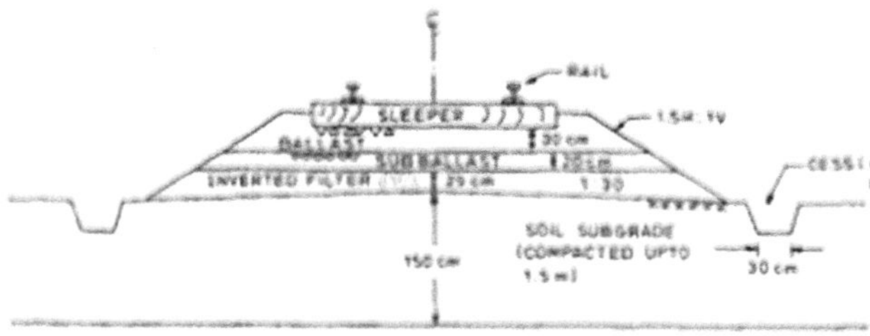

Fig. 7.6 Schematic representation of the railway formation proposed by Shahu et al. (2000)

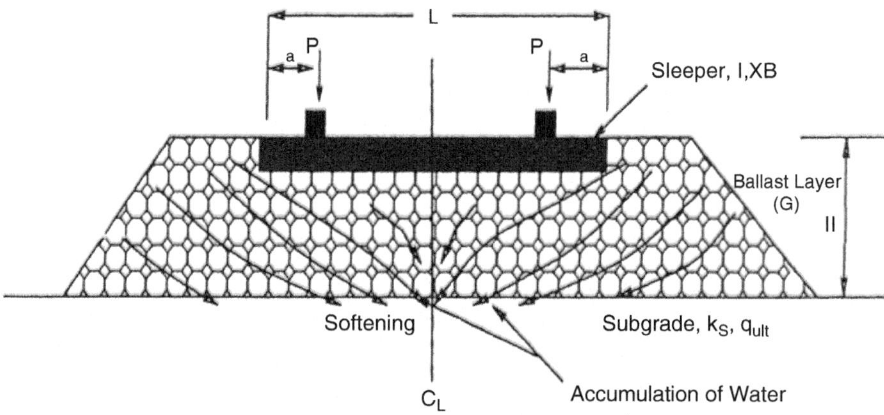

Fig. 7.7 Schematic representation of railway sleeper-ballast-subgrade system modeled by Ghosh (2001)

method, unlike previously existing methods, was applicable to various soil types and different properties of the ballast and sub-ballast layers and was found to be highly suited in cases of high-speed traffic and heavy axle loads.

The study provided a simple procedure for quick evaluation of the resilient modulus and threshold stress in the subgrade soil. The performance of an actual track in the field was studied to develop this method.

Ghosh (2001) studied the effect of softening of the subgrade, due to percolating rainwater, on the response of the railway track. It was known that 95 % of failure or maintenance problems in tracks were due to the poor bearing capacity of the subgrade and gave rise to ballast penetration and mud pumping which affected the track geometry.

The railway sleeper-ballast-subgrade system was idealized as a two-parameter Pasternak model (refer Fig. 7.7). The effect of subgrade softening on the modulus of subgrade reaction was defined as a linear and nonlinear variation. The solution was obtained using a finite difference solution scheme.

The study concluded that subgrade softening affected both deflection and bending moment in the railway sleeper, with maximum deflection being observed at the location of maximum softening, i.e., at the center of the sleeper. The remedial design solution proposed was to introduce a geosynthetic layer at the interface of the sub-ballast and subgrade.

Singh (2002) studied the steady-state response of a uniform beam placed on an elastic foundation and subjected to concentrated moving load at a constant speed.

The authors idealized the system as an infinite Euler-Bernoulli beam of constant cross section resting on an elastic foundation modeled using both one- and two-parameter models. The response of the foundation was attained analytically to get closed-form solution for different cases of speed of moving load and damping present in the beam.

The study presented numerical results of the system at subcritical and supercritical speed in terms of its responses such as beam deflection, bending moment, and shear force in the system for different load speeds, damping, and foundation modeling.

Maheshwari (2004) carried out an extensive study to determine the response of beams resting on reinforced granular soil systems and subjected to static and moving loads. The study was aimed at developing a generalized model applicable to railroad tracks, foundations, etc.

In the first model developed by the author, the foundation was analyzed as a beam of finite length subjected to a static load (refer Fig. 7.8). The geosynthetic reinforcement was idealized as another beam in order to incorporate its bending stiffness. The two beams were considered to be of the same length. The governing equations for the model were solved using appropriate boundary and continuity conditions, and a closed-form analytical solution was obtained for the response of the model. The analysis concluded that the placement of the reinforcement, relative stiffness of the foundation soil, and relative flexural rigidity of the beam affected the model response significantly.

The author developed two more models with the foundation modeled as an infinitely long beam subjected to a load moving at a constant velocity (refer Fig. 7.9). Two conditions were studied in the above setup – first, the foundation beam was assumed to be in contact with the ground surface, and second, the

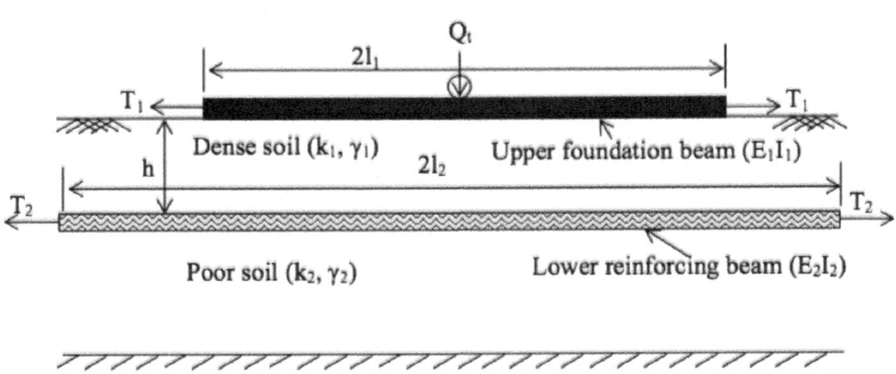

Fig. 7.8 Schematic representation of the two-beam model proposed by Maheshwari (2004)

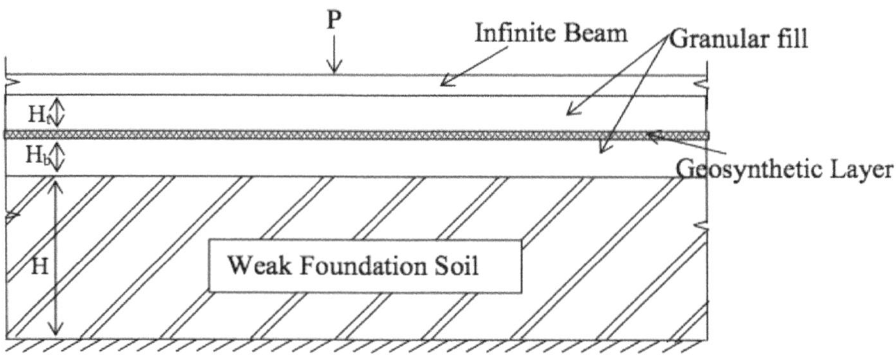

Fig. 7.9 Schematic representation of the model proposed by Maheshwari (2004)

separation between the beam and ground surface was considered. The geosynthetic was modeled as a rough elastic membrane, inextensible in nature in one model, and as an extensible membrane with tensile stiffness, in the other model. The analyses of both models were carried out by the finite difference solution scheme. The responses of both models were found to be significantly affected by the intensity of the applied load, velocity of the load, and relative compressibility of the granular fill.

Vihari (2005) proposed a new design approach to determine the thickness of the granular layer in a railway track, which took into account the effect of shear strength parameters of the foundation soil and the foundation coefficient. The design approach adopted was stress controlled which focused on the strength parameters.

The author adopted a design method that considered the mode of failure of the subgrade soil to be rotational failure about a point just below the end of the sleeper (refer Fig 7.10). A C code was developed to determine the factor of safety of the track and soil, which was then used to evaluate the thickness of the granular soil layer.

The study established that the factor of safety was linearly related to the shear strength parameters of the foundation soil and inversely related to the speed of the train. The depth of the granular fill was found to be inversely related to the shear strength parameters of the soil and highly influenced by the foundation coefficient.

Chaudhari (2012) performed a study to formulate the settlement response in a railway track structure reinforced with a geosynthetic layer and compare it with the unreinforced track model.

The ballast and sub-ballast were modeled as Pasternak shear layers with a prestressed geosynthetic placed at the interface of the two layers. The subbase was idealized as a layer of Winkler springs, and the subgrade was modeled as a viscoelastic medium represented by Burger elements (refer Fig. 7.11). A uniformly distributed load was applied to the system, and the governing equations obtained for the settlement response were solved using a finite difference scheme.

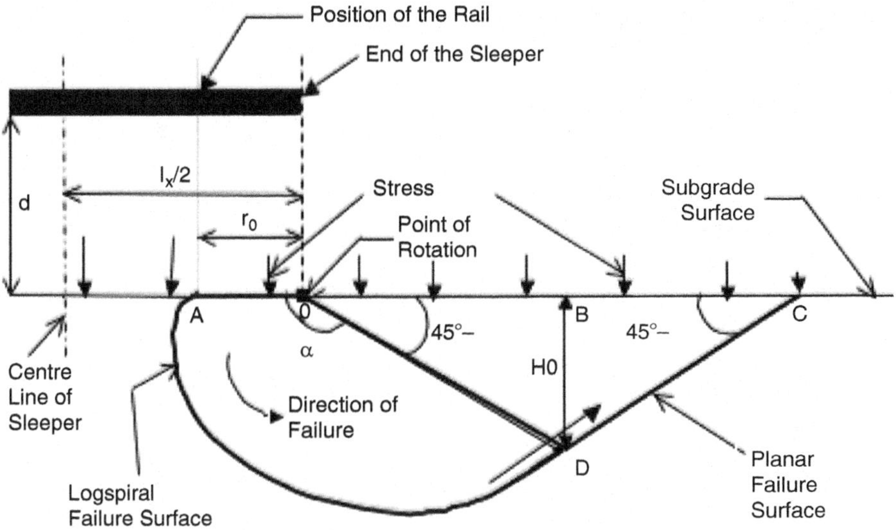

Fig. 7.10 Schematic diagram of failure mode in the subgrade soil proposed by Vihari (2005)

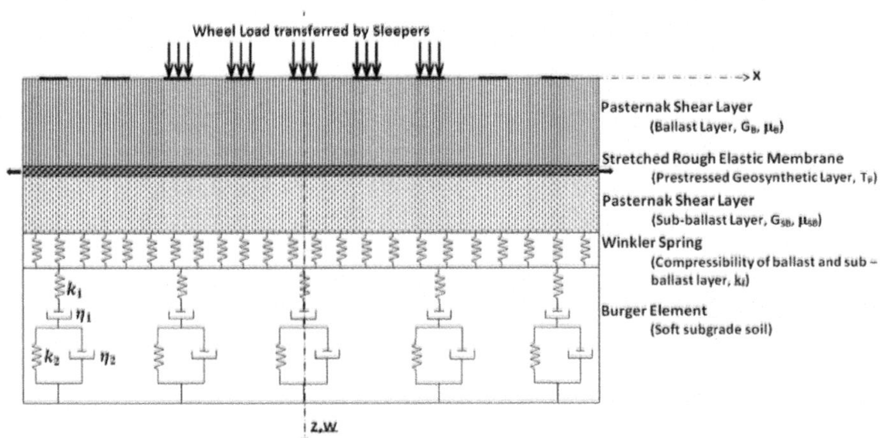

Fig. 7.11 Schematic representation of the model proposed by Chaudhary (2012) for railway track modeling

The study concluded that settlement depended on a number of factors, namely, the presence or absence of reinforcement, the thickness of the ballast and sub-ballast layers, and the shear moduli of the two layers. The presence of geosynthetic reinforcements, greater thickness, and higher shear modulus of the ballast and sub-ballast resulted in lower settlements in the track structure upon application of loads.

Prasad and Chandra (2014) carried out a study to investigate the response of a railway track when subjected to a moving load. The track was modeled as an Euler-Bernoulli beam resting on a two-parameter foundation model with the sub-ballast and subgrade being represented by two layers of Winkler springs (refer Fig. 7.12). The finite difference solution scheme was used to determine the uplift, settlement, and bending moment caused due to variations in parameters such as velocity of the moving load, damping, and stiffness of the subgrade.

It was found that with increase in stiffness of the subgrade layer, the point of maximum deflection shifted behind the applied load. Changes in the stiffness of the subgrade layer were found to have an insignificant effect on the behavior of the beam. The results were found to be consistent with those obtained by Mallik et al. (2006).

Kumar (2013) conducted one of the most recent analytical studies on the railway track structure. The author developed a new model to study the behavior of a traditional ballasted railway track, reinforced with a geosynthetic layer and subjected to a moving load at a constant velocity (refer Fig. 7.13). The moving load was due to a single axle load moving at a constant velocity. The track substructure included the ballast, sub-ballast, and subgrade layer. The rail was idealized as an Euler-Bernoulli beam, while the ballast and sub-ballast were modeled as Pasternak shear layers resting on the subgrade, which was represented as a series of Winkler springs.

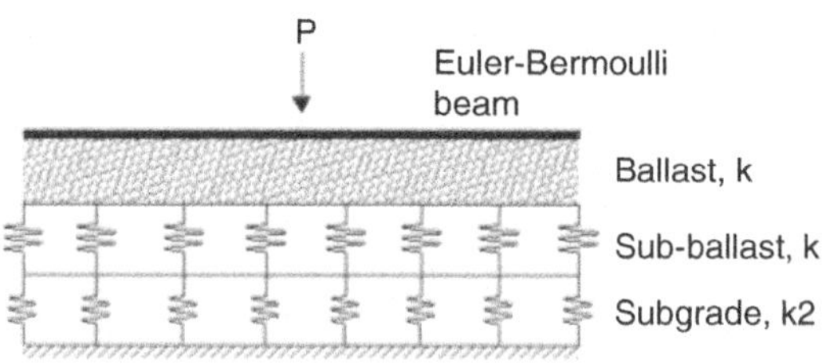

Fig. 7.12 Schematic representation of the model developed by Prasad and Chandra (2014)

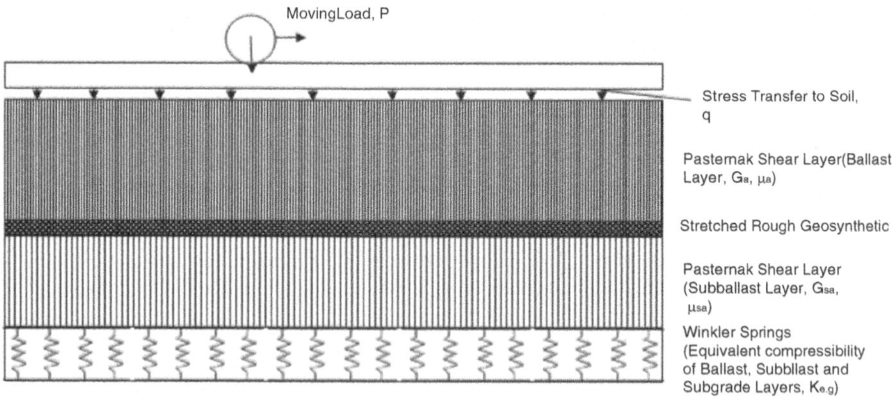

Fig. 7.13 Schematic view of the model proposed by Kumar (2013)

Two governing differential equations were obtained for the system. The first equation was the beam equation modified according to the reinforced model, while the second equation dealt with the development of tension in the geosynthetic membrane. The two equations were solved simultaneously by the finite difference solution scheme using an iterative approach.

The author studied the effects of variation of different parameters on the track system. The system was analyzed with and without a geosynthetic layer. When the geosynthetic was taken into consideration, it was analyzed with and without pretension, and the corresponding results and track responses were compared. The results obtained were validated with those obtained by Singh (2002).

For the parametric study undertaken by the author, axle loads varying from 20 tonnes to 200 tonnes were considered. The range of values was chosen to be wide because even though currently the axle load on Indian railway tracks is approximately 30 tonnes, with the advent of high-speed trains, it is likely to increase up to 200 tonnes and more. Thus, the track responses for high-speed scenarios were also taken into consideration.

The study concluded that the nondimensional displacement varied linearly with the increase in the load intensity for the unreinforced as well as the reinforced case. However, in the case of the track reinforced with a pretensioned geosynthetic, the settlement was observed to be 5–10 % less than that in the unreinforced and without pretensioned cases. It was found that the reinforcement had a more pronounced effect for lower thickness of the ballast and sub-ballast layers, lower values of shear moduli for the ballast and sub-ballast, and higher load values.

Through the author's analysis, it was established that greater pretension in the membrane results in lower values of the nondimensional displacement in the model. It was also shown that application of pretension helps make the design more economical by reducing the required depth of the ballast and sub-ballast. Thus, the study proposed a feasible and relatively economical means of making the existing tracks more suited to high-speed trains and increasing their sustainability in the long run.

7.4 Conclusions

In this paper, the evolution of the railway track structure, from its beginning up to the present-day developments, has been presented. The efforts of various researchers to develop numerical and analytical models that predict the behavior of the railway track system under external loads have been discussed. The studies indicate that the performance of the railway tracks exhibits a substantial improvement, with respect to reduced settlements, increased bearing capacity, and improved drainage, when reinforced with geosynthetics. The simplicity associated with the modeling and analysis of geosynthetics and their economic viability makes them a suitable option. The scope of such studies is expanding, and more efforts are being directed to find viable solutions that increase the track performance and its sustainability.

References

Chang CS, Adegoke CW, Selig ET (1980) GEOTRACK model for railroad track performance. J Geotech Eng Div ASCE 106(GT11):1201–1218

Chaudhary K (2012) Modelling of time dependent response of unreinforced and reinforced railway tracks, M.Tech thesis, Department of Civil Engineering, Indian Institute of Technology, Kanpur, India, 2012

Duffy DG (1990) The response of an infinite railway track to a moving vibration mass. J Appl Mech ASME 57:66–73

Fryba L (1972) Vibration of solids and structures under moving loads. Noordhoff, Groningen

Ghosh R (2001) Analysis of subgrade softening extended ballast/granular layer and reinforcement effects on sleeper/combined footing response, M.Tech thesis, Department of Civil Engineering, Indian Institute of Technology, Kanpur, India, 2001

Kumar S (2013) Analysis of reinforced railway track under moving load at high speed, M.Tech thesis, Department of Civil Engineering, Indian Institute of Technology, Kanpur, India, 2013

Maheshwari P (2004) Response of reinforced granular bed soft soil system to static and moving loads, PhD thesis, Department of Civil Engineering, Indian Institute of Technology, Kanpur, India, 2004

Mallik AK, Chandra S, Singh AB (2006) Steady state response of an elastically supported infinite beam to a moving load. J Sound Vib 291:1148–1169

Mathews PM (1958) Vibrations of a beam on elastic foundation. J Appl Math Mech 38:105–115

Prasad N, Chandra S (2014) Moving load on a railway track- finite difference solution. Int J Civ Struct Eng 1(4):1–4

Selig ET, Waters JM (1994) Track geotechnology and substructure management. Thomas Telford, London

Shahu JT, Kameswara Rao NSV, Yudhbir (1999) Parametric study of resilient response of tracks with a sub-ballast layer. Can Geotech J 36:1137–1150

Shahu JT, Kameswara Rao NSV, Yudhbir (2000) A rational method for design of railroad track foundation. Soils Found Jpn Geotech Soc 40(6):1–10

Singh AB (2002) Steady state response of an infinite beam under moving load for two parameter foundation model, M.Tech thesis, Department of Civil Engineering, Indian Institute of Technology, Kanpur, India, 2002

Vihari S (2005) Design of the granular layer thickness of a railroad track, M.Tech thesis, Department of Civil Engineering, Indian Institute of Technology, Kanpur, India, 2005

Chapter 8
Using System Dynamics to Identify, Evaluate, and Implement Sustainable Policies and Practices in the Road Construction Industry

Rajib B. Mallick and Michael J. Radzicki

Abstract Sustainable road construction is a topic that is currently in vogue among decision makers in the public and private sectors. On the one hand, the need for new roads and a well-maintained transportation infrastructure is necessary to sustain an ever-growing population and to both foster and accommodate economic development. On the other hand, road construction causes more fossil fuel to be burned, more greenhouse gases to be released, and more agricultural and forest land to be destroyed – effects that, in the extreme, go directly against the tenets of sustainable development. How then do we identify, evaluate, and adopt sustainable road construction and maintenance policies? This question is not simple, has no obvious answer, and has no analytical solution because road construction and maintenance is a multidisciplinary, complex, nonlinear feedback process. Solving problems of this type requires the use of simulation modeling. Simulation can be used to understand the problem and develop policies and practices that lead to improved system performance. The purpose of this paper is to demonstrate how system dynamics computer simulation modeling can be used to better understand the challenges associated with sustainable road construction and maintenance and to develop policies that enable these vital activities to be undertaken sustainably.

Keywords Sustainable • Road construction • System dynamics • Policy • Environment

R.B. Mallick (✉)
Department of Civil and Environmental Engineering, Worcester Polytechnic Institute (WPI), Worcester, MA 01609, USA
e-mail: rajib@wpi.edu

M.J. Radzicki
Department of Social Science and Policy Studies, Worcester Polytechnic Institute (WPI), Worcester, MA 01609, USA
e-mail: mjradz@wpi.edu

G.L. Sivakumar Babu et al. (eds.), *Sustainability Issues in Civil Engineering*, Springer Transactions in Civil and Environmental Engineering,
DOI 10.1007/978-981-10-1930-2_8

8.1 Introduction

Road construction and maintenance is a multidisciplinary, dynamic, nonlinear feedback process. A proper understanding of the relationship between this type of system structure and the holistic behavior it generates is critical for engineers who are involved in designing, building, and maintaining infrastructure facilities, as well as for senior-level managers who are in charge of formulating infrastructure policies that must meet multiple goals.

Due to their inherent complexity, nonlinear, dynamic feedback systems are difficult to understand and control. Indeed, their behaviors are often "counterintuitive" and they tend to generate "unintended consequences." Stated differently, it is essentially impossible for human beings to accurately think through the dynamic behavior of complex nonlinear, dynamic feedback systems, and as a result, their behaviors are often surprising. Unfortunately, no analytical solutions to these sorts of systems exist. The good news, however, is that their behavior can be revealed via simulation.

Sustainable road construction has always proved to be a challenging endeavor. On the one hand, new, and well-maintained, roads are necessary for socioeconomic development. On the other hand, road construction and maintenance results in the loss of forest and agricultural lands, the consumption of energy, and the release of greenhouse gases – activities that, in the extreme, go against the tenets of sustainable development. Furthermore, there are many interrelated socioeconomic activities associated with road building, and the whole process is time dependent. A set of equations that can be solved analytically, in closed form, does not exist for this sort of problem. The only way to reach a solution is through simulation modeling.

Computer-assisted modeling and simulation helps us to understand the complexities of problems, experiment with different proposed solutions, and select the structural redesign that will yield the most improved system behavior. Such modeling provides a "very low cost" laboratory where one can *learn* through controlled simulation experiments. At the same time, data collected from laboratories and the field can be used to enhance these models. Computer-based system modeling should be considered as an integral part of experimental work.

Even though we occasionally refer to "systems" in this paper, we emphasize that the intent is to learn how to model a "problem from a systems perspective" and not to "model the system." That is, the goal is to model only that part of the system that is relevant to the problem under consideration and nothing more.

8.1.1 Objective

The objective of this paper is to illustrate the use of a holistic approach to identifying, evaluating, and implementing sustainable policies and practices in the road construction industry. To accomplish this task, a system dynamics model that links the recycling of aggregates to sustainable road construction practices, as well as a web-based simulation tool based on the model, will be presented. Both the model and the tool are available on the web for use and/or modification.

8.2 A Systems Approach and the Use of System Dynamics

A systems or holistic approach has been defined as "a framework for seeing and working with the whole, [while] focusing on interrelationships between parts rather than individual parts" (Senge 1990). Doyle et al. (2008)) have demonstrated that models that are strengthened by systems thinking are more stable, richer in content, and devoid of the sorts of biases that can be caused by environmental factors. Generally speaking, the success of the system dynamics approach in generating improved human decision making can be attributed to the development of tools and techniques for mapping out the mental models of experts and stakeholders and then having a computer accurately trace through the associated dynamic behavior (Forrester 1971). This process leads to learning, system insight, and improved policy implementation from those who participate in the modeling process (Sterman 1994, 2000, 2012). At a more specific level, various studies [Vennix (1990, 1996); Cavaleri and Sterman (1997); Huz et al. (1997))] have shown that the enhanced ability to understand and control complex systems that is generated by the system dynamics approach comes from helping:

1. Decision makers overcome cognitive barriers
2. Students overcome barriers inherent in classroom-type teaching/lectures
3. Decision makers develop better and more complete, complex, and dynamic mental models

A fundamental idea in system dynamics modeling is that structure causes behavior, and, as a consequence, to alter a system's behavior, one must alter (i.e., redesign) its structure. System dynamics models are built by identifying the aspects of a system's structure that are crucial to the problem under study and then mapping them out on the computer.

A system's structure is mapped out in a system dynamics model with stocks, flow feedback loops, and limiting factors. Stocks are represented by rectangular icons that are intended to conjure up the image of a bathtub, while flows are represented by pipe and faucet assemblies that conceptually fill and/or drain the stocks. A second fundamental idea in system dynamics modeling is the "principle of accumulation," which states that all dynamic behavior in the world occurs when flows accumulate in stocks. In the calculus this is the process of integration.

Stocks and flows do not exist in isolation. Rather, they exist as part of feedback loops. Feedback is the transmission and return of information – information on the amount of "stuff" that has accumulated in a system's stocks over time. In a system dynamics model, this information is transmitted on thin "telephone wire arrows" that trace out paths, either directly or indirectly, from the system's stocks back to its flows. The information is used to regulate the flows and hence determine the system's time path.

Two types of feedback loops exist in system dynamics models, positive loops and negative loops. Positive loops represent self-reinforcing processes in which stocks create more of whatever they contain. For example, people make more people, money is used to make more money, knowledge generates more knowledge, and plant and equipment is used to make more plant and equipment. Generally speaking, positive feedback loops are responsible for the growth and decline of systems.

Negative loops, on the other hand, represent goal-seeking processes in which stocks are guided toward targets or desired amounts of "stuff." For example, if a discrepancy exists between a pavement manager's desired number of workers and the number he or she actually has available, a negative loop will motivate corrective action (e.g., employee hiring, transfer, or separation). Generally speaking, negative feedback loops stabilize systems or cause them to oscillate.

Stocks and flows cannot grow infinitely large, and they often should not be allowed to drop below zero (e.g., birthing negative babies makes no sense) or some other value. As a consequence, system dynamics models contain limiting factors that constrain their components. Limiting factors can be physical, financial, or psychological.

As a general rule, nonlinear relationships describe how a system behaves as it approaches its limiting factors. Mathematically speaking there are many ways to add a nonlinear relationship to a system dynamics model. That said, the most important implication of making a dynamic model nonlinear involves the method used in its solution. From a system dynamics point of view, solving a dynamic system means determining how much "stuff" has accumulated in each of its stocks at every point in time. If a system is nonlinear, it has no closed form and analytical solution, and, as a consequence, it must be solved numerically.

With a numerical solution, a computer is asked to take small steps forward through simulated time. At the conclusion of each step, the computer is asked to calculate the amount of "stuff" in each of the system's stocks. The solution is approximate, rather than exact, as a digital computer cannot take infinitesimally small steps forward in simulated time.

8.3 Illustration Through a System Dynamics Model

8.3.1 Problem Description

Generally speaking, new roads are built because there has been population growth and/or advances in economic prosperity. Existing roads, on the other hand, require maintenance and eventually rehabilitation. Road construction activities (new construction, maintenance, and rehabilitation) require aggregates (stones), which may come from local and/or nonlocal sources. By definition nonlocal sources of aggregate are further away from road construction activities than are local sources. If roads are recycled during the maintenance and rehabilitation processes, the demand for new aggregates can be reduced.

Transporting aggregates to job sites requires trips by trucks. Trucks consume diesel fuel, whose cost increases over time due to inflation. Trucks also generate CO_2, which is released into the atmosphere. Although CO_2 can be absorbed by the environment, as more and more trees are cleared to build new roads, the environment's ability to absorb CO_2 falls over time.

If new roads are not built and old roads are not maintained or rehabilitated, then both the quantity and engineering quality of roads suffer. Alternatively, if new roads *are* built and old roads *are* maintained and rehabilitated, rising fuel costs will lead to a deterioration in the cost to benefit ratio associated with infrastructure construction and maintenance. Moreover, rising levels of CO_2 in the atmosphere will lead to a fall in environmental quality.

The salient question to ask at this point is how can we model this problem holistically and determine the *system-wide* costs and benefits of infrastructure construction and maintenance at any point in time? Our intent is to develop a sustainability score and then use it to evaluate the effects of different infrastructure construction and maintenance policies (including a "business-as-usual" policy), as well as to identify any leverage points that can be exploited to maintain sustainable road construction activity.

8.3.2 Step 1: Identify the Key Sectors

Before mapping out the different sectors of the model, each of which consists of several stocks and flows, it is useful to think in terms of a 10,000 m overview of the problem. In fact, this is the most important step in the modeling process as it focuses the modeling effort and defines the criteria for including factors in the model.

In terms of identifying key sectors, it is often useful for the modeler to ask the following questions (to him-/herself and/or to experts and stakeholders):

1. From your knowledge, experience, and literature review, what topics do you feel are most important to this problem?
2. What are the likely causes and effects associated with these topics?

As shown in Table 8.1, the answers to these questions should be recorded. Since modeling is an iterative process, this list is usually not exhaustive, but rather a first cut that can be used to begin construction of the model.

Next, the salient variables for each of the sectors listed in Table 8.1 should be identified. Additionally, if possible each salient variable should be designated as a stock, flow, or auxiliary equation.

Table 8.1 Important sectors

Question	Probable answers	Identified sectors
1. Why do we need to construct new roads?	As population grows, the number of vehicles increase; growth in population leads to a demand of new roads; as a society becomes more affluent, people can afford to own and drive their own vehicles	Population and road demand
2. What happens to new roads over time?	Roads deteriorate due to traffic and the environment; roads have a life, say 10 years, after which they need to be rehabilitated; rehabilitation also requires construction	Road construction and rehabilitation
3. What is the impact of road construction on the economy?	Road construction requires raw materials, energy plant and equipment, and labor; there are many costs associated with road construction, two of which are more important: cost of fuel that is involved in the transportation of aggregates during construction and the cost of production of materials	Economy (consider fuel used during transportation of aggregates only)
4. How does road construction affect the environment?	In many ways. If we consider the transportation of aggregates, the fuel burnt during transportation leads to the emission of CO_2; this CO_2 is dissipated/absorbed by the trees; if trees are cut down due to road or non-road-related work, then the absorption rate also decreases	Environment sector (consider the generation and absorption of CO_2 only)
5. Given the above factors, how can we determine the sustainability of road construction?	By combining the effects of road quality, CO_2 generation and dissipation, and fuel costs, we can create an index that integrates engineering (or quality of transportation or life), environmental, and economic factors	Sustainability index

8.3.3 Population and Road Demand Sector

Figure 8.1 shows a stock-flow-feedback diagram of the population and road demand sector. Mathematically speaking, system dynamics models are nonlinear ordinary differential equations. As such, each icon in Fig. 8.1 represents an equation. Table 8.2 presents the equations that directly correspond to each of the icons in

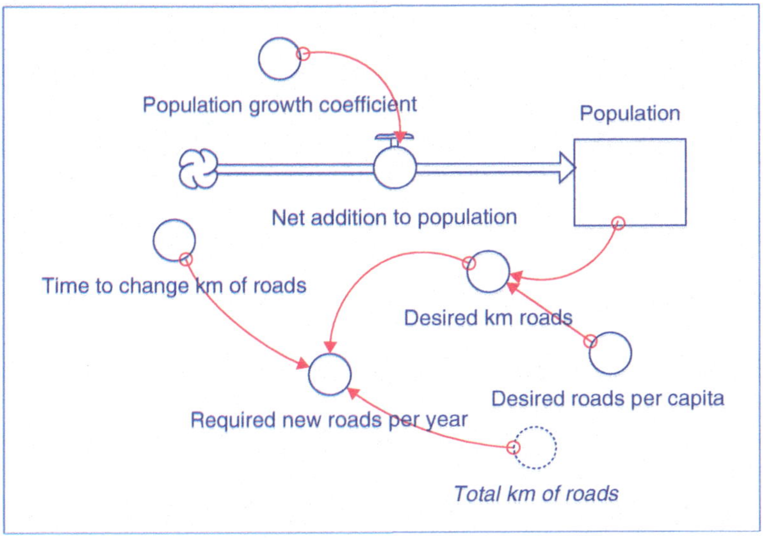

Fig. 8.1 Population and road demand sector

Table 8.2 Equations for the population and road demand sector

Population(t) = Population(t – dt) + (Net_addition_to_population) * dt
INIT Population = 1,000,000 (*consider an initial population of 1,000,000*)
Inflows:
Net_addition_to_population = Population_growth_coefficient * TIME
Desired_km_roads = Population * Desired_roads_per_capita
Desired_roads_per_capita, km = 0.009 (*consider 0.009*)
Population_growth_coefficient, people/(person * year) = 170 (*consider 170*)
Required_new_roads_per_year, km = (Desired_km_roads – Total_km_of_roads)/ (Time_to_change_km_of_roads)
Time_to_change_km_of_roads = 10 years

Fig. 8.1. Commentary is written in bold and italics, while the specific parameter values are assumed or based on literature.

8.3.4 Road Construction and Rehabilitation Sector

Similar to the preceding figure and table, Fig. 8.2 presents a preliminary version of the road construction and rehabilitation sector, and Table 8.3 presents the corresponding equations. Of note is that one or more of the variables in Fig. 8.2 have already been defined in the population and demand sector.

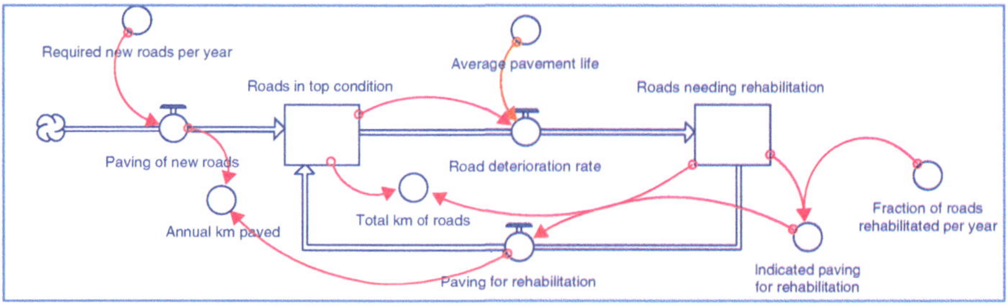

Fig. 8.2 Preliminary road construction and rehabilitation sector

At this point it makes sense to return to the stock-flow-feedback diagram shown in Fig. 8.2. Even though it captures the basic processes of road construction and rehabilitation, it does not include any variables that are important for evaluating any associated economic and environmental impacts, let alone for computing a sustainability score. So what else are needed in this sector? Let's consider the issues one by one.

1. If we wish to consider the effects on the system from transporting aggregate, we need to know the number of trucks and the total distance they travel each year transporting aggregates. Essentially there are two possibilities: (1) transporting aggregates from natural aggregate stocks (e.g., quarries) to construction sites and (2) transporting milled materials from existing roads to plants for recycling aggregates (which can only be done for rehabilitation and not for new road construction). The distance for these two cases will be different. For aggregate transportation from natural aggregate stocks, the model assumes that as more and more stock is exhausted, the transportation distance increases (this is evident from the literature (Mallick et al. 2014)). On the other hand, for recycling, the model assumes an average distance between the job site (i.e., the road to be rehabilitated) and the plant (where the aggregate will be recycled to produce new mixes for rehabilitation). Of course, the more aggregate is recycled, the less new aggregate is taken from the natural aggregate stock. There is one more benefit of recycling and that is the savings in the energy that is required for the production of aggregate and asphalt binder. Based on a literature review, this paper considers the following data regarding energy use and savings (McKetta 1983): 56,000 BTU are required per ton of aggregate, and 587,500 BTU are required per ton of asphalt binder and 36,578 BTU per gallon of diesel; total diesel savings for aggregate per metric ton of recycled aggregate is 1.7 l; considering a 5 % asphalt content, savings for asphalt binder per metric ton of recycled aggregate is 0.9 l; and total diesel savings per metric ton of milled material is 2.6 l.
2. When milling the aggregates from existing roads during rehabilitation, a small fraction (e.g., 5 %) is not recoverable.
3. The total distance traveled for construction of new roads and rehabilitating roads (by recycling or by using new natural aggregate) needs to be considered for the calculation of the total consumption of fuel. This involves the consideration of total aggregates required and tons per trip.

Table 8.3 Equations for the preliminary road construction and rehabilitation sector

Roads_in_top_condition(t), km = Roads_in_top_condition(t – dt) + (Paving_of_new_roads + Paving_for_rehabilitation – Road_deterioration_rate) * dt
INIT Roads_in_top_condition, km = 8000 (*consider 8000*)
Inflows:
Paving_of_new_roads, km per year = Required_new_roads_per_year
Paving_for_rehabilitation, km per year = Indicated_paving_for_rehabilitation
Outflows:
Road_deterioration_rate, k per year = (Roads_in_top_condition)/(Average_pavement_life)
Roads_needing_rehabilitation(t), km = Roads_needing_rehabilitation(t – dt) + (Road_deterioration_rate – Paving_for_rehabilitation) * dt
INIT Roads_needing_rehabilitation, km = 100 (consider an initial value of 100 km)
Inflows:
Road_deterioration_rate = Roads_in_top_condition/Average_pavement_life
Outflows:
Paving_for_rehabilitation, km per year = Indicated_paving_for_rehabilitation
Annual_km_paved = Paving_for_rehabilitation + Paving_of_new_roads
Average_pavement_life, year = 10 (*consider a value of 10*)
Fraction_of_roads_rehabilitated_per_year = 0.7 (*consider a value of 0.7*)
Indicated_paving_for_rehabilitation = Roads_needing_rehabilitation * Fraction_of_roads_rehabilitated_per_year
Required_new_roads_per_year = (Desired_km_roads – Total_km_of_roads)/(Time_to_change_km_of_roads)
Total_km_of_roads = Roads_in_top_condition + Roads_needing_rehabilitation

4. To cope with limited budgets or reduce the impact on the budget and/or the environment, the model can be set up to rehabilitate something less than 100 % of roads that need rehabilitation every year (say 70 %). Similarly, the model can be set up to cancel some projects in order to balance the competing pressures from the intertwined engineering, economic, and environmental objectives.

Including all of the above considerations, we get a modified map and set of equations, as shown in Fig. 8.3 and Table 8.4, respectively.

Next, let us consider the economy sector. We need to determine the cost of fuel used for transportation. Since we have already determined the amount of fuel consumed, in the road construction and rehabilitation sector, we need to consider the unit cost of fuel in the sector. However, over time, the cost of fuel increases due to inflation, and we need to include an inflation factor. Considering all of the above, a good way of expressing the cost of fuel is the cost of fuel per km of road paved (we call it a ratio).

Therefore, the sector map and the equations can be expressed as shown in Fig. 8.4 and Table 8.5, respectively.

Next we consider the environment sector. In this case we need to account for the amount of CO_2 that is generated by the burning of fuel (which is determined in the road construction and rehabilitation sector). This requires considering the amount

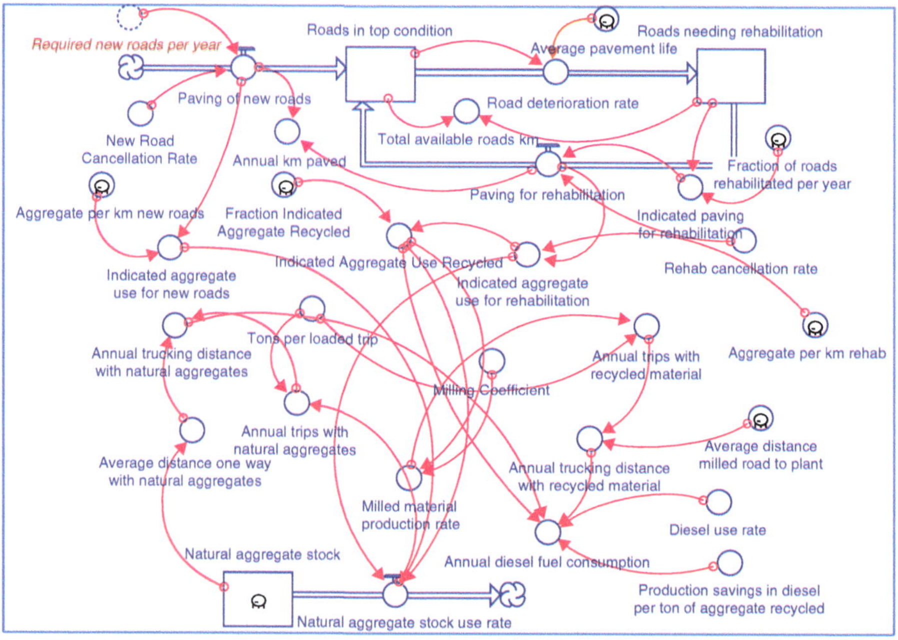

Fig. 8.3 Modified road construction and rehabilitation sector

of CO_2 generated per unit of fuel burned. This calculation must also be tempered by the impact of cleaner technologies that will be developed over time.

Once the CO_2 is released, it can be absorbed by the trees. The area of the forests, however, will be reduced over time due to the need for more area for road construction and other human activities. Thus, the model needs to include the "harvest" rate of forests associated with building one km of road (then we can tie it to total new roads constructed every year), as well as the harvest rate associated with non-road-related activities.

One way to consider the combined effect of CO_2 generation and dissipation will be to utilize a CO_2 flow ratio, which is a ratio of the CO_2 generated due to the burning of fossil fuel to the CO_2 absorbed by trees, per year. Including these factors in the model leads to the stock-flow-feedback diagram and the associated equations shown in Fig. 8.5 and Table 8.6, respectively.

Finally, it is necessary to bring the sustainability sector into the model. In this sector, our intent is to create a composite index that provides an overall score related to the sustainability of road construction. This index will integrate measures of road quality (i.e., an engineering consideration), the CO_2 flow ratio (i.e., an environmental consideration), and costs (i.e., an economic consideration). For road quality the index uses the actual paving of new roads and the actual rehabilitation of existing roads relative to their required (or indicated) values. The difference between the required and the actual values will be a shortfall. An increase in the shortfall will lead to a reduction in road quality and a fall in the "score" of the

Table 8.4 Updated set of equations for the modified road construction and rehabilitation sector

Natural_aggregate_stock(t) = Natural_aggregate_stock(t – dt)+ (-Natural_aggregate_stock_use_rate) * dt
INIT Natural_aggregate_stock = 1000000000
Outflows
Natural_aggregate_stock_use_rate = Indicated_aggregate_use_for_new_roads + Indicated_aggregate_use_for_rehabilitation – Indicated_Aggregate_Use_Recycled
Roads_in_top_condition(t) = Roads_in_top_condition (t – dt) + (Paving_of_new_roads +Paving_for_rehabilitation – Road_deterioration_rate) * dt
INIT Roads_in_top_condition = 8000
Inflows
Paving_of_new_roads = Required_new_roads_per_year-New_Road__Cancellation_Rate
Paving_for_rehabilitation = Indicated_paving_for_rehabilitation-Rehab_cancellation_rate
Outflows
Road_deterioration_rate = Roads_in_top_condition/Average_pavement_life
Roads_needing_rehabilitation(t) = Roads_needing_rehabilitation(t – dt) + (Road_deterioration_rate –
Paving_for_rehabilitation) * dt
INIT Roads_needing_rehabilitation = 100
Inflows
Road_deterioration_rate = Roads_in_top_condition/Average_pavement_life
Outflows
Paving_for_rehabilitation = Indicated_paving_for_rehabilitation-Rehab_cancellation_rate
Aggregate_use_per_km_new_roads = 550
Aggregate_use_per_km_rehabilitated = 300
Annual_diesel_fuel_consumption = (Annual_trucking_distance_with_recycled_material + Annual_trucking_distance_with_natural_aggregates) * Diesel_use_rate
Annual_km_paved = Paving_for_rehabilitation + Paving_of_new_roads
Annual_trips_with_natural_aggregates = (Natural_aggregate_stock_use_rate)/ Tons_per_loaded_trip
Annual_trips_with_recycled_material = Milled_material__production_rate/ Tons_per_loaded_trip
Annual_trucking_distance_with_natural_aggregates = 2 * Annual_trips_with_natural_aggregates * Average_distance_traveled_one_way_with_natural_aggregates
Annual_trucking_distance_with_recycled_material = 2 * Average_distance_from_milled_road_to_recycling_plant * Annual_trips_with_recycled_material
Average_distance_from_milled_road_to_recycling_plant = 4
Average_distance_traveled_one_way_with_natural_aggregates = 126.2 * EXP (-3 * Natural_aggregate_stock/INIT(Natural_aggregate_stock))
Average_pavement_life = 10
Diesel_use_rate = 1/5
Fraction_Indicated_Aggregate_use_Recycled = 0
Fraction_of_roads_rehabilitated_per_year = 0.7

(continued)

Table 8.4 (continued)

Indicated_aggregate_use_for_new_roads = Paving_of_new_roads * Aggregate_use_per_km_new_roads
Indicated_aggregate_use_for_rehabilitation = Aggregate_use_per_km_rehabilitated * Paving_for_rehabilitation
Indicated_Aggregate_Use_Recycled = Fraction_Indicated_Aggregate_use_Recycled * Indicated_aggregate_use_for_rehabilitation
Indicated_paving_for_rehabilitation = Roads_needing_rehabilitation * Fraction_of_roads_rehabilitated_per_year
Milled_material__production_rate = Milling_Coefficient * Indicated_Aggregate_Use_Recycled
Milling_Coefficient = 0.95
New_Road__Cancellation_Rate = 0
Rehab_cancellation_rate = 0
Required_new_roads_per_year = (Desired_km_roads-Total_km_of_roads)/ Time_to_change_km_of_roads
Tons_per_loaded_trip = 25
Total_km_of_roads = Roads_in_top_condition + Roads_needing_rehabilitation

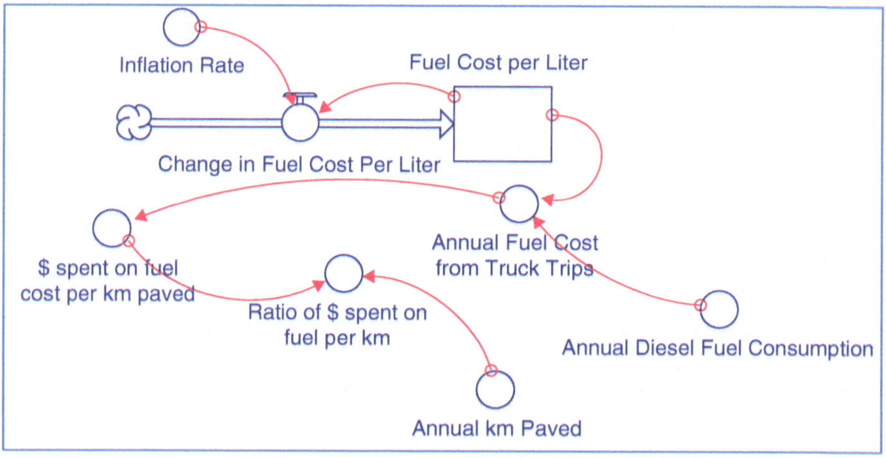

Fig. 8.4 Economy sector

system. Similarly, an increase in the CO_2 flow ratio and $ spent per km of paved roads will lead to a decrease in the score. Then, if we compare the score at any year to the initial score (initial year), we can evaluate the trend in the sustainability of road construction and rehabilitation work. These final elements of the models are shown in Fig. 8.6 and Table 8.7, respectively.

Table 8.5 Equations for economy sector

Fuel_Cost_per_Liter(t) = Fuel_Cost_per_Liter(t − dt) + (Change_in_Fuel_Cost_Per_Liter) * dt
INIT Fuel_Cost_per_Liter = 1
Inflows
Change_in_Fuel_Cost_Per_Liter = Inflation_Rate * Fuel_Cost_per_Liter
Inflation_factor(t) = Inflation_factor(t − dt) + (Change_in_inflation_factor) * dt
INIT Inflation_factor = 1
Inflows
Change_in_inflation_factor = Inflation_factor * Inflation_Rate
$_spent_on_fuel_cost_per_km_paved = Annual_Fuel_Cost_from_Truck_Trips/ Annual_km_paved
Annual_diesel_fuel_consumption = Annual_diesel_fuel_consumption = ((Annual_trucking_distance_with_recycled_material + Annual_trucking_distance_with_natural_aggregates) * Diesel_use_rate)-2.6 * Indicated_Aggregate_Use_Recycled * Production_savings_in_diesel_per_ton_of_aggregate_recycled
Annual_Fuel_Cost_from_Truck_Trips = Annual_diesel_fuel_consumption* Fuel_Cost_per_Liter
Annual_km_paved = Paving_for_rehabilitation + Paving_of_new_roads
Inflation_Rate = 0.01
Percentage_Increase__in_Fuel_Cost = 100*(Annual_Fuel_Cost_from_Truck_Trips-INIT (Annual_Fuel_Cost_from_Truck_Trips))/init(Annual_Fuel_Cost_from_Truck_Trips)
Production_savings_in_diesel_per_ton_of_aggregate_recycled = 10.6
Ratio_of_$_spent_on__fuel_per_km = ($_spent_on_fuel_cost_per_km_paved) / (INIT($_spent_on_fuel_cost_per_km_paved) * Inflation_factor)

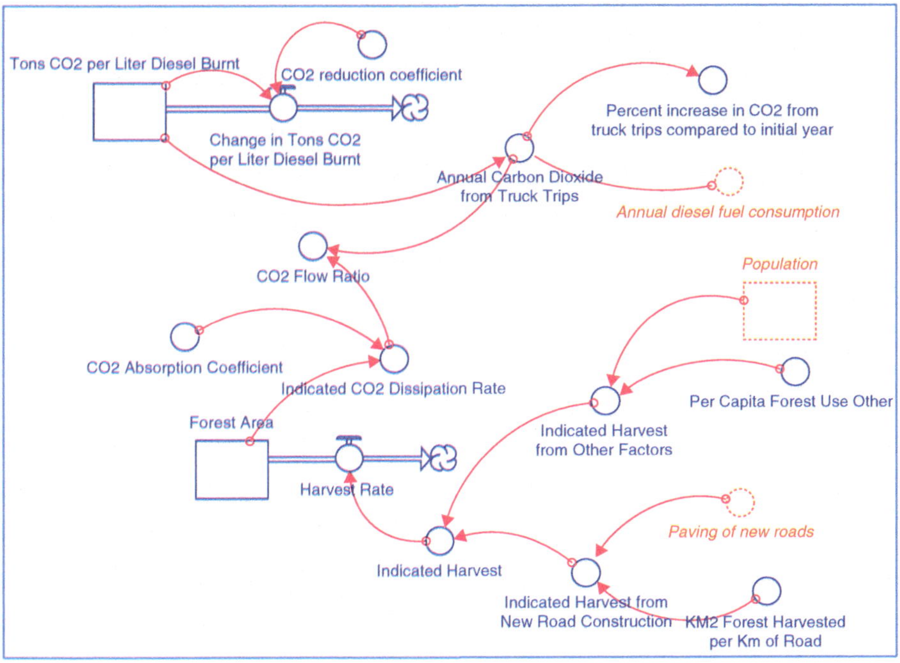

Fig. 8.5 Environment sector

Table 8.6 Equations for the environment sector

Forest_Area(t) = Forest_Area(t – dt) + (-Harvest_Rate) * dt
INIT Forest_Area = 4000000
Outflows
Harvest_Rate = Indicated_Harvest
Population(t) = Population(t – dt)
INIT Population = 1000000
Tons_CO2_per_Liter_Diesel_Burnt(t) = Tons_CO2_per_Liter_Diesel_Burnt(t – dt) + (-Change_in_Tons_CO2_per_Liter_Diesel_Burnt) * dt
INIT Tons_CO2_per_Liter_Diesel_Burnt = 2.65
Outflows:
Change_in_Tons_CO2_per_Liter_Diesel_Burnt = CO2_reduction_coefficient * Tons_CO2_per _Liter_Diesel_Burnt
Paving_of_new_roads = Required_new_roads_per_year-New_Road__Cancellation_Rate
Annual_Carbon_Dioxide_from_Truck_Trips = Tons_CO2_per_Liter_Diesel_Burnt * Annual_diesel_fuel_consumption
Annual_diesel_fuel_consumption = (Annual_trucking_distance_with_recycled_material + Annual_trucking_distance_with_natural_aggregates)*Diesel_use_rate
CO2_Absorption_Coefficient = 439.8
CO2_Flow_Ratio = Annual_Carbon_Dioxide_from_Truck_Trips/ Indicated_CO2_Dissipation_Rate
CO2_reduction_coefficient = 0
Indicated_CO2_Dissipation_Rate = IF (Forest_Area > 0) THEN (Forest_Area * CO2_Absorption_Coefficient) ELSE 0.01
Indicated_Harvest = Indicated_Harvest_from_New_Road_Construction + Indicated_Harvest_from_Other_Factors
Indicated_Harvest_from_New_Road_Construction = Paving_of_new_roads * KM2_Forest_Harvested_per_Km_of_Road
Indicated_Harvest_from_Other_Factors = Population * Per_Capita_Forest_Use_Other
KM2_Forest_Harvested_per_Km_of_Road = 0.01
Percent_increase_in_CO2_from_truck_trips_compared_to_initial_year = 100 * (Annual_Carbon_Dioxide_from_Truck_Trips-INIT (Annual_Carbon_Dioxide_from_Truck_Trips))/INIT (Annual_Carbon_Dioxide_from_Truck_Trips)
Per_Capita_Forest_Use_Other = 0.03

8.4 Combining the Sectors and Running the Model

It is now possible to put all the sectors together and then run the more comprehensive version of the model.

We select a few parameters that are thought to be important for the model and its sustainability score. For example, the fraction of roads rehabilitated, the percentage of aggregates recycled, and the average pavement life are very likely important factors that will affect the sustainability of the system. We would like to see plots of the sustainability score for different combinations of the above factors. Therefore,

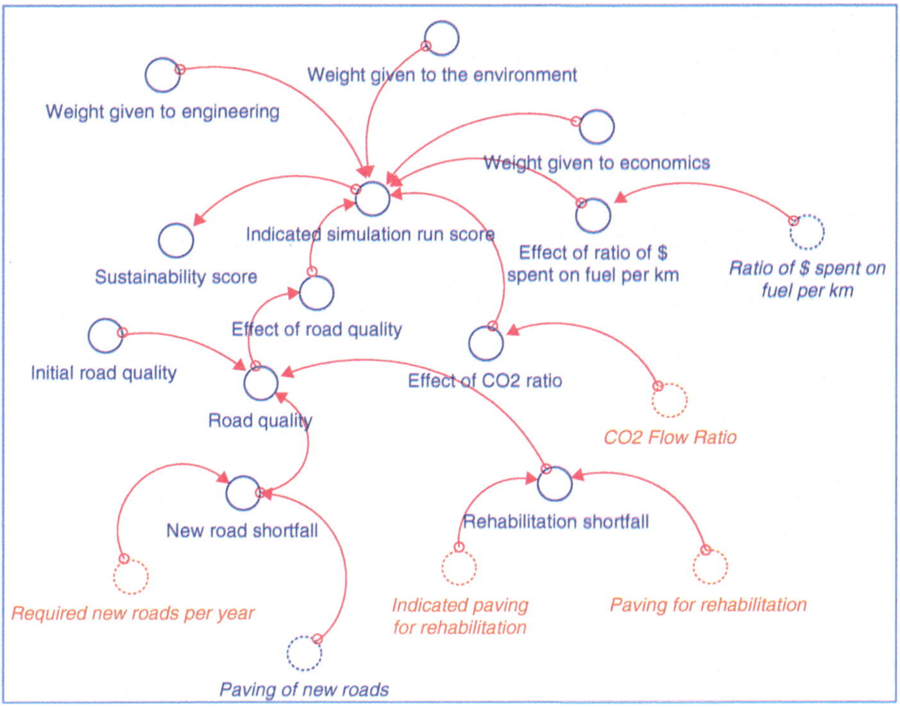

Fig. 8.6 Sustainability sector

we should run the model for the different values of the important parameters and evaluate the results. This is the most important benefit of running the model – to understand which factor(s) is (are) more important than the others. We can, therefore, focus on those factors in our work and policy, such that both are effective. As an example, simulations with 0, 20, 40, and 80 % recycled aggregates were run, and the results are shown in Fig. 8.7. It can be clearly seen that without recycling, the sustainability score decreases over time, first slowly and then at a higher rate, and recycling at higher content can improve the sustainability score over time.

We can call the factors (and the timing of changes in these factors) that can have significant effects on the end result *leverage* factors. Also, since we are considering the effects of a variety of factors (engineering, environment, and economy) on the end result, we are using an *empathetic* view, where we are considering the view point of the economist, the engineer, and the environmentalist. We are considering the fact that road construction leads to a deterioration of the environment, but at the same time, we are also acknowledging the fact that without new roads and properly rehabilitated roads, the quality of life of citizens will fall, and hence, the sustainability score will as well. Policies selected after such a holistic consideration will be more acceptable to different stakeholders and will therefore be easier to implement quickly and effectively.

Table 8.7 Equations for the sustainability sector

Paving_for_rehabilitation = Indicated_paving_for_rehabilitation-Rehab_cancellation_rate
Paving_of_new_roads = Required_new_roads_per_year-New_Road__Cancellation_Rate
CO2_Flow_Ratio = Annual_Carbon_Dioxide_from_Truck_Trips/ Indicated_CO2_Dissipation_Rate
Effect_of_CO2_ratio = 3.3-1.6 * CO2_Flow_Ratio
Effect_of_ratio_of_$_spent_on_fuel_per_km = 2.2-0.08 * Ratio_of_$_spent_on__fuel_per_km
Effect_of_road_quality = -2.1*(Road_quality^2) + 4.26 * Road_quality-0.16
Indicated_paving_for_rehabilitation = Roads_needing_rehabilitation * Fraction_of_roads_rehabilitated_per_year
Indicated_simulation_run_score = ((Effect_of_road_quality ^ Weight_given_to_engineering) * (Effect_of_ratio_of_$_spent_on_fuel_per_km ^ Weight_given_to_economics) * (Effect_of_CO2_ratio^ Weight_given_to_the_environment)) ^ (1/ (Weight_given_to_economics + Weight_given_to_engineering + Weight_given_to_the_environment))
Initial_road_quality = 1
New_road_shortfall = (Required_new_roads_per_year-Paving_of_new_roads)/ Required_new_roads_per_year
Ratio_of_$_spent_on__fuel_per_km = ($_spent_on_fuel_cost_per_km_paved)/ (INIT($_spent_on_fuel_cost_per_km_paved)*Inflation_factor)
Rehabilitation_shortfall = (Indicated_paving_for_rehabilitation-Paving_for_rehabilitation)/ Indicated_paving_for_rehabilitation
Required_new_roads_per_year = (Desired_km_roads-Total_km_of_roads)/ Time_to_change_km_of_roads
Road_quality = IF TIME = 0 THEN Initial_road_quality ELSE(Initial_road_quality * ((1-New_road_shortfall)^0.5) * ((1-Rehabilitation_shortfall)^0.5))
Sustainability_score = 100*Indicated_simulation_run_score/INIT (Indicated_simulation_run_score)
Weight_given_to_economics = 0.33
Weight_given_to_engineering = 0.34
Weight_given_to_the_environment = 0.33

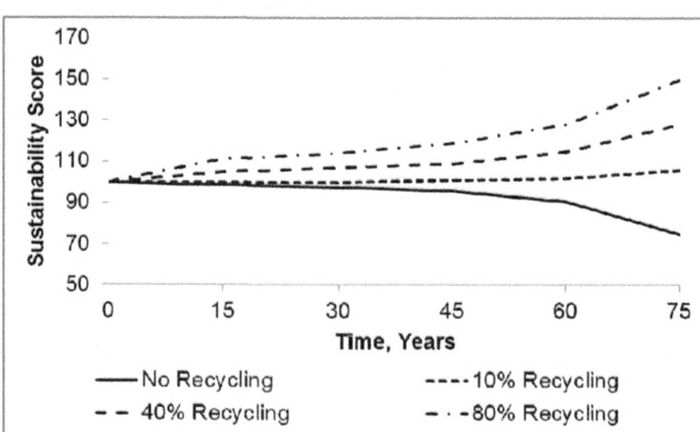

Fig. 8.7 Plots of sustainability score versus time for different recycling rates

The simulation interface and the model developed in this paper are available at http://goo.gl/jy7bU3.

References

Cavaleri S, Sterman JD (1997) Towards evaluation of systems thinking interventions: a case study. Syst Dyn Rev 13(2):171–186

Doyle JK, Radzicki MJ, Trees WS (2008) Chapter 14: measuring change in mental models of complex systems. In: Qudrat-Ullah H, Spector JM, Davidsen PI (eds) Complex decision making: theory and practice. Springer, Berlin, pp 269–294

Forrester JW (1971) World dynamics. Allen Press, Cambridge, MA/Wright

Huz S, Andersen DF, Richardson GP, Boothroyd R (1997) A framework for evaluating systems thinking interventions: an experimental approach to mental health systems change. Syst Dyn Rev 13(2):149–169

Mallick RB, Radzicki MJ, Zaumanis M, Frank R (2014) Use of system dynamics for proper conservation and recycling of aggregates for sustainable road construction. J Resour Conserv Recycl 86:61–73. doi:10.1016/j.resconrec.2014.02.006

McKetta Jr JJ (1983) Encyclopedia of chemical processing and design: volume 19 – energy: costing thermal electric power plants to ethanol. CRC Press, ISBN 9780824724696

Senge PM (1990) The fifth discipline: the art and practice of the learning. Doubleday, New York

Sterman JD (1994) Learning in and about complex systems. Syst Dyn Rev 10(2/3):291–330

Sterman JD (2000) Business dynamics: systems thinking and modeling for a complex world. McGraw-Hill, Boston

Sterman JD (2012) Sustaining sustainability: creating a systems science in a fragmented academy and polarized world. In: Sustainability science: the emerging paradigm and the urban environment. Springer, pp 21–58. doi:10.1007/978-1-4614-3188-6_2

Vennix JAC (1990) Mental models and computer models: design and evaluation of a computer-based learning environment for policy-making. CIP-Gegevens KoninklijkeBibliotheek, Den Haag

Vennix JAM (1996) Group model building: facilitating team learning using system dynamics. Wiley, New York

Part II
Sustainable Geosystems

Chapter 9
Resilient and Sustainable Geotechnical Solution: Lessons Learned from the 2011 Great East Japan Disaster

Hemanta Hazarika, Tadashi Hara, and Yasuhide Fukumoto

Abstract This paper first describes the field tests conducted on a tire retaining wall that miraculously survived in the 2011 Great East Japan disaster. The paper then proposes a new concept of using waste tires behind the seawall in order to protect such coastal structures from the damage due to impact force of tsunami. A physical model for tsunami impact force simulation was developed and described to evaluate the reduction effect of tsunami impact force by the tire structures. Finally, from aesthetic point of view, cultivation of suitable plants inside the tires was proposed. Field tests on planting trees that can grow in saline soil conditions were performed to see whether tire structures can preserve the greenery of the area. The results of this research, thus, can go a long way toward providing a sustainable solution for infrastructure development in the future.

Keywords Earthquake • Resiliency • Sustainability • Tsunami • Waste tires

9.1 Introduction

The 2011 off the Pacific Coast of Tohoku Earthquake, the record-breaking tsunami that easily overtopped many coastal structures, brought devastating damage to many geotechnical structures in the eastern coast of Tohoku area, Japan (Hazarika 2011; Hazarika et al. 2012b, c). The overtopping tsunami caused sliding failures in many land development sites (Hara 2011). Long-term solutions and reconstruction of the Tohoku area are still remaining the challenging issues for the geotechnical engineers and researchers.

H. Hazarika (✉) • Y. Fukumoto
Kyushu University, 744 Motooka, Nishi ku, Fukuoka 819-0395, Japan
e-mail: hazarika@civil.kyushu-u.ac.jp

T. Hara
Kochi University, 200 Monobeotsu, Nankoku, Kochi 783-8502, Japan

G.L. Sivakumar Babu et al. (eds.), *Sustainability Issues in Civil Engineering*,
Springer Transactions in Civil and Environmental Engineering,
DOI 10.1007/978-981-10-1930-2_9

While surveying the tsunami disaster areas immediately after the disaster, the authors were amazed to discover a retaining wall (made of recycled tires) that miraculously survived the disaster in Okirai area, Iwate Prefecture, Japan (Fig. 9.1). Ironically, this tire retaining wall is located just about 150 m away (toward the land) from a completely collapsed seawall (Fig. 9.2). The factory building situated on the backfill ground of the retaining wall was damaged by the tsunami, and a natural slope nearby this tire retaining wall was eroded by the tsunami. Why this tire retaining wall was neither damaged by the earthquake nor by the inundation and scouring due to the tsunami was the source of inspiration of research presented in this paper.

A Japanese government panel is estimating that a 9.0 magnitude earthquake in the Nankai Trough region will bring damage worth $2.2 billion, a figure that is much higher than the $177 million from the Great East Japan disaster of 2011. Scientists are predicting that the earthquake is due to "the not-too-distant future," based on historical rough calculations. The report of central disaster mitigation council of the ministry of Japan (Central Disaster Mitigation Council 2003) states that about 2 m of tectonic subsidence is expected in the Kochi area of Shikoku island of Japan, by that earthquake. In order to mitigate the damage from such future devastating earthquakes, it is necessary to take appropriate measures that can protect the infrastructures from the compound disasters instigated by the combined effect of events such as earthquake, liquefaction, and tsunami. To realize that, what is needed most at this moment is how to make use of the lessons that we civil engineers learned from the 2011 Great East Japan disaster.

According to damage mechanism due to earthquake and tsunami (Hazarika et al. 2013a, b), most of the damage of coastal dikes was mainly due to scouring at the back of the structures by tsunami impact force and backrush force of tsunami. The goal of this research is to develop an earthquake-resistant and tsunami-resistant

Fig. 9.1 Tire retaining wall in Okirai, Iwate Prefecture (picture taken after the disaster)

Fig. 9.2 Collapsed seawall in Okirai, Iwate Prefecture

low-cost technique for coastal structures that can prevent scouring of foundation soils and at the same time is environmentally friendly. It uses waste tires as a tsunami-resistant reinforcing measure of civil engineering structures to prevent compound damage of such structures during tsunami. Considering the aesthetic, the tires are also planted to preserve the greenery. Such reinforcing technique could mitigate the damage to structures during earthquake as well as tsunami and at the same time could also be an effective method of recycling the large number of scrap tires generated each year, which are posing great environmental degradation (Hazarika 2013).

This paper first reports on the site investigations of the non-damaged retaining wall made of tires shown in Fig. 9.1, focusing on the structural and geotechnical aspects. The investigations include seismological, geotechnical, and structural aspects of the retaining wall due to the tsunami using both the field survey and laboratory testing. As a part of the field survey, in situ density test, dynamic cone penetration test, micro tremor measurement, and surface wave exploration were conducted.

Secondly, in order to protect coastal structures such as seawall from impact force of tsunami, a new concept of using waste tire structure behind the seawall is introduced in this paper. A new model for tsunami impact force simulation was developed to evaluate the reduction effect of tsunami impact force by using tire.

On the other hand, from aesthetic point of view, tire structure on the back of the seawall along with filled soil does not look good. Therefore, cultivation of suitable plants inside the tire is suggested. Field tests on planting trees that can grow in saline soil conditions were performed to see whether such structure can preserve the greenery of the area.

9.2 Investigations on Tire Retaining Wall

9.2.1 Overview of the Damage and the Types of Investigations

The whole of Okirai area, where the tire retaining wall is located, was completely inundated by the tsunami (Fig. 9.3). The casualties in this area include 66 deaths and 30 missing people. As seen in Fig. 9.2, the concrete seawall along the coastal line completely collapsed. As a result, the tsunami run-up washed away most of the wooden buildings in the area. The tsunami run-up was up to the third floor of the Okirai Elementary School, which is located 200 m from the coastline. The tsunami run-up height in this location was found to be 16 m by the RTK-GPS survey conducted by Hazarika et al. (2012a). According to previous records, the tsunami run-up height in Sugishita area of Okirai was 11.6 m during the Showa Sanriku Tsunami in 1933 and was 7.8 m during the Meiji Sanriku Tsunami in 1896 (Shuto 2011). Based on our surveying data using the total station, the tsunami run-up height this time was estimated to be 16.79 m, which is much higher than the past tsunamis that inundated this area.

As a part of the site investigation, GPS surveying, portable dynamic cone penetration test (PDCP), surface wave exploration, and micro tremor measurement were conducted.

PDCP is recognized widely as a standard method for obtaining dynamic characteristics of soils at the site by the Japanese Geotechnical Society (JGS 1433). In PDCP, a drop hammer weighing 5 kg is allowed to fall through a rod from 50 cm

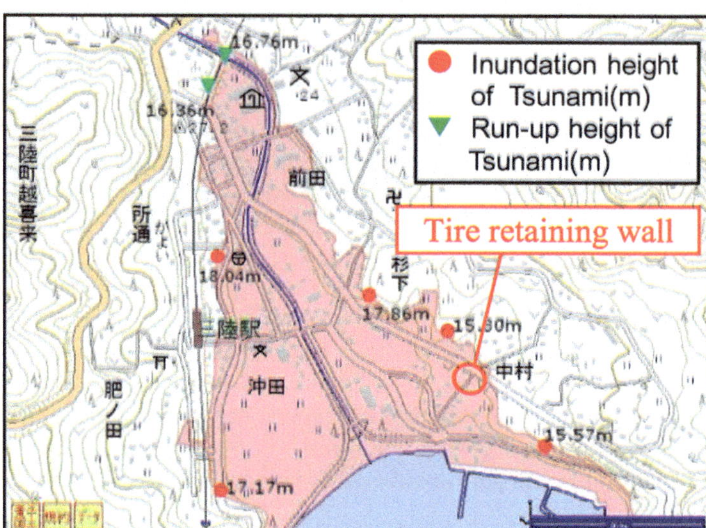

Fig. 9.3 State of tsunami inundation in the surveyed area

height, which enables the cone attached at the toe of the rod to penetrate into the ground. The number of blows (N_d) to penetrate every 10 cm of the ground is measured. N_d is related to the *N*-value of the standard penetration test. In this study, using the relationship proposed by Okada et al. (1992) for sandy soils, N_d values were converted to *N*-values. The location of the groundwater table was judged from the wet condition of the rod immediately after termination of the test.

The surface wave exploration is a convenient method to obtain S-wave velocity distribution (within the ground up to a depth of 10 m), which can measure and analyze the transmission of the surface wave (Rayleigh wave) that transmits near the ground surface. In this method, a wave is generated by striking the ground surface with a hammer. The generated wave propagates according to the surface and subsurface material conditions. During our investigation, in order to obtain the characteristics of the ground layer indirectly from the surface, the surface wave exploration was carried out together with the PDCP.

Micro tremor measurement has become a powerful tool to estimate the ground motion characteristics, amplification of ground motion in the soil deposits, microzonation, and dynamic behavior of existing service structures. The micro tremor observations described in this study were carried out by portable micro tremor equipment (type New PIC). Measurements (two horizontal and one vertical components) were conducted in velocity mode, which were recorded by the sensors. H/V ratio is then calculated based on the smoothened ratio of horizontal to vertical Fourier spectra of the micro tremor data. The amplitude ratio calculated in this study was based on the method proposed by Nakamura (1989). The value corresponding to the peak represents the predominant frequency of the motion.

Figure 9.4 shows the plane view of the tire retaining wall in Okirai. The figure also shows the locations of the various field surveys (in situ density, PDCP test, surface wave exploration, and micro tremor measurement) that were conducted. Disturbed soil samples were also collected from the three locations (No. 1, No. 2, soil within the tire), and laboratory investigations were carried out.

9.2.2 *Results of Field Investigations*

The distribution of S-wave velocity by surface wave exploration method conducted on the backfill soils is shown in Fig. 9.5. As seen in the figure, beyond the depth of 10 m, S-wave velocity is greater than 220 m/s implying a hard stratum near the sloping side. Within the depth of less than 10 m, stratum with 150–200 m/s of S-wave velocity exists. Since the average height of the retaining wall was 3.2 m with a maximum height of about 4 m, it can be said that as a whole, the backfill soil was in the loose state. Figure 9.15 also shows converted *N*-value obtained from the PDCP test. Converted *N*-value and S-wave velocity in general are showing the similar trend. The converted *N*-values within the depth of 0–70 cm are high. Therefore, it can be inferred that near the surface, the backfill soil has a very high density. The higher density near the surface may be the result of influence of cyclic load experienced by the backfill soils due to parked cars, since the yard was used as

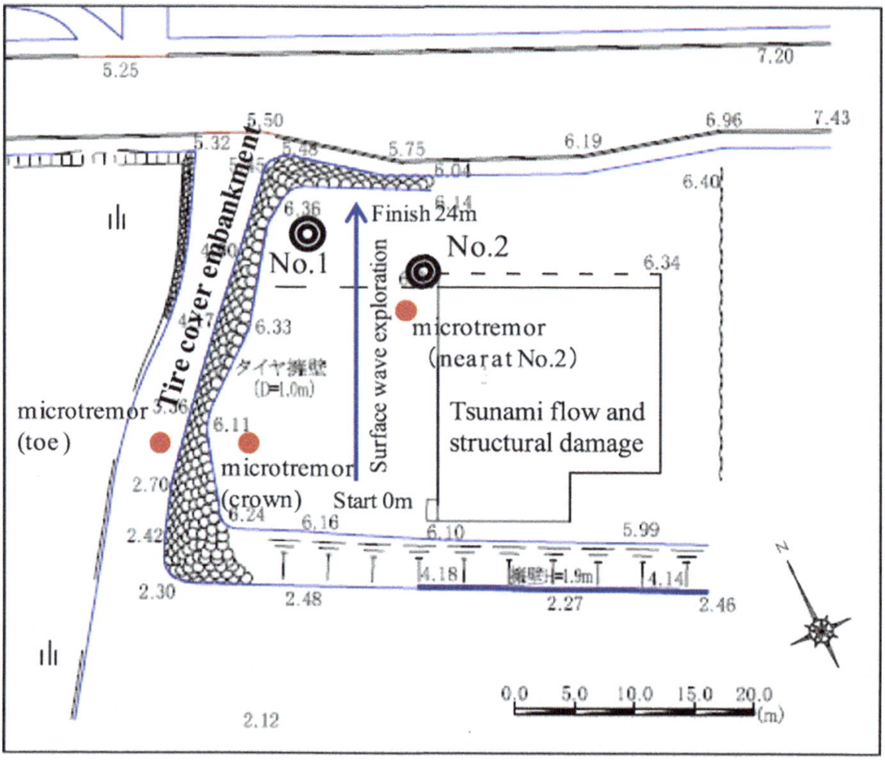

Fig. 9.4 Plan view of the retaining wall and locations of the field survey

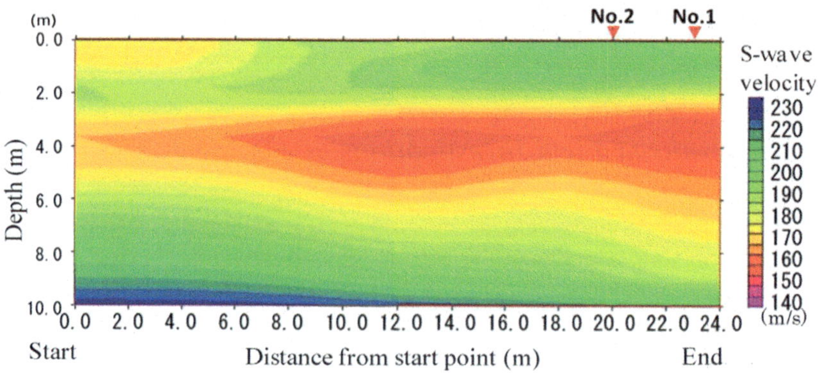

Fig. 9.5 Distribution of shear wave velocity

a parking lot. On the other hand, converted N-value within the depth of 1–1.5 m is about 5. Therefore, if we consider that the backfill consists only of sandy soil, it can be said that in general, the backfill soil was in loose state.

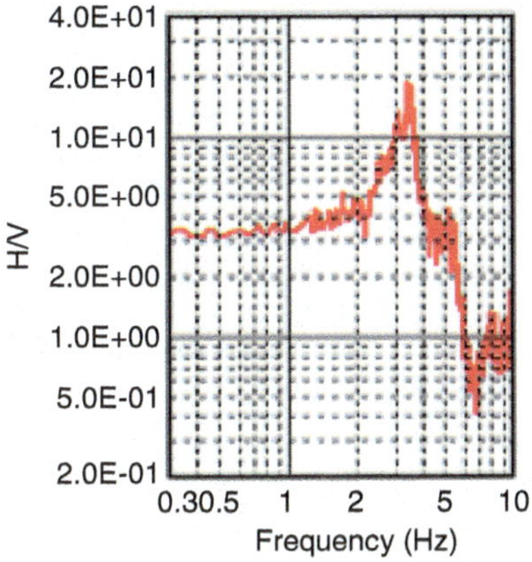

(a) H/V-ratio at the surface

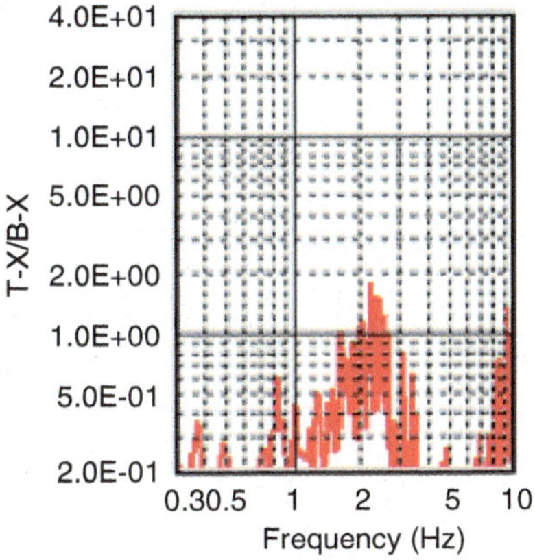

(b) Transfer function

Fig. 9.6 Results from micro tremor observations. (**a**) H/V-ratio at the surface. (**b**) Transfer function

Figure 9.6 shows the results of the micro tremor measurements conducted at the top as well as at the bottom of the wall. At location No. 2, the predominant frequency of H/V ratio of the soil deposits is 3.1 Hz (Fig. 9.6a). Spectral ratio between the top and the bottom of the sloping side was found to be 2.1 Hz (shown in Fig. 9.6b). The shear wave velocity from Fig. 9.5 was found to be approximately 200 m/s beyond 10 m depth. The predominant frequency of the site is 3.8 Hz. Therefore, it can be said that the results of the micro tremor measurements and the surface wave exploration are in good agreement.

9.2.3 Laboratory Investigations

Laboratory tests were conducted for soil samples that were collected at the sites. The in situ density of the backfill soil was measured using the core-cut method (JGS1613-2003). The values of in situ wet density ρ_t at the two locations (No. 1 and No. 2 in Fig. 9.4) are almost the same (a little more than 1.5 g/cm^3). From the laboratory testing, the void ratio of the soil was found to be less than 1.1. The grain size distribution showed that the soils were well grained (Fig. 9.7). It can also be said that the soils are well compacted. On the other hand, it is found that the filled soils inside the tires are found to be of higher density and thus can be said that they were compacted better. The angle of internal friction of the sample soil (collected within 30 cm from the surface) of the backfill ground is 45.7°, while it is 48.2° for the soil within the tires. Therefore, it can be said that near the surface, both soils have high value of internal friction, because the wet density ρ_t is quite high.

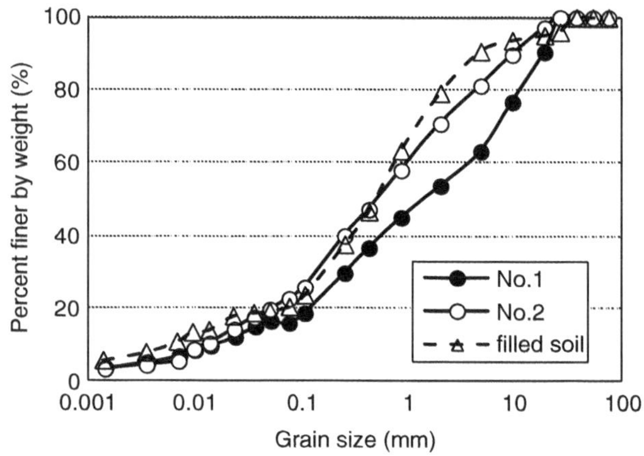

Fig. 9.7 Grain-size distribution of soil samples

9.2.4 *The Reasons for Survival of the Wall*

Figure 9.8 shows the state of the tire retaining wall at the time of earthquake (Fig. 9.8a) and tsunami (Fig. 9.8b). The tire retaining wall has the confining effect which can make the wall strong against earthquake and could prevent any sliding failure or surface failure of the backfill soils. On the other hand, permeable and flexible structures like the tire retaining wall can reduce the earth pressures and water pressures during earthquakes and tsunamis. In addition, the earthquake motion could be attenuated because of the seismic isolation characteristics of the tire and that prevented the failure of the backfill ground. On the other hand, as shown in Fig. 9.8b, tsunami entering from the backfill side damaged the building situated on the backfill ground and also eroded the natural slope opposite to the road parallel to the building. However, the tire wall could prevent damage to the backfill as well as the sloping side of the backfill from erosion. The flexible structure also could prevent any scouring of the wall at the bottom, a phenomenon which was prevalent in almost any seawalls, sea dikes, breakwaters, and quay walls in many parts of Tohoku area due to the tsunami this time.

If we calculate the coefficient of permeability, k using the relationship proposed by Creager et al. (1945) based on the D_{20} of Fig. 9.7, the permeability of the No. 1 and No. 2 samples was found to be $k = 1 \times 10^{-5}$~10^{-6} m/s, which is rather small. The fine contents F_c of the two samples (16 % for No. 1 and 23 % for No. 2) are rather high. Therefore, it is very unlikely that the water level inside the embankment could increase or decrease according to the tsunami water level.

The fact that tires are strong against scouring is evident from the fact that even during the backrush, the tsunami could not do any damage to the retaining wall. The hoop tension due to confinement, isolation effect, and the anti-scouring effect is due to the high strength attributable to the hoop stress in each individual tire and the resiliency of the tire retaining wall (Fukutake and Horiuchi 2007; Yasuhara

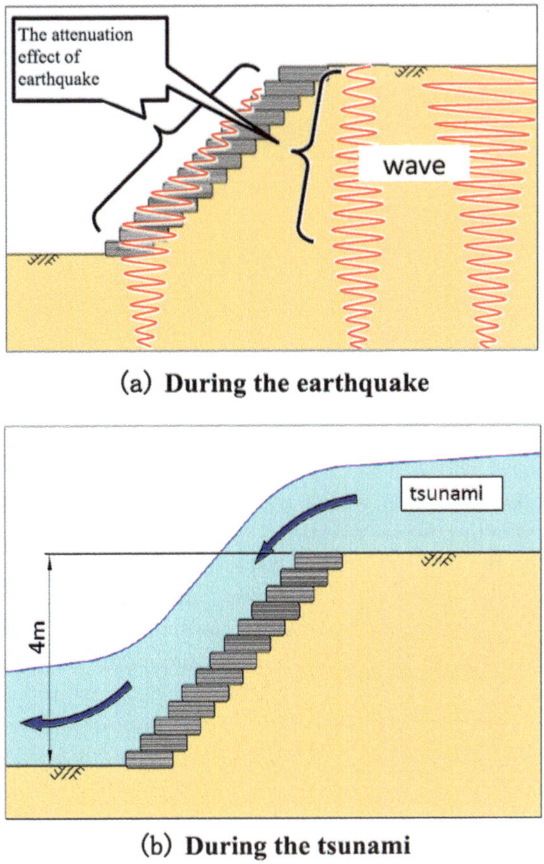

Fig. 9.8 Behavior of the tire retaining wall at the time of earthquake and tsunami. (**a**) During the earthquake. (**b**) During the tsunami

et al. 2006). However, to arrive at a very definite conclusion, detailed study is required.

The performance of the wall during the tsunami can be explained using the classic Newton's law, which is explained using Eq. (9.1) and Fig. 9.9:

$$F = \frac{mv}{\Delta t} \tag{9.1}$$

where F is the impact force acting on the structure, m is the approaching mass, v is the velocity with which the mass approaches the structure, and Δt is the time interval within which the mass acts on the structure.

The force of impact will be less when the time interval will be large. Due to flexible nature of tires, the impact force will be acting for a comparatively longer time, which resulted in a lesser force acting on the structure. In addition, permeable nature of the structure allows the force to dodge rather than bounce, which resulted in the good performance of the structure. A classic example of this technique is found in sports like sumo wrestling or judo.

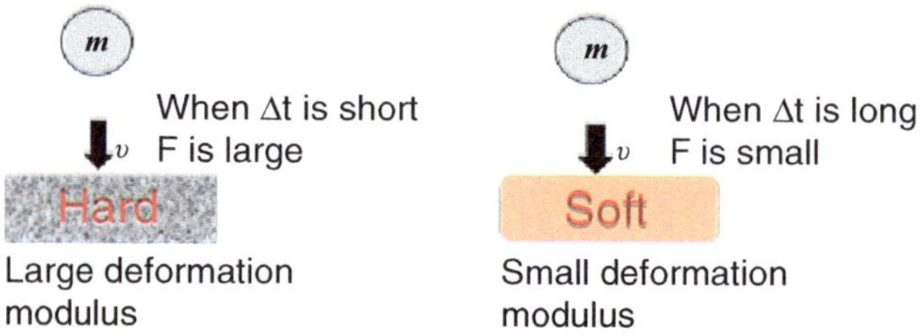

Fig. 9.9 Impact force on structure depending on the structural materials

9.3 Countermeasure Against Tsunami-Induced Scouring

Before discussing about any countermeasures to protect seawalls from future damage during tsunami, damage mechanism of seawalls due to tsunami should be understood well. The damage mechanism is mainly due to scouring at the back of the structures which can be divided into two stages. In the first stage, tsunami (leading wave) overflows the seawall, and in the second stage, tsunami returns back to the landside (called backrush). When tsunami overflows the seawall, it collides with the seawall foundation, and due to rapid overflow, the soils of the seawall embankment is scoured due to the impact of water resulting in outflow of concrete panel covering of seawall embankment. On the other hand, when tsunami returns back to the landside, the force of backrush hits the wall with a huge impact force, and as a result, the seawall collapses completely due to reduced strength of the foundation and the structure. Examples of failures of many seawalls and breakwaters during the 2011 Great East Japan disaster are reported in Hazarika et al (2012b).

Therefore, to protect the seawall from damage due to earthquake and tsunami, we need to provide two measures simultaneously: (a) protection of soils behind the seawall due to tsunami impact force and (b) protection of concrete cover behind the seawall due to force of backrush. In this research, to protect the seawall from impact force of tsunami, a new concept of using waste tire (a resilient material) structures behind the seawall was developed as shown in Fig. 9.10.

The main purpose of this research is to validate the above technique, which aims to prevent scouring of seawall foundation and resulting damage during tsunami by placing tires behind the seawall. In order to achieve the aim of this research, the following model tests and field experiments were performed:

1. Tsunami overflow test (TOT)

To evaluate the tsunami impact force absorption property and the dispersion effect of the tire structures, tsunami overflow tests (TOT) were conducted by placing the model tires behind the seawall in different laying style and different filling conditions of tires.

2. Field planting test (field test)

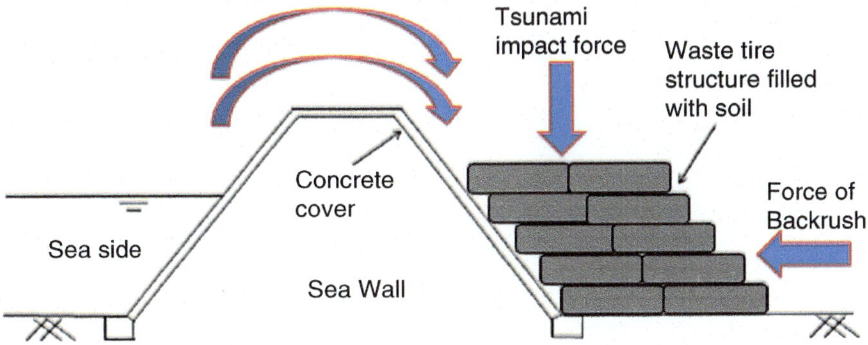

Fig. 9.10 New concept of protecting the seawall by using waste tire structure

To know the growth rates of plants inside the tire, field tests were performed in which various types of plants were cultivated inside tires filled with soils as well as tire chips mixed soils. Tire chips are sustainable materials, which are finding increasing application in Geotechnical Engineering, due to their cost-performance benefit (Hazarika 2013; Hazarika et al. 2008, 2010).

9.4 Tsunami Overflow Test (Tot)

The ability of the tire to reduce the impact force of tsunami will be different according to the number layers of tire, type of filling materials, and pattern of placing of tires. Therefore, to observe and evaluate the optimum condition that can maximize the reduction of tsunami impact force, tsunami overflow test (referred hereafter as TOT) was conducted using the various options (number and laying style of tires, filling materials).

9.4.1 Description of the Test Model

A new apparatus for TOT was developed in Geotechnical Engineering Laboratory of Kyushu University. In this apparatus, a model soil box made of acryl (1200 mm in length, 300 mm in width, and 1000 mm in height) was used to reproduce the overflow phenomenon of tsunami. The schematic diagram of the apparatus developed in this research is shown in Fig. 9.11. The model seawall (height 60 cm and crown width 12 cm) is placed inside the soil box. There is a hinged gate above the model seawall that stores water and can reproduce the overflow phenomena of tsunami. A constant hydraulic pressure is maintained by storing a fixed amount of water, which is pooled on the landside to reproduce the tsunami overflow. In order

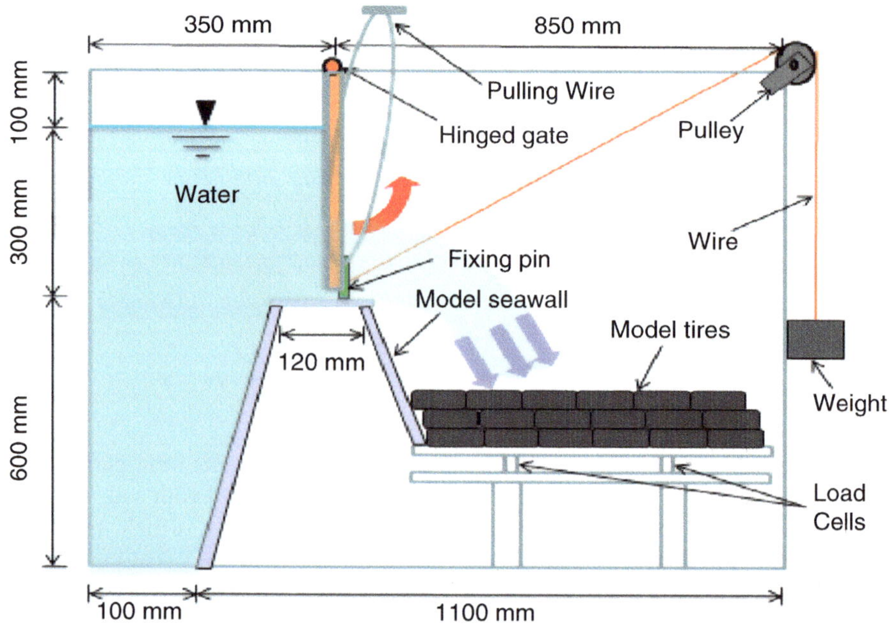

Fig. 9.11 Cross-sectional view of the Tsunami Overflow Test (TOT) apparatus

to allow overflow at a constant rate, a hinged gate was constructed as shown in Fig. 9.11. The hinged gate is connected to a weight (1527 g) by means of a wire and a pulley. The gate can be kept closed by fixing two pins in front of the gate as shown in the figure so that the water to be overflowed can be stored.

By pulling up the pin, the gate can be opened, resulting in the overflow of stored water behind the gate. A series of tests were performed with different test conditions under constant overflow condition. The reduction of water impact force due to model tires was recorded using load cells.

9.4.2 Construction of Model and Data Acquisition System

In this study, model tires with outer diameter 85 mm, inner diameter 45 mm, and thickness 21 mm were used as shown in Fig. 9.12. In order to provide sufficient rigidity, tires should be filled with some suitable materials. Soils and tire chips (size less than 2 mm) were selected as the filling materials as they are easily available in the desired quantity. At the same time, the stiffness of tire structure should be determined depending on the number of tire layers and laying pattern of tires so that the stiffness values can be compared and correlated (Hazarika et al. 2013a, 2014; Pradhan et al. 2014).

Fig. 9.12 Dimension of model tire

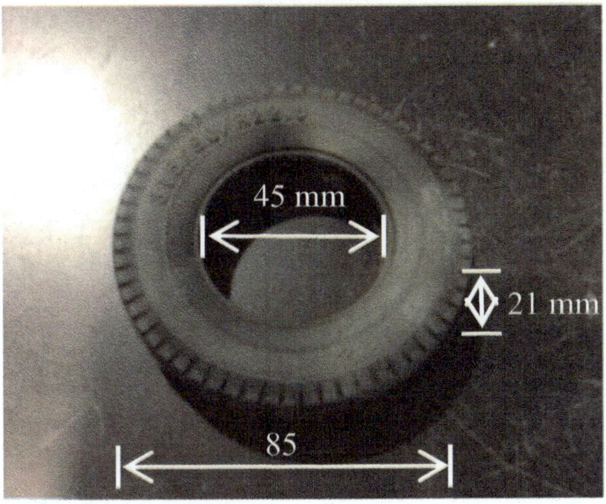

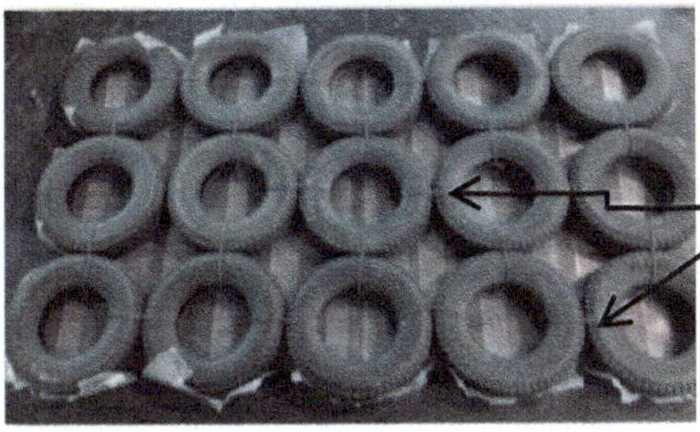

Fig. 9.13 Connections of the tires

Model tires were connected together using thin binding wires so that they cannot spread out due to the impact force of water (Fig. 9.13). In addition, the whole tire assembly was fixed to prevent it from horizontal displacement during the experiment.

Arrangements of the load cells are shown in Figs. 9.14 and 9.15, respectively. Five load cell tables (named as Table 1, Table 2, Table 3, Table 4, and Table 5) were constructed which were made of acrylic plate. Each table was connected with two load cells as shown in Fig. 9.14. The average value of the impact force was calculated using the values recorded by the two load cells in each table.

9.4.3 Test Conditions

In order to determine the maximum reduction of water impact force, the tire structure was constructed using different laying patterns as well as different filling

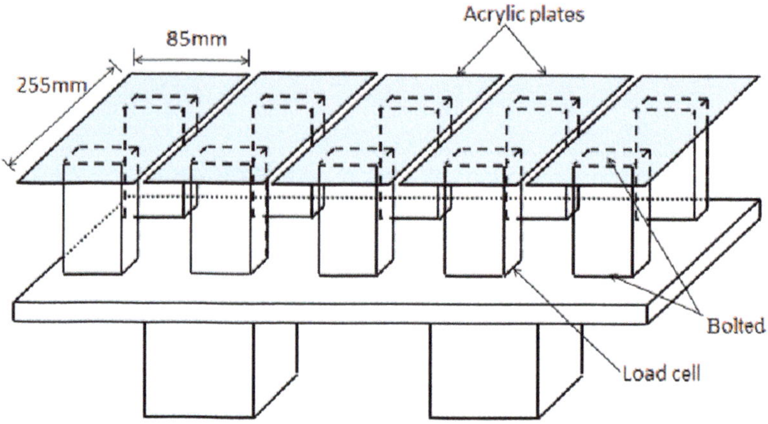

Fig. 9.14 Load cells and load cell table

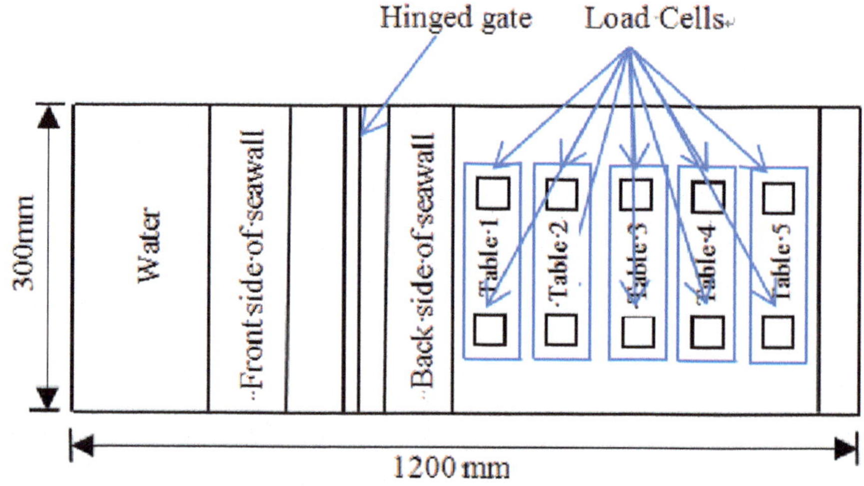

Fig. 9.15 Arrangement of load cells inside of TOT apparatus

conditions (materials). The following laying patterns of model tires were used (Fig. 9.16): Case 2 (two layers vertical), Case 3 (three layers vertical), and Case 5 (slopped three layers). Two-layer and three-layer tires were used to observe whether the reduction of water impact force will be affected by the depth of the filling material. Slopped three-layer tires were used because this will be more practical in the actual seawall construction.

Tires were filled with two kinds of filling materials: pure soils and tire chips (size less than 2 mm). The dry density (ρ_d) of soil filling was 1.43 g/cm^3, while that of tire chips was 0.47 g/cm ^{3}. Hollow tires (without any filling) were also used in the test. The conditions of filling are displayed in Fig. 9.17.

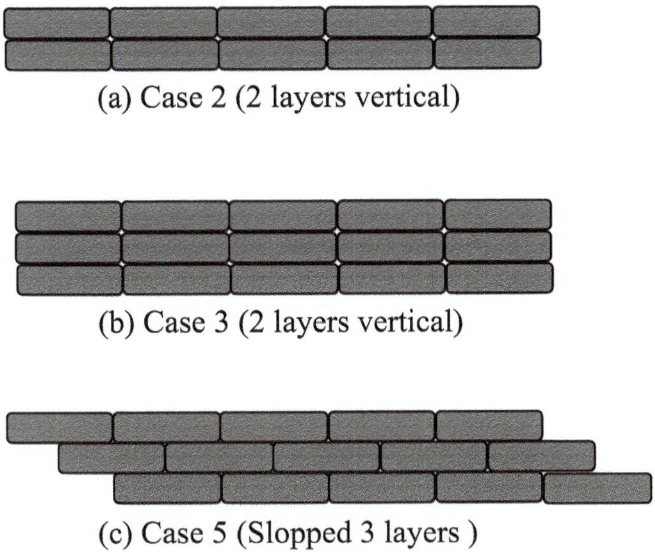

Fig. 9.16 Different laying patterns of tires. (**a**) Case 2 (2 layers vertical). (**b**) Case 3 (2 layers vertical). (**c**) Case 5 (Slopped 3 layers)

Fig. 9.17 Conditions of tire specimen according to filling materials. (**a**) Hollow tire. (**b**) Tire filled with soils ($\rho_d = 1.43$ g/cm^3). (**c**) Tire filled with tire chips (Size <2 mm), ($\rho_d = 0.47$ g/cm^3)

9.4.4 Test Results and Discussion

For each test condition, the average values of the two load cells in each load table were calculated, and they were plotted against time. From these plots, the maximum value in each load table can be determined for different tire laying patterns.

Figure 9.18 shows the typical variation of the water impact force with time when no tire was not placed behind the seawall. The graph indicates that after 1.6 s, a maximum impact load of 50 N was recorded.

Figure 9.19 shows the results using hollow tires under the different laying conditions. The graph indicates that by changing the laying style of tires from Case 2 to Case 5, the impact force of water could be decreased from 39 N to 26 N.

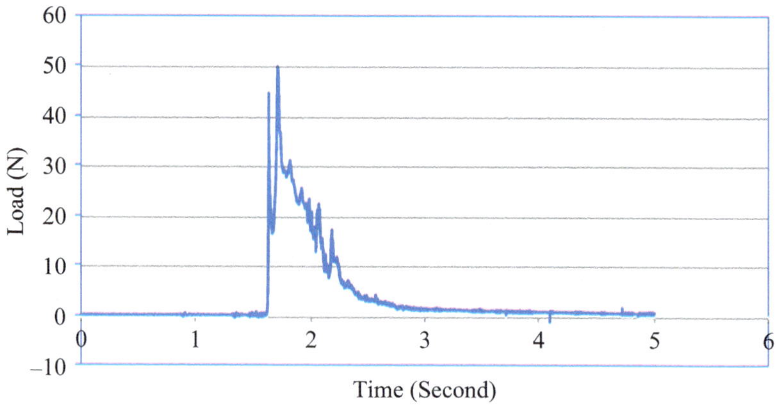

Fig. 9.18 Impact load recorded by Table 3 without using tire

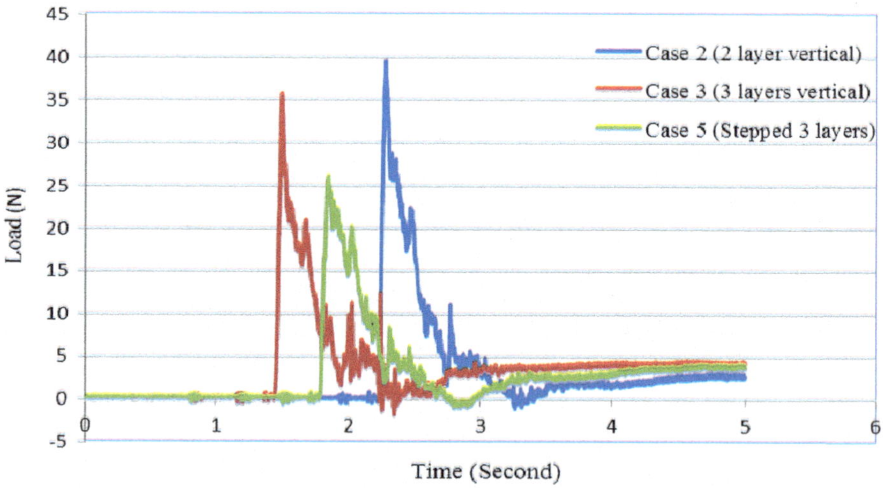

Fig. 9.19 Reduction of water impact force in case of hollow tire

Figure 9.20 shows the performance of soil-filled tires under the different laying conditions. The graph clearly indicates that by changing the laying style of tires from Case 2 to Case 5, the impact force of water could be decreased from 26 N to 18 N.

Similarly, Figure 9.21 shows the performance of tires filled with tire chips under the different laying styles of tires. In this case, the impact force of water could be decreased only up to 22 N. In this case, the effect of laying patterns on the reduction performance is not significant as compared to the conditions when they are filled with soils.

The trends in the measured data in Figs. 9.19, 9.20, and 9.21 indicate that by changing the laying style of tires from Case 2 to Case 5, the impact force could be

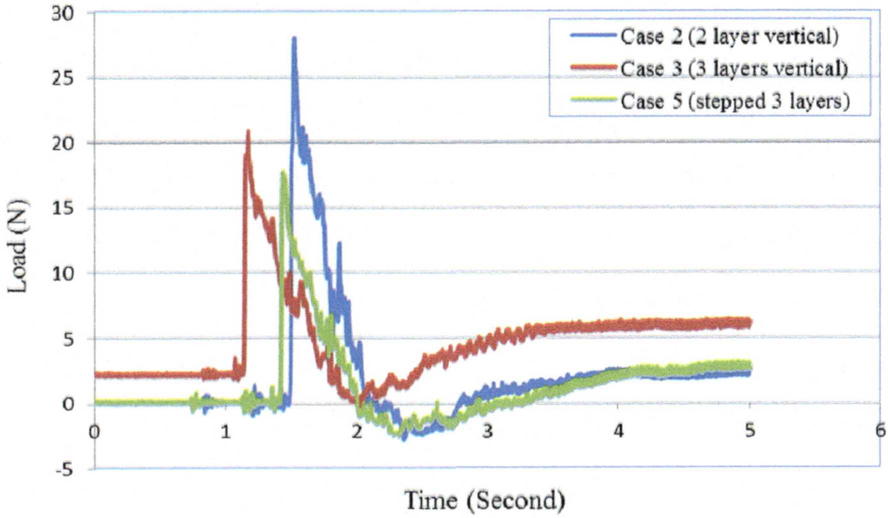

Fig. 9.20 Reduction of water impact force in case of tire filled with soil

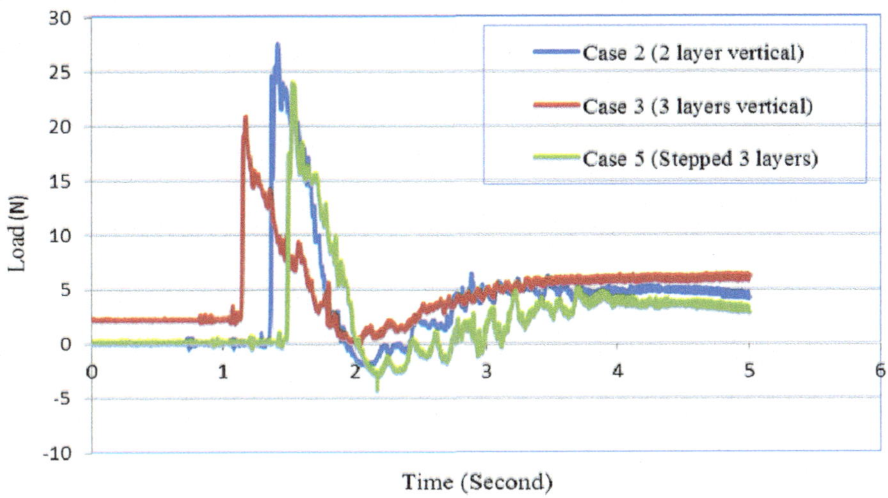

Fig. 9.21 Reduction of water impact force in case of tire filled with tire chips

gradually decreased in all the filling conditions of the tire. Also, tires filled with soils lead to the significant decrease of the impact load as compared to tires filled with pure tire chips. Therefore, it can be said that, instead of using pure tire chips, mixing tire chips with soils can be a viable alternative.

Since the highest value of impact load was recorded in Table no. 3 (Fig. 9.15), therefore, the discussions that follow will be in relation to the data recorded in that load table. Also, since the stepped arrangement of tires leads to the maximum reduction of impact load, the discussion will be limited to only that particular case (Case 5 of Fig. 9.16).

Figures 9.22, 9.23, and 9.24 show the results for the three different filling conditions of tires. The maximum value of impact load without any tires behind the seawall was recorded to be 50 N. In case of hollow tires, the impact load could be reduced to 25 N, implying 50 % reduction of load. In case of tires filled with soils, the impact load could be reduced to 18 N, implying 64 % reduction of load. In case of tires filled with tire chips, the impact load could be reduced to 23 N,

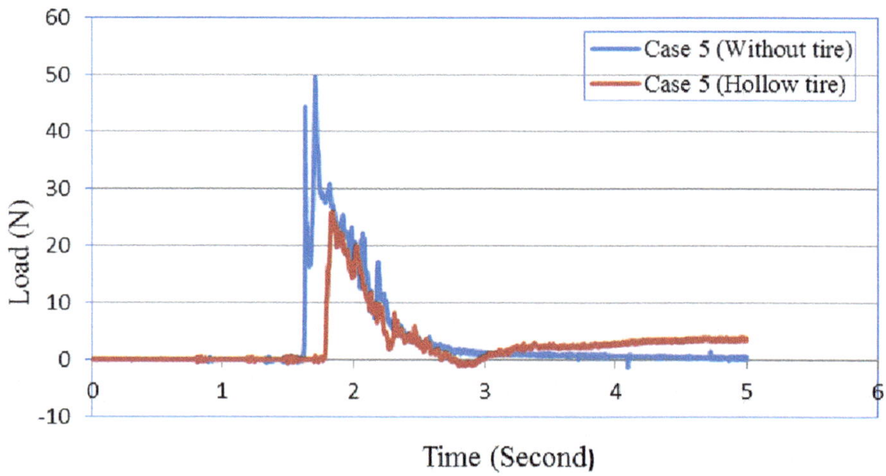

Fig. 9.22 Impact load recorded by Table 3 for slopped three layers (without using tires and hollow tires)

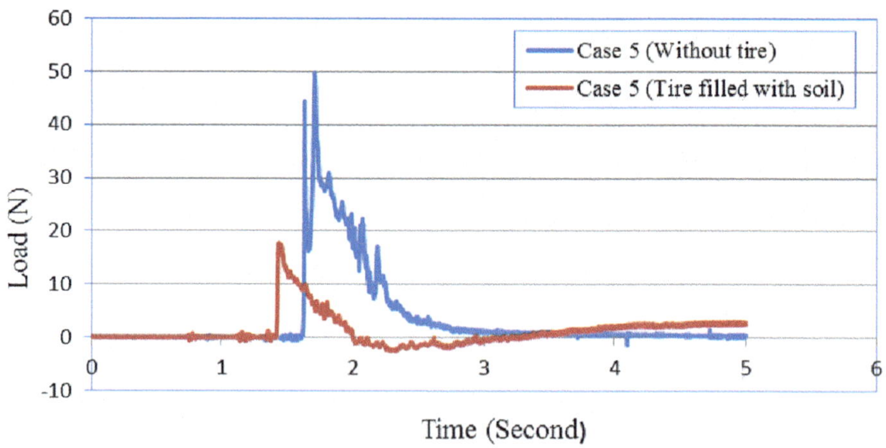

Fig. 9.23 Impact load recorded Table 3 for slopped three layers (without using tires and soil-filled tires)

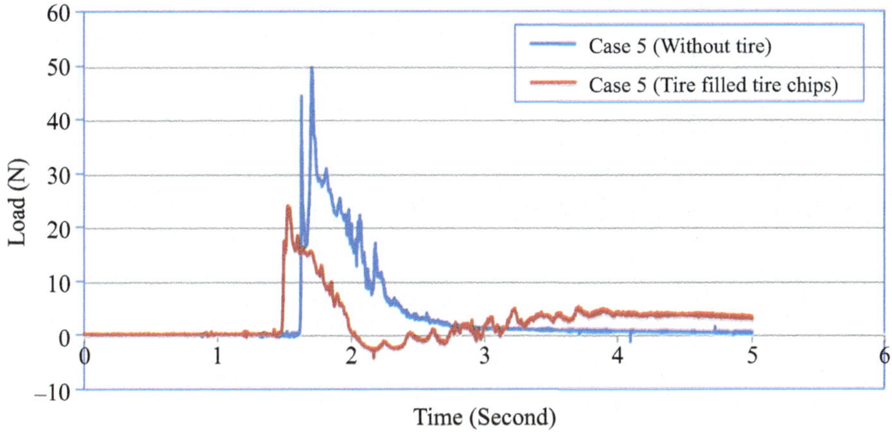

Fig. 9.24 Impact load recorded Table 3 for slopped three layers (without using tires and tire chips filled tires)

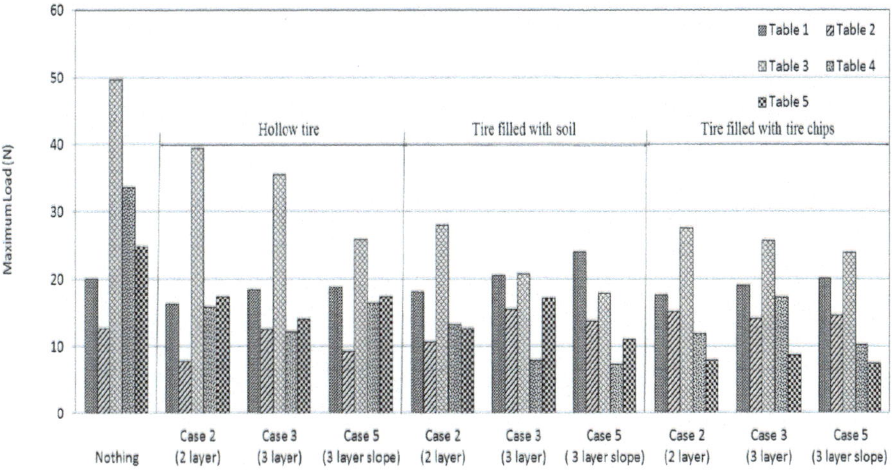

Fig. 9.25 Maximum load recorded in each load table for all the test conditions

implying 54 % reduction of load. Therefore, it can be concluded that a tire filled with soils is the most effective solution. The above results, thus, indicate that the reduction of impact force using a tire is sensitive to the laying patterns of tires and the filling materials of the tires.

Figure 9.25 shows the result of maximum value of impact loads recorded by each load table (Table 1 to Table 5) due to overflowed water. In the figure, the peak values of each load table with respect to the different conditions of tire placement are indicated. It can be observed that the largest value of load, which is expected to cause the most serious damage limited to this model apparatus, gradually decreases with the introduction of tire layers behind the seawall. Paying attention only to load

Table 3 and Case 5, it can be seen that as compared to the case without tires, the case of tires filled with soil shows the smallest value of impact force. On the other hand, the case with hollow tires shows the largest value of impact force.

9.5 Field Test on Plant Cultivation Inside Tires

From aesthetic point of view, tire structures with filled soil on the back of a seawall are not appealing. Therefore, cultivation of suitable plants inside actual tires (used passenger car tire) was conducted in the field using a yard inside the Ito campus of Kyushu University. It is to be noted that plants are to be grown inside the tires near the coastal area, where the soil conditions will be generally saline. Concentrated component of salt (Na) can damage plant tissue whether it contacts above or below the ground. High salinity can reduce plant growth and may even cause plant death. Saline soils generally occur along the coastline and barrier islands where seawater may enter and collect in the soil (Appleton et al. 2009). Therefore, plants to be selected are such that they can sustain certain saline condition.

9.5.1 Selection of Plants

The availability of plants is dependent on the location and season of the particular place. Furthermore before selecting the plants, cold hardiness zone and heat tolerance zone should be checked well. Therefore, we tried to find plants in our location that can grow easily in a wide range of weather condition so that it can be cultivated inside the tires in coastal area. After careful observation of the characteristics of different available plants, *Kirinsou* and *Dichondra* (Fig. 9.26) were

(a) Kirinsou (b) Dechondra

Fig. 9.26 Selection of plants. (**a**) Kirinsou. (**b**) Dechondra

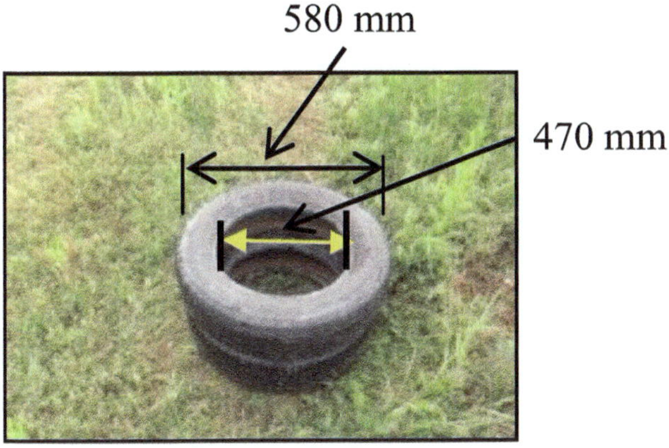

Fig. 9.27 Passenger car tire used in the field test

finally selected to be grown inside actual tires (Fig. 9.27), since these plants can grow in the wide range of climate as well as soil condition.

9.5.2 *Field Test Preparations for Cultivation of Plants*

Tires were filled with soil with a dry density of 1.426 g/cm^3 (same density, which was used to fill the tires of the retaining wall in Okirai, Iwate Prefecture (Fig. 9.1)). Tires also were filled with soil and tire chips (10 % by volume) with a dry density of 0.47 g/cm^3. The idea was to observe the effect of tire chips as a filling material in the growth of plants. According to Cetin et al. (2006), up to 20 % coarse-grained tire chips and 30 % fine-grained tire chips can be used as a lightweight filling material to increase the shear strengths. In this study 10 % of tire chips (size $<$ 2 mm) were mixed with soil to maintain an optimum balance between the strength and growth condition of plants, as high tire chips content may affect the growth of the plants.

9.5.3 *Test Conditions*

Field tests were conducted for three cases: Case a (two layers vertical), Case b (three layers vertical), and Case c (stepped three layers) as shown in Fig. 9.28. Two layers and three layers of tire were used to observe how the plants will grow inside the tires filled with different thickness of soil. In two layers and three layers vertically placed tires, large surface area will be available for the plants to grow, whereas when slopped tires are placed, the surface area will be reduced. Therefore, to observe the growing condition of plants in reduced soil surface area, slopped three-layer tires were used in the test. In addition, to study the effect of tire chips in

Fig. 9.28 Different condition of tire placement for cultivation. (**a**) Case a (Tire 2 layer). (**b**) Case b (Tire 3 layer) (**c**) Case c (3 layer in slope)

the growth of plants, 10 % of tire chips (size < 2 mm) were mixed with soil and used inside the two-layer vertical tire as a filling material.

To maintain the optimum dry density, each tire was filled with soils in three layers, and each layer was compacted well. Weights of the soil and 10 % tire chips to be added were calculated based on the total inner volume of the tire. The growths of the two different plants (*Dichondra* and *Kirinsou*) were observed in the field from July 2013 (beginning of plantation) to September 2013.

9.5.4 Test Observations

Figures 9.29 and 9.30 are the photographs showing the state of growth of *Kirinsou* and *Dichondra*, respectively, for Case a, Case b, and Case c. These photographs were taken at the end of July 2013 and beginning of September 2013. It can be seen that the two plants grow well during the period (July 2013 to September 2013). Also, the growth rate of these plants does not depend on the filling materials and the thickness of the soil layers.

Even though the growth of the two plants from the July 2013 to the September 2013 was very well, the harsh winter made the plants vanish. With the advent of spring (April 2014), *Kirinsou* was found to grow again naturally. However, the growth rate of *Dichondra* was observed to be rather slow. Furthermore, it was also observed that the growth rates of these plants in two different filling materials (10 % tire chips mixed with soil and only soil) were the same. Therefore, after the winter season, observations were made only for Case a (two layers vertical) and Case b (stepped three layers) as shown in Figs. 9.31 and 9.32.

9.6 Conclusions

The following are some of the main conclusions derived based on the different tests, which were performed in this research:

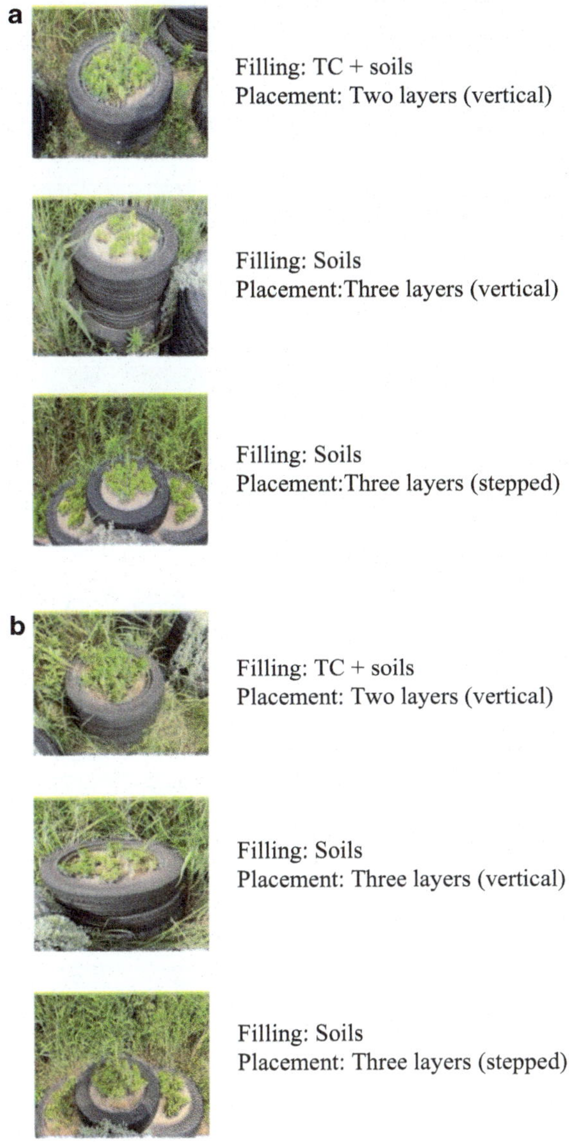

Fig. 9.29 State of growth of *Kirinsou*. (**a**) In the beginning of planting (July 2013). (**b**) At the end of September, 2013

Field Survey of Tire Retaining Wall

1. The tire retaining wall was not damaged in spite of the loose state of the backfill. Damage to the tire retaining wall could be prevented by the earthquake due to confinement effect of each individual tire.

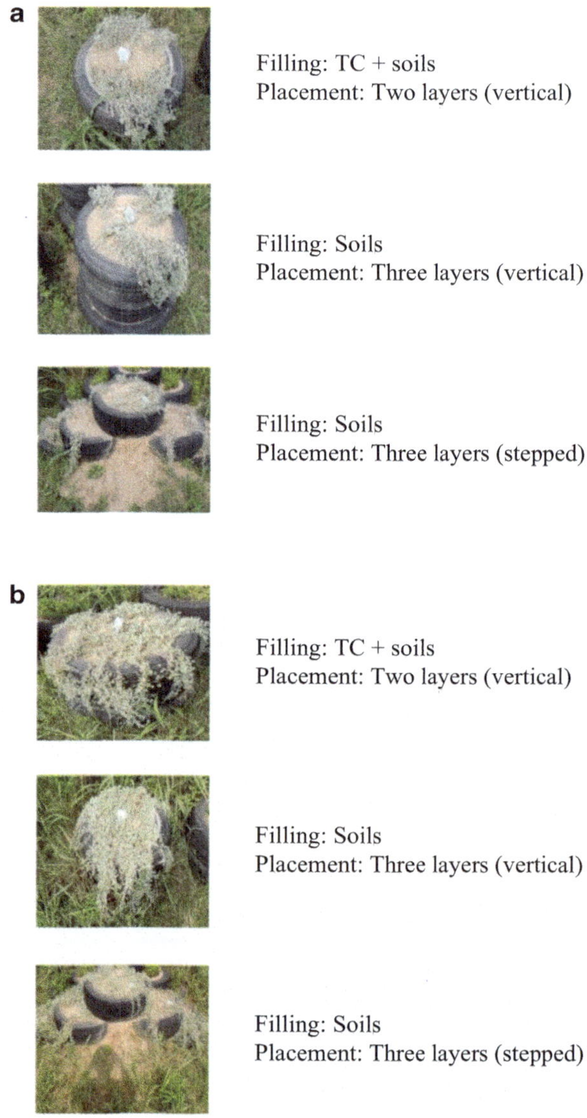

Fig. 9.30 State of growth of *Dichondra*. (**a**) In the beginning of planting (July 2013). (**b**) At the end of September, 2013

2. The flexibility inherent in tires made the earth pressure and water pressure reduce during the earthquake and the tsunami. The seismic isolation effect was the reason behind such good performance.

(a) 2 layers vertical (b) Stepped 3 layers

Fig. 9.31 Growth of *Kirinsou* after the winter season (Case a and Case b) in April 2014. (**a**) 2 layers vertical. (**b**) Stepped 3 layers

(a) 2 layers vertical (b) Stepped 3 layers

Fig. 9.32 Growth of *Dichondra* after the winter season (Case a and Case b) in April 2014. (**a**) 2 layers vertical. (**b**) Stepped 3 layers

Tsunami Overflow Test of the Seawall

3. The tsunami impact force can be reduced considerably by placing filled tires (with a suitable filling material) behind seawalls to protect the damage of such structures from impact force and resulting scouring and erosion.
4. The stepped arrangement of tires results in the maximum reduction of tsunami impact force as compared to the other arrangement of tires.
5. The tires must be filled with soils and with other materials to have certain stiffness, so that they would be able to absorb the impact force of tsunami.

Field Test on Cultivation of Plants Inside Tires

6. The greening effect could be maintained by planting trees inside the tires and could be one of the effective methods for recycling of waste tires.
7. Appropriate plants are to be selected so that the plants can survive in the saline conditions of soils that exist in coastal areas. Depending on the location, geography, and geomorphology, the selection of plants will, however, differ.
8. The filling materials of tires do not have any significant effect on the growth of plants. However, an appropriate percentage of tire mixed in the soils is to be

selected so that the strength of the tire structures is not affected and at the same time plant's growth is not endangered.

In this research, the effect of impact force only due to the leading wave of tsunami was considered. In the future, the effect of the backrush of tsunami is also necessary to investigate. In addition, the impact force reducing property of tires is going to be verified by conducting theoretical analysis based on the theory of conservation of thermal energy (Fukumoto and Izuyama 1992).

The tsunami disaster mitigation technique presented in this paper combines three important factors that we modern-day engineers must deal with cost reduction, environmental protection, and disaster mitigation (Yasuhara 2010). Establishment of the technique is expected to provide a new guideline for not only the scientific and technological fields but also the environmental and cost-performance aspects.

Acknowledgments The financial assistance for a part of this research was provided by Kyushu University under university's P&P research grant for Education and Research in Asian Region. The first author gratefully acknowledges this support. The authors also would like to extend their heartfelt thanks and gratitude to Mr. Kenichiro Yagi, the manager of the land on which the tire retaining wall is located, for generously allowing us to use the area during our field survey. Special thanks go to Dr. Tsuneo Ohsumi of NIED, Tsukuba; Dr. Minoru Yamanaka of Kagawa University, Japan; and Mr. Hideo Furuichi of Giken Co. Ltd., Tokyo, for their support in conducting the field tests in Tohoku area and drafting some of the figures. Last but not least, the authors acknowledge the contributions of Mr. Kiran Hari Pradhan and Mr. Nozomu Hirayu, former students of Geotechnical Engineering Laboratory, Kyushu University, in conducting the tests and data processing.

References

Appleton B, Green V, Smith A, French S, Kane B, Fox L, Downing A, Gilland T (2009) Trees and shrubs that tolerate saline soils and salt spray drift. Virginia cooperative extension, Publication no. 430-031

Central Disaster Mitigation Council (2003) Report of the 16th committee meeting, Investigation committee on Tonankai earthquake and Nankai earthquake. Cabinet Office, Government of Japan (in Japanese)

Cetin H, Fener M, Gunaydin O (2006) Geotechnical properties of tire-cohesive clayey soil mixtures as a fill material. Eng Geol 88:110–120

Creager WP, Justin JD, Hinds J (1945) Engineer for dams. Earth Rock-Fill Steel Timber Dams 3:645–650

Fukumoto Y, Izuyama T (1992) Thermal attenuation and dispersion of sound in a periodic emulsion. J Am Phys Soc 46(8):4905–4921

Fukutake K, Horiuchi S (2007) Stacks of tires for earth reinforcement using their resistance to hoop tension and land reclamation methods. In: Hazarika H, Yasuhara K (eds) Scrap tire derived geomaterials. Taylor and Francis, London, pp 205–214

Hara T (2011) Relationship between geotechnical disaster and building. Wood Ind 66 (11):492–497 (in Japanese)

Hazarika H (2011) Historic tsunami and associated compound disaster triggered by the 2011 great east Japan earthquake – a reconnaissance report. Keynote paper, Geosynthetics India' 2011, Chennai, India, KN57–KN75

Hazarika H (2013) Paradigm shift in earthquake induced geohazards mitigation -emergence of nondilatant geomaterials. In: Keynote lecture for the annual conference of Indian Geotechnical Society. CD-ROM, Roorkee

Hazarika H, Kohama E, Sugano T (2008) Underwater shaking table tests on waterfront structures protected with tire chips cushion. J Geotech Geo Environ Eng ASCE 134(12):1706–1719

Hazarika H, Yasuhara K, Kikuchi Y, Karmokar AK, Mitarai Y (2010) Multifaceted potentials of tire-derived three dimensional geosynthetics in geotechnical applications and their evaluation. Geotext Geomembr 28:303–315

Hazarika H, Okada H, Hara T, Ueno M, Ohsumi T, Yamanaka M, Yamazaki T, Kosaka N, Minowa H, Furuichi H (2012a) Case studies of geotechnical damage by the 2011 off the Pacific Coast of Tohoku Earthquake and tsunami in Japan. In: Proceedings of the 15th world conference on earthquake engineering, Lisbon, Portugal, Paper no 4796

Hazarika H, Kasama K, Suetsugu D, Kataoka S, Yasufuku N (2012b) Damage to geotechnical structures in waterfront areas of northern Tohoku due to the March 11, 2011 tsunami disaster. Indian Geotech J Indian Geotech Soc 43(2):137–152

Hazarika H, Kataoka S, Kasama K, Kaneko K, Suetsugu D (2012c) Compound geotechnical disaster in Aomori prefecture and northern Iwate prefecture due to the earthquake and tsunami. Jpn Geotech J Jpn Geotech Soc 7(11):13–23 (in Japanese)

Hazarika H, Eto I, Yasufuku N, Ishikura R, Fukumoto Y (2013a) Fundamental research on tsunami countermeasures for seawalls using low carbon material. In: Proceedings of the 10th national symposium on environmental geotechnology, Tokyo, Japan, pp 351–354

Hazarika H, Hara T, Furuichi H (2013b) Soil-Structure interaction during earthquake and tsunami – two case studies from the latest disaster in Japan. In: Proceedings of the eighteenth international conference on soil mechanics and geotechnical engineering, Paris, France, pp 131–142

Hazarika H, Pradhan K, Hirayu N, Fukumoto Y, Yasufuku N, Ishikura R (2014) Protection of seawall against earthquake and tsunami using flexible material. In: Proceedings of the 14th international conference of the International Association for Computer Methods and Advances in Geomechanics (IACMAG), Kyoto, pp 1897–1902

Nakamura Y (1989) A method for dynamic characteristics estimation of subsurface using micro tremor on the ground surface. Q Rep RTRI 30(1):25–33

Okada K, Sugiyama T, Noguchi T, Muraishi N (1992) A correlation of soil strength between sounding tests on embankment surface. Tsuchi to kiso 40(4):11–16 (in Japanese)

Pradhan K, Hazarika H, Hirayu N, Fukumoto Y, Yasufuku N, Ishikura R (2014) Experimental evaluation of tsunami impact force absorption by scrap tires. In: Proceedings of the Japanese Geotechnical Society special symposium on the great East Japan earthquake, Tokyo, pp 782–787

Shuto N (2011) History of tsunami in Sanriku area (Part 1): information on tsunami due to 2011 off the Pacific Coast of Tohoku earthquake. Joint investigation report of the 2011 off the Pacific Coast of Tohoku earthquake http://www.coastal.jp/ttjt/index.php (in Japanese)

Yasuhara K (2010) Global warming induced climate change adaptation measures from geotechnical perspec-tives. In: Proceedings of the 65th national conference of Japan Society of Civil Engineers, pp 953–954 (in Japanese)

Yasuhara K, Hazarika H, Fukutake K (2006) Technology for applications of waste tires in geotechnical structures. Doboku Gijutsu 61(10):83–85

Chapter 10
Reinforced Piled Embankments for Sustainable Infrastructure Development

K. Rajagopal and Anjana Bhasi

Abstract The Geosynthetic Reinforced Piled Embankment Systems provide cost economic solutions for the construction of high embankments even on very soft clay soils. They are fast to construct and enable the usage of marginal locally available soils to help in sustainable growth of infrastructure. These systems consist of cement concrete columns acting as pile columns and basal reinforcement at the ground level below the embankment. The design of these systems relies on empirical methods leading to uneconomical solutions. Especially, the design of such embankments resting on floating piles in thick soft clay soils is extremely conservative. No detailed guidelines are available for such cases. Through numerical simulation studies, this paper brings out the performance of these embankments under progressive consolidation of the foundation soil. More accurate arching coefficients are proposed for the design of these systems.

Keywords Geosynthetics • Arching • Pile platforms • Embankments • Load transfer • Finite element modeling

10.1 Introduction

The construction of high embankments for roadways and railway lines through soft clay soils is a challenging task because of several problems caused by large total and differential settlements, difficulty of moving the construction equipment through the construction area, etc. Several techniques ranging from simple replacement methods to pre-consolidation and geosynthetic reinforcement techniques are employed to construct in such difficult conditions as illustrated in Table 10.1. The replacement methods are feasible if the thickness of the soft clay soil strata is very

---

K. Rajagopal (✉)
Department of Civil Engineering, IIT Madras, Chennai 600 036, India
e-mail: gopalkr@iitm.ac.in

A. Bhasi
Department of Civil Engineering, NIT Karnataka, Surathkal, Karnataka 575025, India
e-mail: anjanabhasi@yahoo.com

G.L. Sivakumar Babu et al. (eds.), *Sustainability Issues in Civil Engineering*,
Springer Transactions in Civil and Environmental Engineering,
DOI 10.1007/978-981-10-1930-2_10

Table 10.1 Different methods of ground treatment for soft clay foundations

Method	Advantages	Disadvantages
Replacement	Fast and easy	The disposal of extracted soil and transportation of the new fill material increase the cost
		It is an expensive method for a large area and/or deep deposits of soft clay
		Issues like stability of the cut edge, dewatering if ground water is encountered
Preloading	Easiest method	Time for consolidation very high (2–10 years)
	Does not require extra materials except for preloading weight (cost effective)	Transportation and disposal of the preloading weights could increase the total cost
	It is very efficient if enough consolidation time is allowed	
Preloading with vertical drain	Faster than preloading	Costlier than preloading
	Compared with sand drains, the installation of PVDs is much easier and faster	Monitoring is required to determine the degree of consolidation
Vacuum preloading	Compared with the two preceding preloading techniques, it saves time and cost of transporting preloading weights	Efficient pumping equipment and impervious membrane are vital
		Limited experiences are available
Lightweight material	Fast and easy to handle	The lightweight material with reasonable strength is expensive
	Least disturbance on in situ soils	Protective cover is needed
Jet-grouting	Fast and reliable method	Expensive
	It is a very popular ground improvement technique	Complicated
Stone column, sand column	Fast	They are expensive
		Cannot be used in very soft soil situation since stones and sand need some confinement to sustain the strength
		Geosynthetic encased columns can overcome the above limitation
Piled raft	Fast, effective, and reliable	It is almost the most expensive technique among the ones discussed
		It is typically used for bridge approach embankment
Geosynthetic reinforced	Faster construction	Theories have not been well established
Column-supported system	Eliminates differential settlement especially for large height embankments	Limited studies or guidelines available on the use of floating piles
	Low long-term maintenance cost	
	Slope stability	

Several researchers have worked on the above topics, e.g., Bhasi (2013), Bhasi and Rajagopal (2014), etc low. The pre-consolidation techniques are suitable if there is sufficient lead time before opening the road for traffic due to large total and differential settlements. Probably the best option is to use reinforced concrete piles with a raft. However, this option is very expensive and not feasible for large areas of treatment. The use of stone columns or jet columns may not work or become quite expensive in case of very soft clay soils with cohesive strength of 10 kPa or less. Flexible solutions using geosynthetic materials are promising as they allow for use of local materials, and the construction technologies are simple.

10.2 Geosynthetic Reinforced Piled Embankment Systems

The Geosynthetic Reinforced Piled Embankment Systems (GRPES) consist of cement concrete piles and one or more layers of horizontal geosynthetic reinforcement layers at the base to help in load transfer from the soil to the piles. Figure 10.1 illustrates the concept of GRPES.

The diameter, length, and center to center (c/c) spacing of piles and the basal reinforcement layers are designed to minimize the total and differential settlements within the allowable limits. Due to the arching mechanism within the embankment, higher loads are transferred to the geosynthetic reinforcement layers and the piles.

The GRPES support systems are reported to be economical for both the initial construction and their long-term maintenance (Alexiew and Gartung 1999).

10.3 Load Transfer Mechanism in GRPES

Load transfer in the geosynthetic reinforced piled embankments is mainly due to two mechanisms. Firstly soil arching develops as the embankment fill mass between piles has a tendency to move downward due to the presence of soft foundation soil. This movement is partially restrained by shear resistance from the fill above the piles. The shear resistance reduces the pressure acting on the reinforcement but increases the load applied onto the pile caps (Fig. 10.2). This load transfer mechanism is the classical soil arching (Terzaghi 1943). The arches span

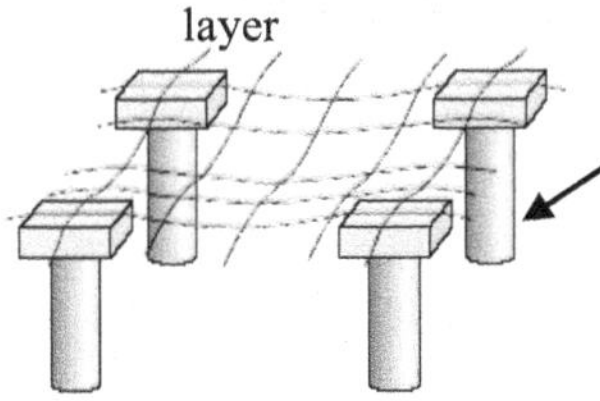

Fig. 10.1 Concept of geosynthetic reinforced piles embankment (After Kempton et al. 1998)

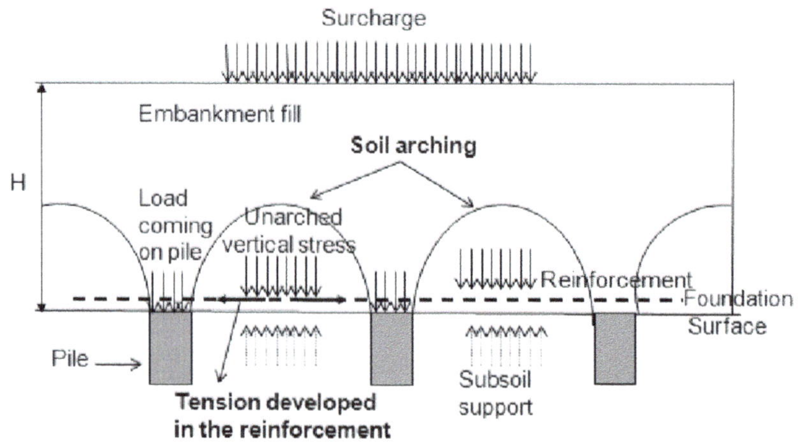

Fig. 10.2 Load transfer mechanism in geosynthetic reinforced pile supported embankments (After Lawson 2012)

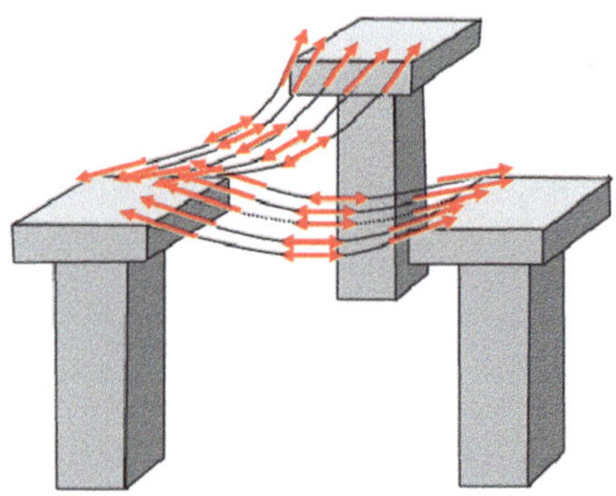

Fig. 10.3 Tension developed in the reinforcement creating membrane effect (www.delftcluster.nl 2008)

the soft soil and the applied load is transferred onto the piles and then to the firm bearing stratum (Kempfert et al. 2004).

Secondly the unarched vertical stress between the columns is assumed to be carried by the geosynthetic reinforcement. This load applied normal to the surface of the reinforcement creates tension in the membrane, creating the membrane effect (Zhan and Yin 2001). Then this load is transferred to the columns through the vertical component of the tensile forces in the membrane (Han and Gabr 2002). Figure 10.3 shows the direction of tensile force developed in the reinforcement.

The tensile load generated in the reinforcement occurs in two directions along the length of the embankment and across its width. The transfer of the unarched portion of the embankment loading occurs in the two directions, while the transfer of the horizontal outward thrust in the fill region occurs across the width of the embankment only. Thus in the direction along the length of the embankment, the tensile load requirements are fairly minimal (T_{min}). For optimal reinforcement layout, the reinforcement should be laid across the width of the embankment (T_{crit}) in one continuous length.

Some reported applications of GRPES are listed below:

- Bridge approach embankments (Reid and Buchanan 1984)
- Airport runway projects (Hossain and Rao 2006)
- Subgrade improvement (Han 1975)
- To minimize differential settlement under storage tanks (Alzamora et al. 2000)
- Foundation for segmental retaining walls (Alzamora et al. 2000)
- Widening of the existing roadway embankment (Han and Gabr 2002)
- To construct confined embankment structures (Lawson 2012)
- Embankment supported over soft ground where only minimal settlement can be tolerated, e.g., highway construction adjacent to a rigid railway structure (www.tensar.co.uk 2012)

10.4 Numerical Analyses of GRPES

Finite element analyses of the GRPES resting on soft clay soil are performed using the program ABAQUS (SIMULIA 2009). As the analysis involves both fluid flow and soil stresses, the analyses were performed using special coupled displacement/pore pressure elements. The finite element equations for the coupled analyses are written as

$$\begin{bmatrix} K & C \\ C^T & -(E + \theta \Delta t H) \end{bmatrix} \begin{Bmatrix} v_{t+\Delta t} \\ u_{t+\Delta t} \end{Bmatrix} = \begin{Bmatrix} F_{t+\Delta t} \\ Q_{t+\Delta t} \end{Bmatrix} \tag{10.1}$$

where $[K]$ is the stiffness matrix of the soil, $[C]$ is the matrix of interaction terms between the soil and the pore fluid, $[E]$ is the compressibility matrix of the pore fluid, θ is the convolution integration factor which varies from 0 to 1, $[H]$ is the matrix governing the dissipation of pore fluid, $v_{t+\Delta t}$ is the nodal displacement vector, $u_{t+\Delta t}$ is the nodal pore pressure vector, $F_{t+\Delta t}$ consists of the incremental applied load terms, and $Q_{t+\Delta t}$ consists of the seepage forces on the boundary.

Full three-dimensional (3-D) analysis is required to accurately represent the actual arching phenomenon and the load transfer mechanism. However, performing 3-D analysis is time-consuming and needs high computing power (Smith and Filz 2007; Ariyarathne et al. 2012). Therefore, two-dimensional (2-D) analysis (Hewlett and Randolph 1988; Russell and Pierpoint 1997; Kempton et al. 1998; Han and Gabr 2002; Ariyarathne et al. 2012) and 3-D column analysis (Yoo and Kim 2009; Bhasi and Rajagopal 2013) are usually employed extensively for numerical analyses. According to Bhasi and Rajagopal (2014) and Yoo and Kim (2009), results obtained from the 3-D column models of geosynthetic reinforced piled embankment and geosynthetic encased stone column reinforced ground shows good agreement with the results obtained from the respective full three-dimensional models.

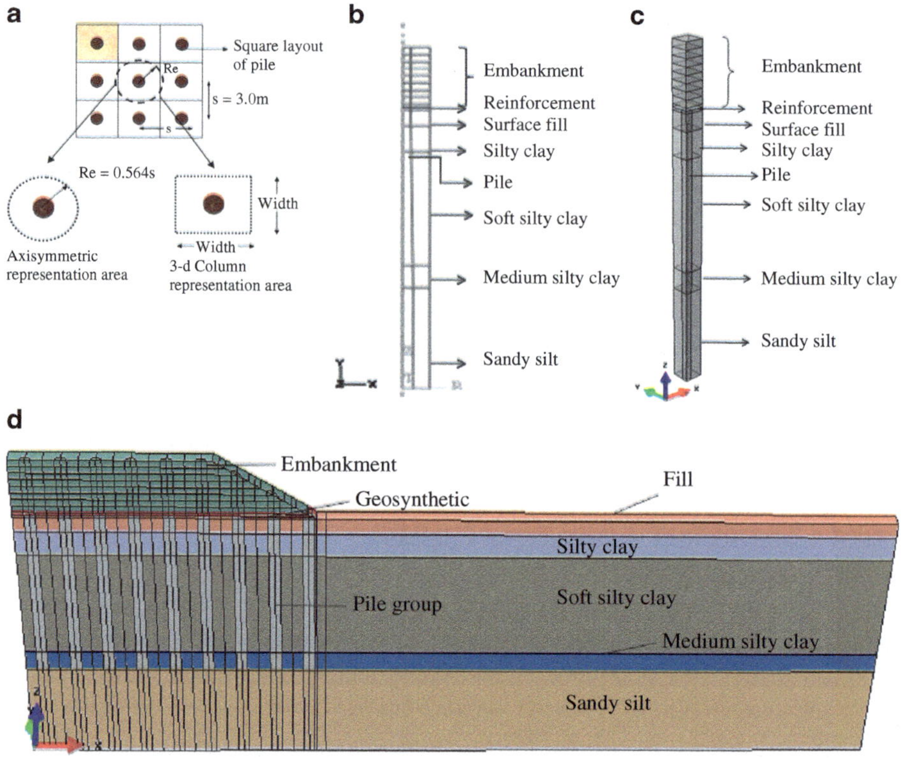

Fig. 10.4 Influence areas and the different modeling approaches used for the analyses. (**a**) Axisymmetric and 3-D column representation regions; (**b**) axisymmetric model; (**c**) 3-D column model; (**d**) full three-dimensional model

The three numerical approaches for simulating the behavior of GRPES are illustrated in Fig. 10.4. A comparison of typical CPU times needed for solving different types of numerical models is illustrated in Table 10.2. All calculations shown are performed on a high-performance computing facility having a processor of Intel Xenon E5472, 3GHz, 1600 MHz FSB with 16 GB RAM.

As all the computations are performed in a shared environment, the actual clock-time required to perform the calculations is much larger than the CPU times shown in the above table.

The diameter of the axisymmetric model was selected so as to produce exactly the same plan area as the three-dimensional column model corresponding to the area replacement ratio used in the field construction by Liu et al. (2007). Due to the symmetry, only one quarter of the pile and its tributary area was modeled (Yoo and Kim 2009). On the vertical planes of symmetry, horizontal displacements were constrained, and no shear stresses were allowed to develop by using smooth roller boundary conditions on these surfaces. The bottom surface of the mesh was made rough and rigid by setting all the displacement components to zero. The water table

Table 10.2 Comparison of different types of numerical models

Model	Embankment		Reinforcement		Pile		Foundation		CPU time
	dof	Elements	dof	Elements	dof	Elements	dof	Elements	
Axisymmetric	2	234	2	9	2	184	3	891	15 min
3-D column	3	1860	3	25	3	452	4	9439	45 min
Full 3-D	3	22,814	3	274	3	6074	4	47,906	90 h

was at a depth of 1.5 m below the ground level, and the initial pore pressures prior to the embankment construction were taken to be hydrostatic. The vertical and bottom boundaries were treated as impermeable surfaces.

In the case of axisymmetric model, eight node stress–pore pressure-coupled axisymmetric elements with reduced integration (CAX8RP) were used to represent the clay soil layer. The embankment fill, reinforced gravel layer, the coarse-grained soil fill above the piles, and the piles were represented using conventional CAX8R elements. The clay soil in the three-dimensional models was discretized using 20-node stress–pore pressure-coupled brick elements with reduced integration (C3D20RP), whereas 20-node brick elements with reduced integration (C3D20R) were used for piles, embankment fill, reinforced gravel layer, and the coarse-grained fill above the piles. The reinforcement layer was modeled using the membrane elements MAX2 (axisymmetric) and M3D8R (3-D models). A linear elastic perfectly plastic model with the Mohr–Coulomb failure criterion was used to model the embankment fill, gravel, and the coarse-grained surface fill. The soft clay soil was modeled using the Modified Cam Clay model. The pile was modeled as an isotropic linear elastic material with a Young's modulus of 20 GPa and a Poisson's ratio of 0.2. The geogrid was modeled as an isotropic linear elastic material with a modulus of 1180 kN/m and a Poisson's ratio of 0.3. Surface to surface contact was used to model the interaction between the reinforcement and the surrounding soil. More details on the material properties assumed can be found in Bhasi and Rajagopal (2014).

10.5 Methodology of Analyses

Initially the foundation soil layer was constructed in one step and its effective self-weight was applied using the geostatic option. To model pile installation, the soft soil elements were removed and replaced by the piles. The embankment construction was then simulated in stages by adding successive layers of elements that form the different embankment layers. Overall height of 5.6 m was achieved linearly over a period of 55 days. After full placement of the embankment, consolidation analysis was carried out for a period of 650 days until the excess pore pressures fell to near zero. Both the responses at the end of construction and at the end of consolidation were analyzed.

Figure 10.5 shows the comparison between the foundation soil settlements predicted by different numerical models with the measured values by Liu et al. (2007). It can be observed that the unit cell models which are computationally inexpensive are able to accurately simulate the general trend of the measured data. The result demonstrates that the axisymmetric, 3-D column and full 3-D models are reasonable for simulating the settlement behavior of geosynthetic reinforced piled embankments. Further calibration of the numerical models was performed for both end-bearing and floating piles against the data reported by Lee et al. (2002). More details of these analyses can be found in Bhasi (2013).

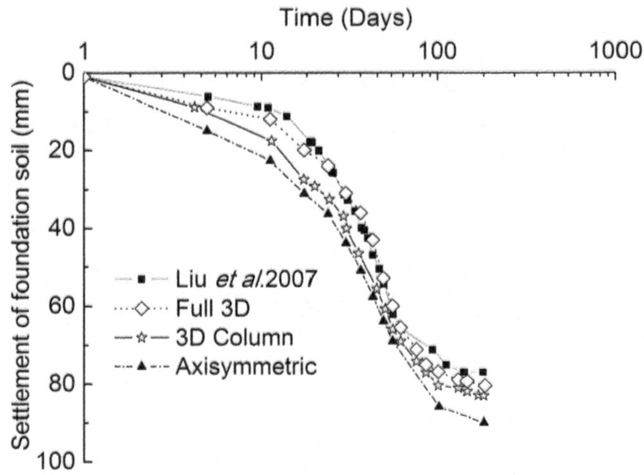

Fig. 10.5 Time vs. settlement behavior predicted by three different numerical models

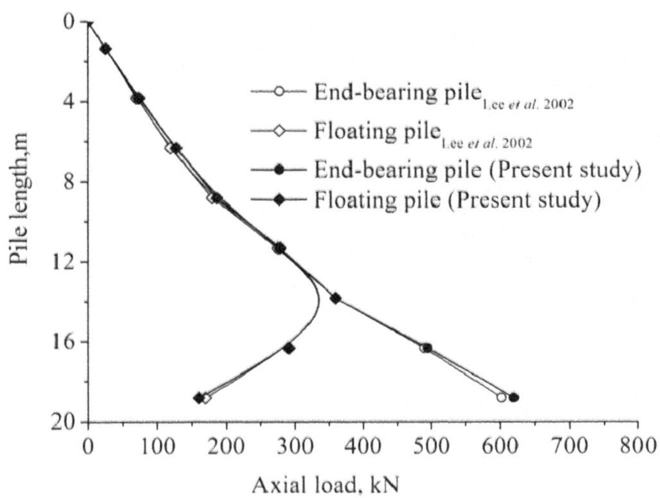

Fig. 10.6 Axial load variation in pile for both end-bearing and floating piles

The results presented in Figs. 10.5, 10.6, and 10.7 show that the numerical models developed in the current research work are able to capture the essential behavioral aspects of both end-bearing and floating piles. More details of these analysis could be found in Bhasi (2013) and Bhasi and Rajagopal (2014).

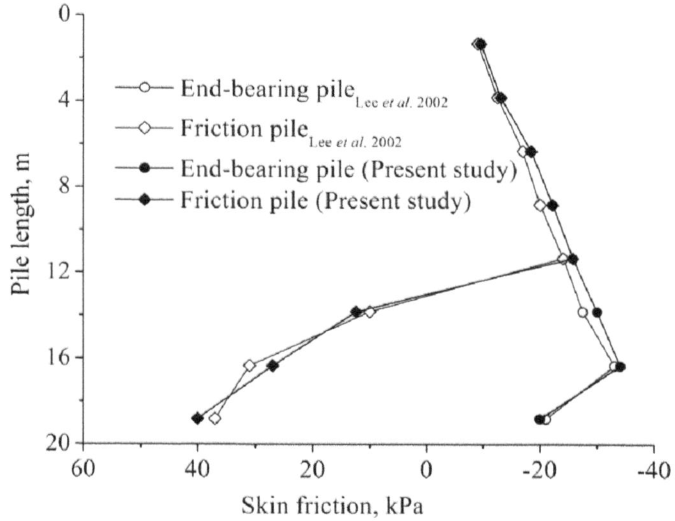

Fig. 10.7 Variation of skin friction stress along the length of the pile

10.6 Parametric Studies

A hypothetical construction of geosynthetic reinforced embankment supported on piles is considered for the study. A 4 m high embankment of crest width 35 m and having side slopes of 1 V:1.5H, comprising sandy soil, is constructed on a 35 m thick soft clay layer below which a rigid layer exists. A working platform of 1 m is provided for laying of the piles and the basal reinforcement. Piles of 0.6 m diameter (a) are assumed to be placed in a square grid pattern at 2.2 m c/c spacing (s). The area replacement ratio corresponding to this spacing is approximately 6 %. Detailed material parameters and the constitutive models used are listed in Table 10.3. The reinforcement is modeled as an isotropic linear elastic three-dimensional membrane element having tensile modulus of 1200 kN/m and a Poisson's ratio of 0.3. The pile was modeled as an isotropic linear elastic material with a Young's modulus of 20 GPa and a Poisson's ratio of 0.15.

Satibi (2009) has developed an analytical equation considering the vertical force equilibrium of a soil element to find the critical length (ℓ_{crit}) of a floating pile. Critical length is defined as the pile length at which the entire embankment load is supported by the floating pile of length ℓ_{crit}. Critical length is calculated using the equation

$$q_{\text{emb}} + \gamma'_{\text{soft}} l_{\text{crit}} + \frac{\gamma'_{\text{soft}}}{\beta} = \frac{\gamma'_{\text{soft}} e^{\beta l_{\text{crit}}}}{\beta} \text{ where } \beta = \frac{2\text{r}}{\text{R}^2 - r^2} K_0 \sin \phi'_{\text{softsoil}} \qquad (10.2)$$

where $\gamma'_{\text{soft}t}$ is the submerged unit weight of the foundation soft, R is the radius of the unit cell, r is the radius of the pile, ϕ'_{softsoil} is the effective friction angle of the foundation soil, and K_0 is the coefficient of lateral earth pressure at rest.

Table 10.3 Material properties used in parametric studies

Property	Embankment fill (Mohr – Coulomb)	Soft clay (Modified Cam Clay) I	II	III	Working platform (Mohr – Coulomb)
Unit weight (kN/m^3)	19	17	17	18	18.5
Young's modulus (kPa)	20,000	–	–	–	7000
Poisson's ratio	0.3	0.35	0.35	0.35	0.3
Cohesion (kPa)	10				15
Friction angle (°)	30				28
Dilation angle (°)	0				0
Critical state stress ratio, M	–	1.0	1.0	1.2	
Logarithmic hardening constant for plasticity, λ	–	0.2	0.15	0.08	
Logarithmic bulk modulus for elastic material behavior, k	–	0.02	0.03	0.008	
Initial void ratio, e_0	–	1.1	1.3	0.87	
Permeability, k(m/day)	–	1.14×10^{-4}	1.1×10^{-4}	1.1×10^{-4}	

For the embankment under consideration, the critical length was estimated to be 14.5 m using Eq. 10.2. Based on the critical length of 14.5 m, four lengths of pile are chosen for parametric studies, viz., 15 m, 18 m, 22 m, and 36 m. The thickness of the soft clay layer is taken as 36 m and hence the case of 36 m long pile is the case of end-bearing pile.

Two modeling approaches of varying complexities and computational requirements were considered for the analyses, viz., 3-D column model and full three-dimensional (3-D) model (Fig. 10.4). In the 3-D column models, only one quarter of the column and tributary area is considered for analysis by considering symmetry around the pile. Most analyses were performed with 4 m high embankment, while some analyses were performed with a maximum height of 10 m.

10.7 Results and Discussion

The negative skin friction (or negative drag) is developed when the soft soil adjacent to a pile undergoes larger relative settlements leading to transfer of additional loads into the pile. It is important to estimate the negative drag loads accurately while designing the structures supported on floating piles as they lead to total and differential settlements leading to failures (Lee et al. 2002; Leung et al. 1991).

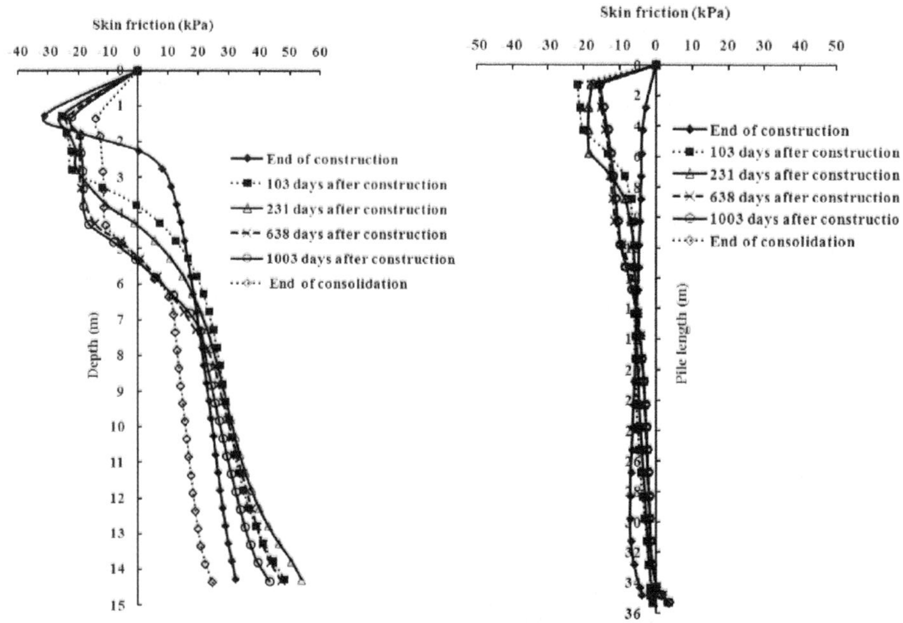

a.15 m long floating piles b. 36 m long pile

Fig. 10.8 Variation of skin friction force along the length of pile with time. (**a**) 15 m long floating piles. (**b**) 36 m long pile

Figure 10.8 shows the distribution of skin friction for 15 m long floating pile and 36 m long end-bearing pile. It is clear that irrespective of the length of the pile, the neutral plane is located at a depth of 2.4 m at the end of construction of the embankment. The location of this plane moves down as the consolidation progresses. The neutral plane location is near the tip of the pile for end-bearing piles as shown in Fig. 10.8b.

The locations of the depth of neutral plane from the ground surface for different lengths of piles are shown in Table 10.4. It is clear that the neutral plane depth at the end of consolidation is about 38 % of the length of the pile. This shows that for floating piles of length greater than the critical length, the depth of neutral plane where the change of negative skin friction to positive skin resistance takes place is the same percentage of length of the pile.

10.7.1 Axial Force Distributions with Depth

Axial force (P_a) was calculated from the vertical stresses developed in the pile using the equation

Table 10.4 Depth of neutral plane for different lengths of pile

	End of construction	End of consolidation	
Length of pile, *l* (*m*)	Neutral plane depth from ground surface (m)	Neutral plane depth from ground surface (m)	Percentage pile length
15 m	2.4	5.7	38.0
18 m	2.4	7.0	38.8
22 m	2.4	8.3	37.7

$$P_a = \sigma_v \left(\pi d^2/4\right) \tag{10.3}$$

where a is the diameter of the pile and σ_v is the vertical stress in the pile element averaged at the same elevation. Figure 10.9 shows the distribution of axial forces along the length of the pile for two cases. The data for all pile lengths is presented in Table 10.5. The axial force has initially increased with depth due to the negative drag forces and reached a maximum value at the neutral plane. For the end-bearing piles, as the neutral plane is near the toe, the axial force showed an increasing trend up to the toe of the pile. The axial force in each case is found to increase as the consolidation progressed and reached the maximum value near the end of consolidation as illustrated in Table 10.5 and Fig. 10.9.

10.7.2 Development of Arching

Distribution of embankment load between different components of a GRPES depends on the amount of soil arching developed in the embankment and membrane action of the geosynthetic reinforcement. As a result of differential movements in the embankment soil, shear resistance (τ) is developed in the fill mass on the vertical interface between the two soil masses. The developed shear resistance reduces the pressure acting on the reinforcement and the foundation soil and correspondingly increases the load transferred to the pile caps. Figures 10.10 and 10.11 illustrate the formation of arches along the width of embankment and show the reorientation of principal stresses in the embankment soil for floating and end-bearing piles. It could be seen that the stresses acting on the foundation soil reduce due to soil arching (indicated by smaller length arrows).

It is clear from these figures that the arching is not an instantaneous phenomenon and that the stresses reorient themselves during the consolidation and the soil arching develops gradually as the consolidation of the foundation soil progresses. The gradual transfer of stresses due to soil arching could be understood from the results presented in Table 10.6 as shown by the reduction of stresses on soil surface.

Fig. 10.9 Distribution of axial force in pile with time and depth below ground surface. (**a**) 15 m long pile. (**b**) 36 m long pile

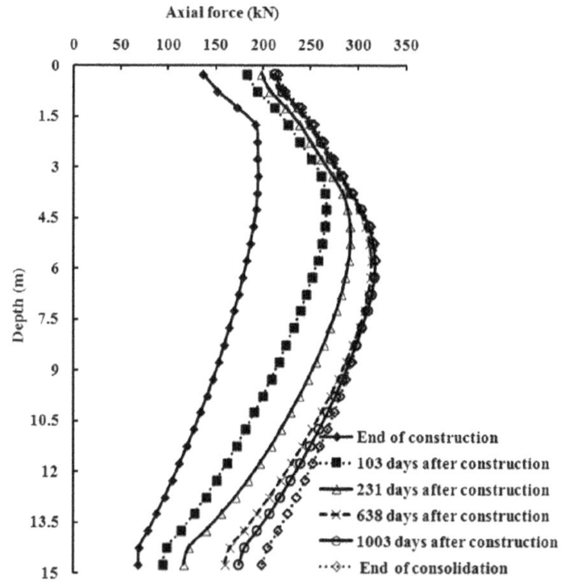

a. 15 m long pile

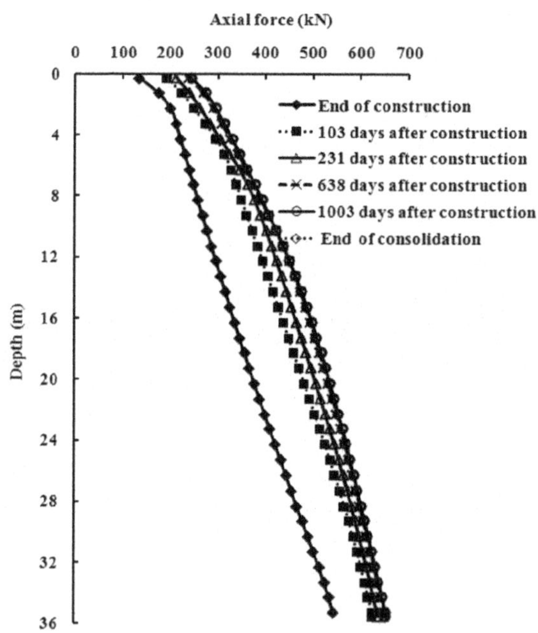

b. 36 m long pile

Table 10.5 Maximum axial force developed for different lengths of pile

Pile length, ℓ (*m*)	End of construction	End of consolidation	
	Maximum axial force developed (kN)	Maximum axial force developed (kN)	% increase in axial force compared to 15 m long pile
15 m	194.4	316	
18 m	195.5	341.3	8.0
22 m	197.8	370.4	17.2
36 m	543	651.1	106.0

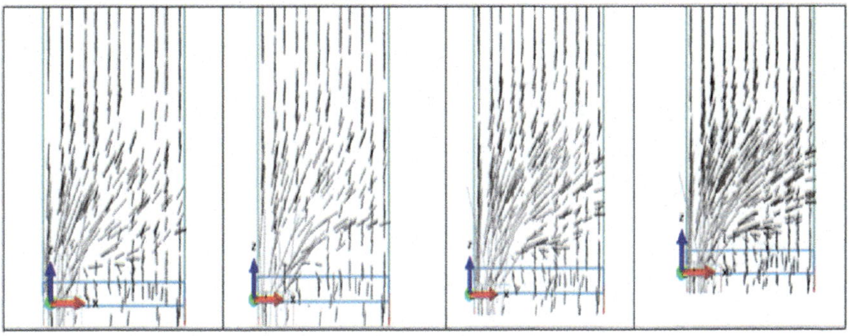

a.during construction (H=8.5 m)   b. during construction (H=10 m)   c. 30 days after construction   d. 100 days after construction

Fig. 10.10 Development of arching in the case of floating piles (**a**) during construction (H = 8.5 m), (**b**) during construction (H = 10 m), (**c**) 30 days after construction, (**d**) 100 days after construction

10.7.3 Stress Concentration Ratio

The stress concentration ratio is defined as the ratio of stress at pile cap level to the static pressure due to the height of soil above the pile cap. Due to soil arching, higher stresses are transferred to the piles. This is reflected in higher stress concentration ratios as the arching develops. Figure 10.12a and b shows the stress concentration ratio for both reinforced and unreinforced cases. The modulus of the basal reinforcement layer is 4000 kN/m. It is clear that the influence of basal reinforcement is to help in transfer of larger percentage of loads into the piles. It is interesting to note that even short-length floating piles are able to attract large amounts of loads due to the interaction with basal reinforcement. Figure 10.13 shows the gradual increase in stress concentration ratio with time. As higher loads are transferred to the piles, the stresses in the foundation soil decrease as illustrated in Fig. 10.14.

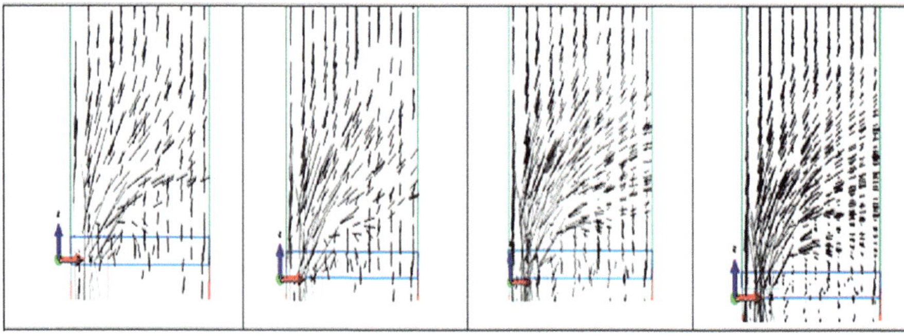

Fig. 10.11 Development of arching in the case of end-bearing piles (**a**) during construction (H = 8.5 m), (**b**) during construction (H = 10 m), (**c**) 30 days after construction, (**d**) 100 days after construction

Table 10.6 Stresses on the pile cap and soil surface

	End of construction		End of consolidation	
Length of pile (m)	Pile head (kPa)	Foundation soil surface (kPa)	Pile head (kPa)	Foundation soil surface (kPa)
15 m	841.2	58.7	1201.2	43.9
18 m	862.9	57.2	1362.5	40.3
22 m	872.8	55.8	1444.1	37.6
36 m	873.1	53.5	1676.3	29.2

10.7.4 Comparison of the Results with BS8006 (2010)

From the work done by previous researchers, it is observed that detailed studies are required for embankments supported on floating piles. Among the design codes, only the British Standard BS8006 (2010) gives consideration for floating piles. When the embankment height H exceeds *1.4(s-a)*, BS8006 assumes that the full arching is developed, and the basal reinforcement or the foundation soil is not affected by the surcharge or any weight of embankment soil beyond that height. Due to this assumption, the predicted tensile force in the reinforcement shows a dip at the transition between partial and full arching, and the tensile forces remain constant with further increase in embankment height. The transition height of embankment for this case is 2.33 m. For the partial- to full-arching assumption (Case1), both BS8006 (2010) and Modified BS8006 proposed by Van Eekelen et al. (2011) show a slight reduction in the reinforcement tensile forces after reaching the full-arching height, and these forces remain constant with further increase in the height of embankment.

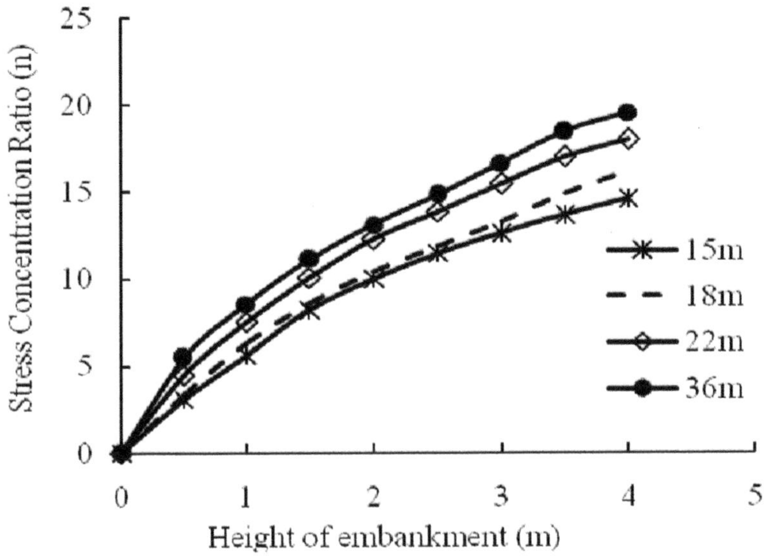

a. with basal reinforcement

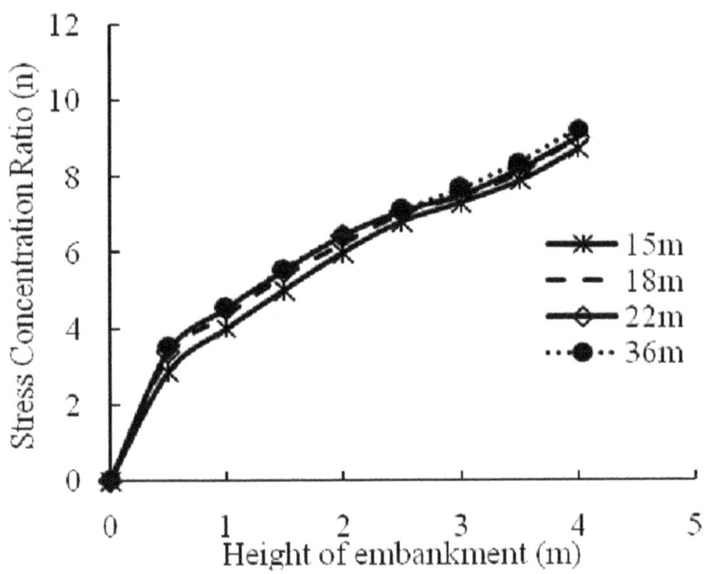

b. without basal reinforcement

Fig. 10.12 Stress concentration ratio on the pile cap (**a**) with basal reinforcement, (**b**) without basal reinforcement

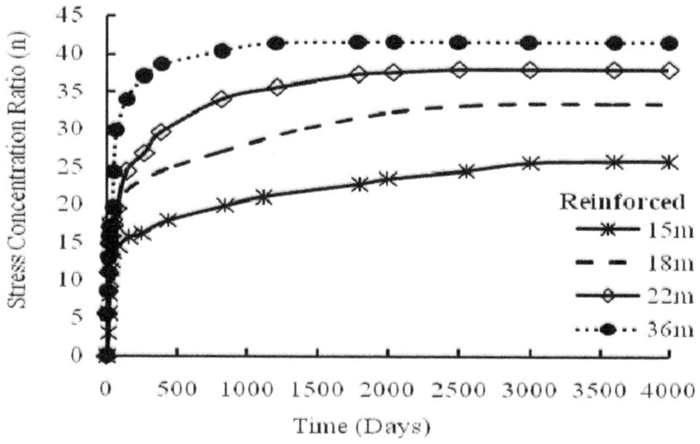

Fig. 10.13 Variation of stress concentration with time

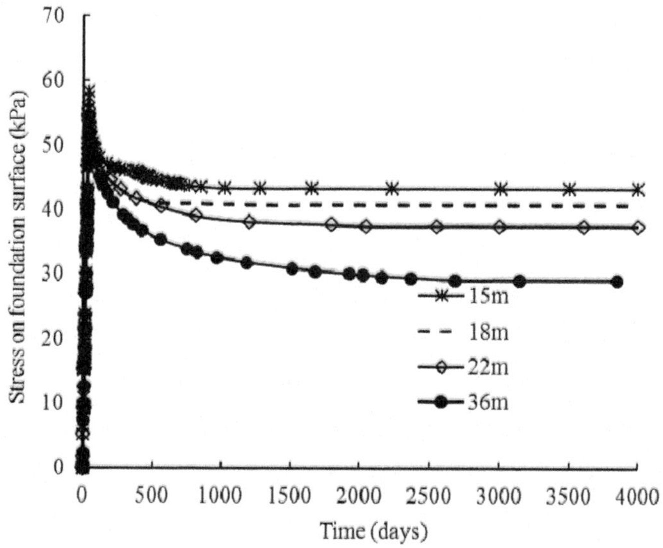

Fig. 10.14 Variation of foundation soil pressures with time

Full three-dimensional analyses were carried out with (R) and without (U) reinforcement layer for different pile lengths of 15 m, 18 m, 22 m, and 36 m. The load acting on the pile without reinforcement caters to the case of load on the pile due to arching alone given by the BS8006. From the equations in BS8006 (2010) and Modified BS8006 (Van Eekelen et al. 2011) used for calculating the stresses acting on the pile, it is clear that the equations do not depend on the length of the pile. However, the results from the numerical analyses clearly indicate that

Table 10.7 Numerically predicted pile loads for different cases

	Pile Load (kN)							
	Floating piles						End-bearing piles	
	15 m		18 m		22 m		36 m	
Embankment height (m)	R	U	R	U	R	U	R	U
1	81.4	41.7	80.6	42.3	81.2	43.1	81.9	43.1
2	140.3	67.1	139.5	69.3	140.5	69.6	141.9	69.7
2.5	169.0	81.6	167.8	84.6	169.0	84.7	170.9	84.8
3	197.7	96.8	196.1	99.4	198.3	99.5	200.1	100.1
4	235.1	122.0	245.7	123.7	250.8	123.8	256.2	125.5

the portion of embankment load transferred into the floating piles depends very much on its length (Table 10.7). Table 10.7 shows the load acting on the pile for different lengths of pile with and without reinforcement.

The numerically predicted pile loads are nearly the same for all the floating pile cases below the arching height of 2.33 m. The pile loads are seen to increase when the height of embankment is increased beyond the arching height. The corresponding pile loads as per the BS code are very low.

From the data presented in Table 10.7, it could be observed that the load acting on the piles showed considerable increase with increase in the embankment height. The rate of increase is higher for a reinforced case than for the unreinforced case. This shows that basal reinforcement is effective in transferring larger loads to the piles.

10.7.5 *Modified Arching Coefficient* (C_c)

Based on the estimated loads for different cases of floating piles and the end-bearing piles, the following modified equations are proposed for the arching coefficient (C_c). As the load transferred to the floating piles are found to be dependent on their length and the location of the neutral plane, these two lengths are incorporated in the equation for arching coefficient of floating piles.

10.7.5.1 End of Construction

$C_c = 1.8\left(1.95\frac{H}{a} - 0.18\right)$ for end-bearing piles.

$$C_c = 2.2\left(1.5\frac{H}{a} - 0.07\right) = 5.8\left(\frac{\overline{x}}{l}\right)\left(1.5\frac{H}{a} - 0.07\right) \text{ for floating piles.} \qquad (10.4)$$

10.7.5.2 End of Consolidation

$C_c = 2.5\left(1.95\frac{H}{a} - 0.18\right)$ for end-bearing piles.

$$C_c = 2.8\left(1.5\frac{H}{a} - 0.07\right) = 7.4\left(\frac{\overline{x}}{l}\right)\left(1.5\frac{H}{a} - 0.07\right) \text{ for floating piles.} \quad (10.5)$$

where $\overline{x}$ is the depth of neutral plane at the end of consolidation, H is the height of embankment, a is the pile diameter, and ℓ is the length of the floating pile.

10.8 Conclusions

The 3-D column models are found to be accurate and fast for the numerical modeling of the behavior of GRPES. The load distribution between the piles and the basal reinforcement is found to be much different from that of the recommendations given in design codes. Based on the studies carried out, the following conclusions are drawn:

1. Formation of arching is not an instantaneous phenomenon and the development of full arching takes place during the consolidation phase.
2. The depth of neutral plane is at a shallow depth at the end of construction and increases with the progress in consolidation.
3. The numerical results do not support the partial- and full-arching assumptions made in BS 8006 (2010).
4. The arching coefficient strongly dependent on the length of the floating piles.
5. The pressure on the foundation soil decreases and the loads transferred to the piles increase as the consolidation of foundation soil progresses.

References

Alexiew D, Gartung E (1999) Geogrid reinforced railway embankments on piles: performance monitoring 1994–1998. In: Proceedings of the 1st South American symposium on geosynthetics/3rd Brazilian Symposium on Geosynthetics, Rio de Janerio, pp 403–411

Alzamora D, Wayne MH, Han J (2000) Performance of SRW supported by geogrids and jet grout columns. In: Proceedings of the ASCE Specialty Conference on Performance Confirmation of Constructed Geotechnical Facilities, vol 94. Geotechnical Special Publication, pp 456–466

Ariyarathne P, Liyanapathirana D, Leo C (2012) A comparison of different two dimensional idealizations for a geosynthetic reinforced pile-supported embankment. Int J Geomech 13:754–768

Bhasi A, Rajagopal K (2013) Study of the effect of pile type for supporting basal reinforced embankments constructed on soft clay soil. Indian Geotech J 43(4):344–353

Bhasi A, Rajagopal K (2014) Geosynthetic reinforced piled embankments: comparison of numerical and analytical methods. ASCE Int J Geomech. doi:10.1061/(ASCE)GM.1943-5622.0000414

Bhasi, A (2013) Performance analysis of geosynthetic reinforced embankments supported on piles, Thesis submitted to Indian Institute of Technology Madras, Chennai for the award of Ph.D. Degree

BS8006 (2010) Code of practice for strengthened/reinforced soils and other fills. British Standard Institution, London

Han J, Gabr M (2002) Numerical analysis of geosynthetic-reinforced and pile-supported earth platforms over soft soil. J Geotech Geoenviron 128(1):44–53

Hewlett WJ, Randolph MF (1988) Analysis of piled embankments. Ground Eng 21(3):12–18

Hossain S, Rao KN (2006) Performance evaluation and numerical modeling of embankment over soft clayey soil improved with chemico-pile. Transp Res Rec, USA, Issue Number: 1952:80–89

Kempton G, Russel D, Pierpoint ND, Jones CJFP (1998) Two and three dimensional numerical analysis of the performance of piled embankments. 6th international conference on geosynthetics. Atlanta, USA, pp 767–772

Kempfert HG, Gobel C, Alexiew D and Heitz C (2004) German recommendations for reinforced embankments on pile-similar elements. Proceedings of the EuroGeo3, Munich DGGT, 279–284

Lawson CR (2012) Role of modelling in the development of design methods for basal reinforced piled embankments. In: Proceedings of Eurofuge 2012. Delf, Netherland

Lee CJ, Bolton MD, Al-Tabbaa A (2002) Numerical modeling of group effects on the distribution of drag loads in pile foundations. Geotechnique 52(5):325–335

Leung C, Radhakrishnan R, Tan S (1991) Performance of precast driven piles in marine clay. J Geotech Eng 117(4):637–657

Liu H, Ng C, Fei K (2007) Performance of a geogrid-reinforced and pile-supported highway embankment over soft clay: case study. J Geotech Geoenviron 133(12):1483–1493

Reid WM, Buchanan NW (1984) Bridge approach support piling. In: Proceedings of the conference on piling and ground treatment. Institution of Civil Engineers, London, pp 267–274

Russell D, Pierpoint N (1997) An assessment of design methods for piled embankments. Ground Eng 30(10):39–44

Satibi S (2009) Numerical analysis and design criteria of embankments on floating piles. PhD thesis, Universität of Stuttgart, Stuttgart, Germany

SIMULIA (2009) ABAQUS/CAE user's manual, Pawtucket, Rhode Island, United States

Smith M, Filz G (2007) Axisymmetric numerical modeling of a unit cell in geosynthetic-reinforced, column-supported embankments. Geosynth Int 14(1):13–22

Terzaghi K (1943) Theoretical soil mechanics. Wiley, New York

Van Eekelen SJM, Bezuijen A, van Tol AF (2011) Analysis and modification of the British standard BS8006 for the design of piled embankments. Geotext Geomembr 29(3):345–359

Yoo C, Kim S-B (2009) Numerical modeling of geosynthetic-encased stone column-reinforced ground. Geosynth Int 16(3):116–126

Zhan C, Yin JH (2001) Elastic analysis of soil-geosynthetic interaction. Geosynth Int 8:27–48

Chapter 11
Experimental Modeling of Unsaturated Intermediate Geomaterials for Sustainable Design of Geotechnical Infrastructure

Laureano R. Hoyos and Anand J. Puppala

Abstract Over the last few decades, intensive and sustained experimental efforts have been undertaken worldwide that have defined the threshold of our understanding of unsaturated soil behavior. The adoption of matric suction and the excess of total stress over air pressure, that is, net normal stress, as the relevant stress state variables, have facilitated the investigation of key features of unsaturated soil behavior via either axis-translation or vapor transfer technique. The present paper documents some of the most recent advances in experimental modeling of intermediate geomaterials over a whole range of suction-controlled paths and modes of deformation. Its main sections describe the test protocols and corresponding results from suction-controlled resonant column, biaxial, triaxial, and ring shear testing programs recently accomplished at the Advanced Geomechanics Laboratory (AGL) of the University of Texas at Arlington, via either axis-translation or vapor transfer technique.

Keywords Unsaturated soil • Matric suction • Total suction • Axis-translation • Relative humidity

11.1 Introduction

Over the last few decades, intensive research efforts have been undertaken worldwide that have defined the threshold of our understanding of unsaturated soil behavior. The adoption of matric suction $(u_a–u_w)$ and the excess of total stress over air pressure, that is, net normal stress, $(\sigma–u_a)$, as the relevant stress state variables, have facilitated the investigation of essential features of unsaturated

L.R. Hoyos (✉) • A.J. Puppala
University of Texas at Arlington, Arlington, TX 76019, USA
e-mail: lhoyos@uta.edu; anand@uta.edu

G.L. Sivakumar Babu et al. (eds.), *Sustainability Issues in Civil Engineering*,
Springer Transactions in Civil and Environmental Engineering,
DOI 10.1007/978-981-10-1930-2_11

soil behavior, via either the axis-translation or the vapor transfer technique, for a wide range of matric and total suction states (Fredlund and Rahardjo 1993; Lu and Likos 2004). It is the relative success of these techniques that has prompted researchers in the discipline to devote countless hours to fine-tuning myriad details of existing and standardized soil testing devices and hence keep the focus of their efforts on expanding their testing capabilities.

The present keynote paper documents some of the most recent advances in experimental modeling of unsaturated intermediate geomaterials over a whole range of suction-controlled paths and modes of deformation. Its main sections summarize the test protocols and corresponding results from suction-controlled resonant column, biaxial, triaxial, and ring shear testing programs recently undertaken at the Advanced Geomechanics Laboratory (AGL) of the University of Texas at Arlington, via either the axis-translation or the vapor transfer technique. The experimental data and related analyses are expected to be of extreme interest to both geotechnical and geological engineering communities worldwide, facilitating the incorporation of more reliable material properties in the analysis and design models of geotechnical infrastructure made of compacted soil or resting on unsaturated ground.

11.2 Suction-Controlled Resonant Column Testing

Dynamic properties of unsaturated soils, particularly shear-wave velocity V_s, shear modulus G, and material damping D, play a critical role in the analysis and design of geotechnical infrastructure resting on unsaturated ground or made of compacted soil, as they are subjected to cyclic loading (Fig. 11.1). Several power law relationships have been developed, mainly for shear modulus of unsaturated soil, for different choices of state variables, including void ratio, K_O, OCR, degree of saturation, volumetric water content, soil suction, and, more recently, the role of hardening and hydraulic hysteresis (Cho and Santamarina 2001; Mancuso et al. 2002; Sawangsuriya et al. 2008; Ng and Yung 2008; Khosravi and McCartney 2012). Very few attempts, however, have been made to develop a unified

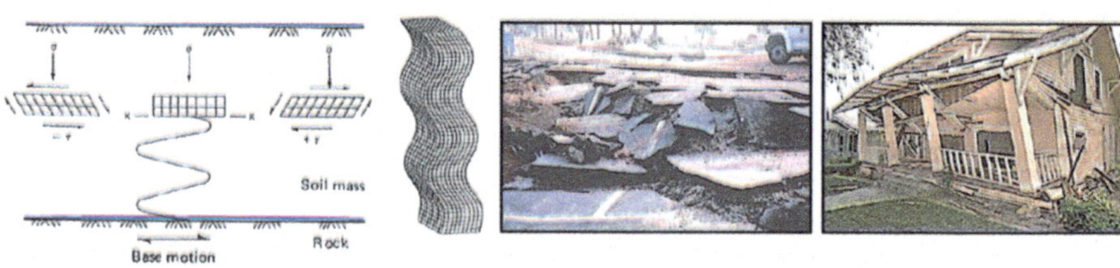

Fig. 11.1 Failures caused by cyclic loading on shallow geotechnical infrastructure (Adapted from Prakash 1981 Photos courtesy of USUCGER)

framework for unsaturated soil stiffness that also embraces material damping as a fundamental property.

This paper introduces a suction-controlled, proximitor-based resonant column device with self-contained bender elements which is referred to as the RC/BE device. The model THS-100 cell features a reinforced acrylic chamber of 1000-kPa pressure capacity. The bottom pedestal, for specimens of 70-mm diameter, features a full set of 5-bar ceramics and one bender element crystal (receiver) for V_s readings. The top cap features a full set of coarse porous stones, for uniform pore-air pressure application, and one bender element transmitter. An electrical servo-motor actuator is used for the application of torsional loads with ±23.3-kN-m capacity, and 300-Hz frequency range. Mounted on an internal floating frame to allow for large vertical deformations, the actuator includes a servo amplifier for closed-loop control of torsional loads and one proximitor mounting acting as internal angular displacement transducer. A model PCP-15U pressure panel is used for direct control of pore-air pressure u_a through the top cap, with dual-pressure regulators/gauges for precise control and measurement of matric suction, $s = u_a$ (i.e., $u_w = 0$).

In the present work, special attention was given to the effects of matric suction over damped free-vibration cycles, threshold shear strain amplitudes, and normalized shear moduli G/G_{max} and damping ratio D/D_{min}. The soil classifies as silty sand (SM) according to the USCS, with 70 % sand and 30 % silt, specific gravity, $G_s = 2.71$, and coarse-fraction particle sizes between 0.5 and 1.2 mm. The passing No. 40 sieve fraction has liquid limit, LL = 26 %, and plastic limit, PL = 21 %. All specimens were statically compacted into a 70-mm diameter, 130-mm height, split mold via a triaxial loading frame. Specimens were prepared in three lifts, under a constant displacement rate of 1.0 mm/min, to target void ratio, $e = 0.85–0.95$, and dry unit weight, $\gamma_d = 13.0$ kN/m^3, with an average water content of 26 %, which corresponds to a degree of saturation of 72 % and initial matric suction of about 20 kPa. (The soil-water retention curve was obtained via pressure plate extracting technique.)

The performance of the RC/BE apparatus, featuring an electrical servomotor actuator, was first experimentally verified against a fully standardized, accelerometer-based RC/TS apparatus, which uses an electromagnetic torque driver. Figure 11.2 shows the change in normalized shear modulus G/G_{max} and normalized damping ratio D/D_{min} with shear strain γ as determined from both devices in identically prepared specimens of SM soil under 276-kPa (40 psi) confinement. Experimental data from both devices have been best fitted altogether with Ramberg-Osgood model parameters (Borden et al. 1996). Values of damping ratio D were calculated via the half-power bandwidth method from frequency response curves. The trends from both devices are reasonably similar, yielding a threshold strain amplitude, $\gamma_{th} = 0.000018$ cm/cm (0.0018 %), for a value of $G/G_{max} = 0.90$, that is, using a 10 % modulus degradation criterion.

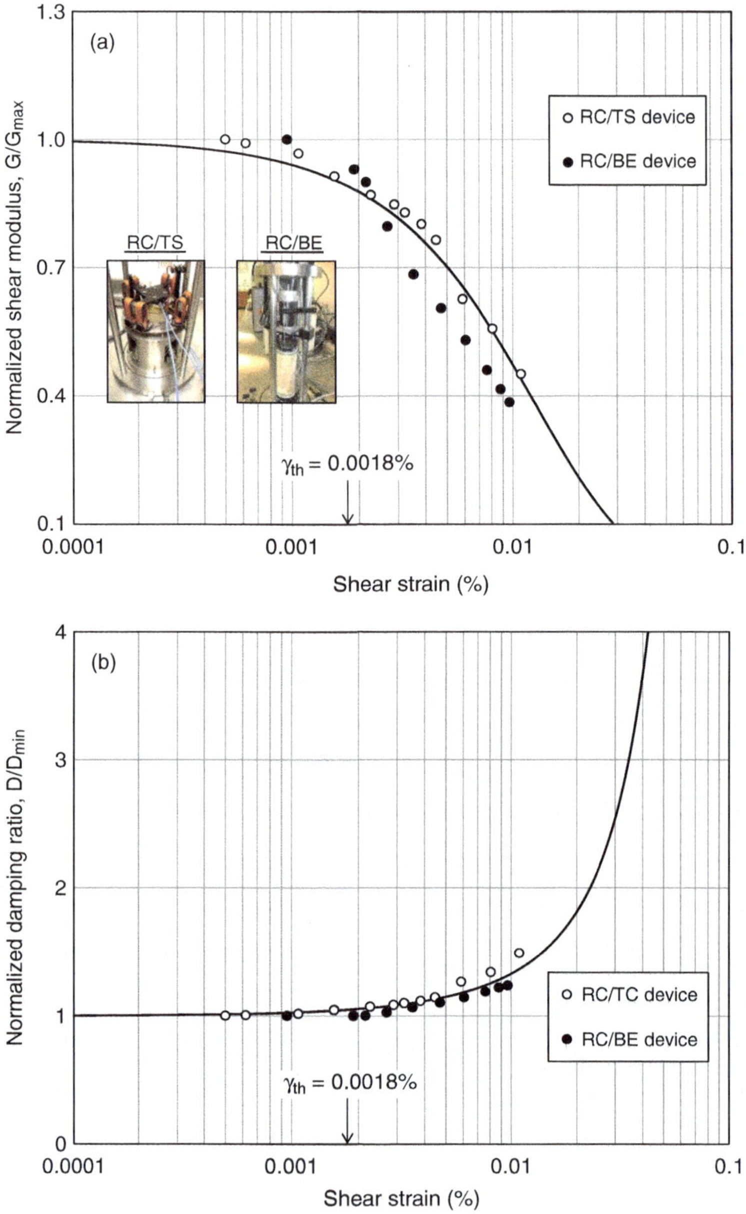

Fig. 11.2 Response of SM soil from RC/TS and RC/BE devices (276-kPa confinement)

A series of constant suction tests were then performed on identically prepared specimens of SM soil. Each compacted soil specimen was tested for a different matric suction state, $s = 50$, 100, 200, or 400 kPa, attained via axis-translation technique. The main intent was to induce suction values beyond the air-entry value of the test soil, approximately 10–20 kPa, and well into the drying loop of the SWCC. RC/BE tests were performed on each specimen under a constant suction state and four different net confining pressures, $(p–u_a) = 50$, 100, 200, and 400 kPa.

Figure 11.3 shows a full set of damped free-vibration cycles generated from two SM soil specimens under constant matric suctions, $s = 50$ kPa (curves on left) and 400 kPa (curves on right), respectively. Free-vibration cycles for net confining pressures, $(p–u_a) = 50$, 100, 200, and 400 kPa, are shown for each suction value. The initial shear strain amplitude at time $t = 0$, that is, right after the torsional vibration is cut off, tends to be lower at higher suctions, with almost negligible shear strain amplitudes for the highest stress state induced, that is, $s = 400$ kPa and $(p–u_a) = 400$ kPa. Most free-vibration cycles, regardless of the initial magnitude of shear strain amplitude, appear to be fully attenuated after 0.08 s. Since the initial shear strain amplitude is lowest under the highest suction state, $s = 400$ kPa, this is strongly indicative of lower material damping at higher suction values, which is directly related to the increase in stiffness generally observed with increasing suction.

Nonlinear seismic site response analyses of unsaturated soil deposits require a thorough assessment of seasonal changes in shear modulus reduction function G/G_{max} as well as normalized damping ratio D/D_{min}. Alterations in these functions may lead to changes in near-surface wave propagation mechanisms, acceleration amplification patterns, and seismically induced settlements. A series of suction-controlled RC tests were also conducted at small- to mid-shear strain amplitude levels (0.0001–0.01 %) in order to assess the effect of matric suction on threshold strain limits of SM soil.

Figure 11.4 shows the change in normalized shear modulus G/G_{max} and damping ratio D/D_{min} with increasing shear strain γ for two identically prepared specimens of SM soil at matric suctions, $s = 25$ kPa and 200 kPa. The experimental data have been best fitted with Ramberg-Osgood model parameters. Each specimen was subjected to increasing cyclic-torque magnitudes ranging from 0.1 to 1.0 kN-m. It can be readily observed that the shear strain amplitudes induced by the largest torque decrease with increasing suction.

Figure 11.5 shows the effect of matric suction on threshold shear strain values for additional SM soil specimens tested under constant matric suction, $s = 50$ kPa or 100 kPa, and increasing net confining pressures, $(p–u_a) = 25$, 100, and 200 kPa. Serving as further substantiation of those observed in Fig. 11.4, this figure shows a reasonably well-defined pattern of decreasing threshold shear strain limit with increasing suction, with a more pronounced effect of suction under the lowest net confinement of 25 kPa.

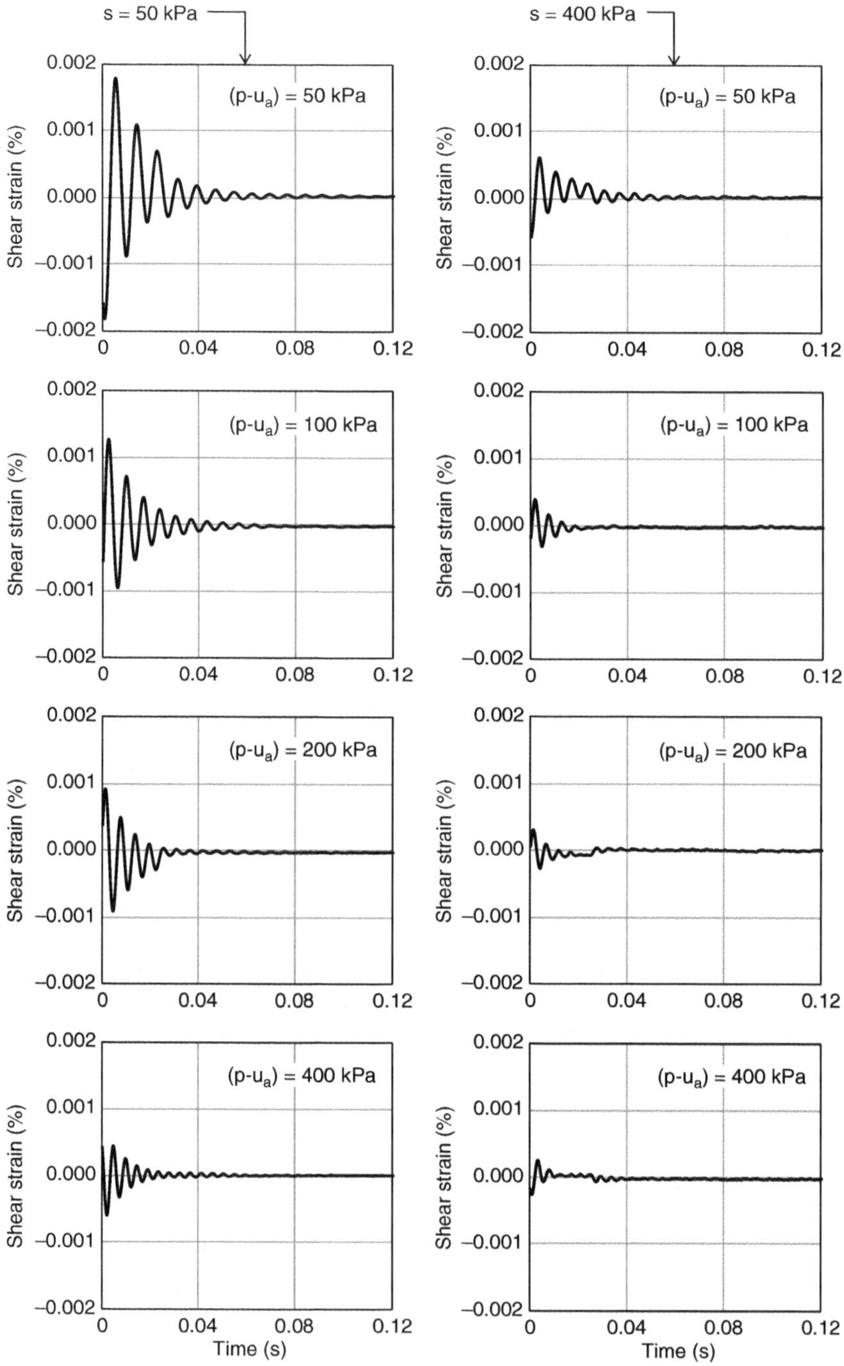

Fig. 11.3 Damped/suction-controlled free-vibration cycles from SM soil

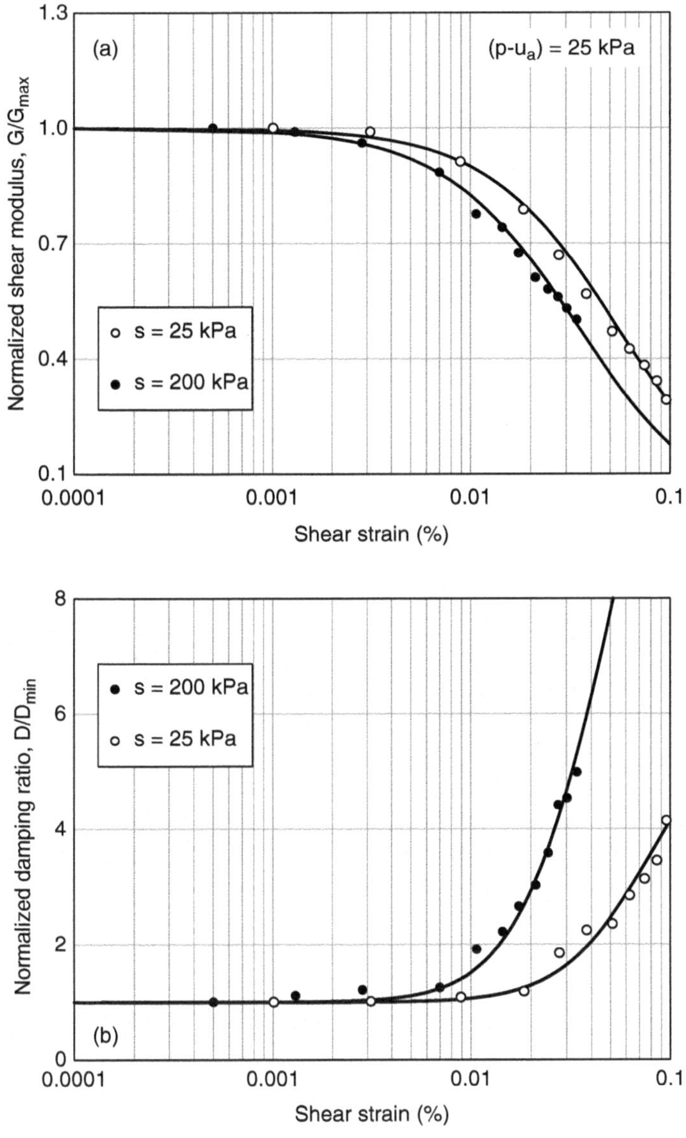

Fig. 11.4 Effect of matric suction on normalized shear modulus and damping of SM soil

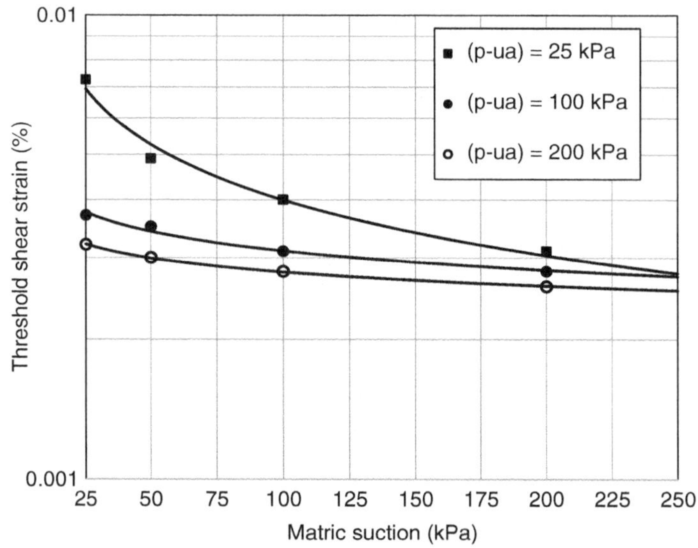

Fig. 11.5 Effect of matric suction on threshold shear strain of SM soil

11.3 Suction-Controlled Plane Strain Testing

The performance of a vast majority of geotechnical infrastructure, such as pavements, tunnels, embankments, earth slopes, and earth retaining systems, may be more suitably modeled via plane strain analyses that account for the particular geometries, boundary conditions, and stress paths that such geosystems normally feature or experience in the field: Fig. 11.6. Biaxial devices allow for reliable testing of soil materials under truly plane strain conditions, facilitating accurate assessments of shear-banding phenomena and stress-strain-strength parameters under these conditions.

This paper introduces a suction-controlled biaxial apparatus that is suitable for unsaturated soil testing via axis-translation technique. The design of its core system is based upon the original Vardoulakis type of apparatus (Vardoulakis and Goldscheider 1981), thus allowing for direct measurements of normal stresses generated between a soil specimen and two lateral rigid walls, as well as volumetric strain changes during suction-controlled plane strain shearing. Detailed descriptions of its original design, components, and performance verification testing are presented by Cruz et al. (2014). Plane strain condition is imposed on the soil specimens by means of two 8-mm thick rigid walls made of Type 304 stainless steel. These walls prevent the specimen from deforming along the intermediate principal axis X_2 (Fig. 11.6). An orderly step-by-step assembling process of the apparatus is illustrated in Fig. 11.7 and can be summarized as follows:

Step 1 – The base plate supporting the apparatus is fitted with three Sensotec-type miniature load cells (A, B, C) located right underneath the bottom pedestal. The

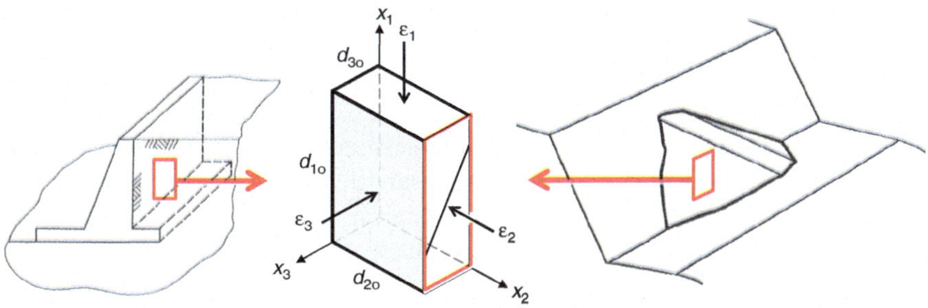

Fig. 11.6 Typical geotechnical infrastructure subjected to plane strain conditions

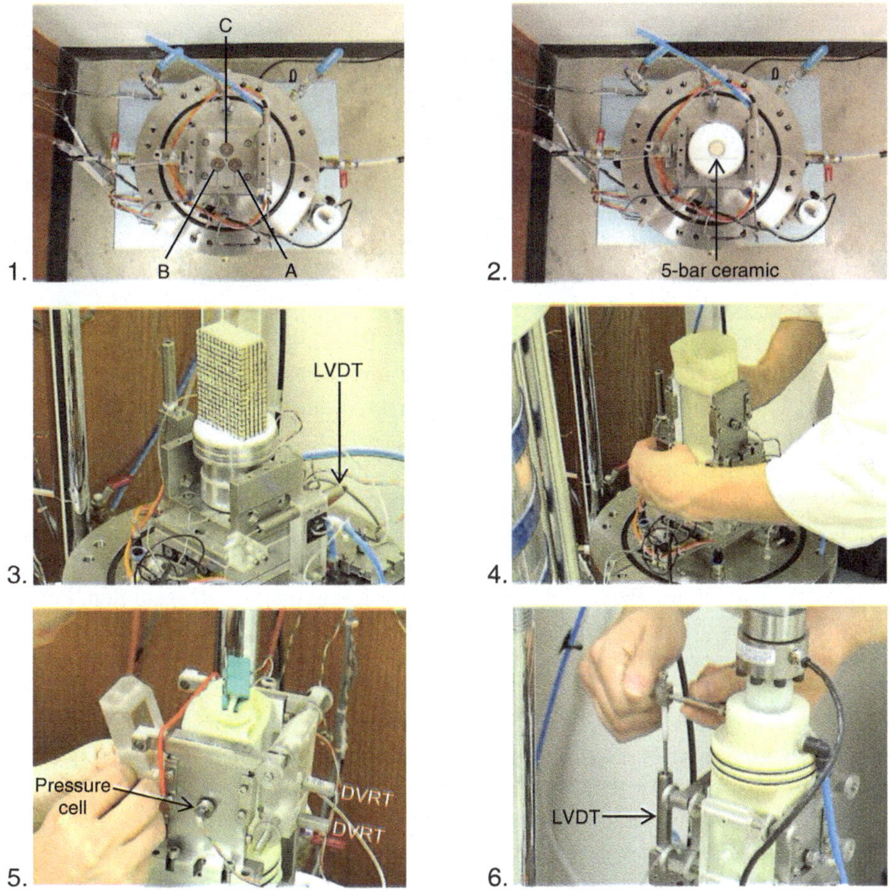

Fig. 11.7 Step-by-step assembling of servo/suction-controlled biaxial apparatus

applied normal stresses can be simultaneously measured in the upper and lower sides of the prismatic specimen, allowing for assessment of the level of frictional resistance generated between the lateral walls and the specimen during plane strain shearing.

Step 2 – The bottom pedestal, made of stainless steel, receives a 3-bar ceramic for control/measurement of pore-water pressure u_w during suction-controlled testing. The pedestal rests on a U-shaped stainless steel base frame, which couples both the pedestal and the rigid lateral walls to a model NKL 6-110 sliding table.

Step 3 – The Schneeberger-type sliding table, which provides an additional degree of freedom to expedite the formation of the failure surface during shearing, is fitted with a horizontal LVDT (linear variable differential transformer) affixed to the base plate of the apparatus. The eventual formation of a failure surface or a shear band(s), perpendicular to the minor principal axis X_3 (Fig. 11.6), can be readily detected by sensing the start of motion of the sliding table with this LDVT.

Step 4 – A latex membrane is gently placed around the compacted specimen and tightly O-ring-secured onto the bottom pedestal. The rigid walls are then set into place. Four stainless steel tie rods are then tightly secured onto each of the rigid walls to prevent the walls from bending during shearing.

Step 5 – Lateral displacements experienced by the soil specimen in the minor principal axis X_3 during plane strain shearing are measured by four Microstrain-type DVRTs (differential variable reluctance transducers) securely attached to the tie rods with an acrylic plate. The intermediate principal stress along axis X_2 (Fig. 11.6) is measured via Sensotec-type pressure cells, one installed on each of the rigid lateral walls. Readings from the cells, along with the measurements of displacement in the remaining principal axes, allow for a complete definition of stress and strain tensors during suction-controlled plane strain testing.

Step 6 – Vertical displacement of the soil specimen in the major principal axis X_1 (Fig. 11.6) during plane strain shearing is measured by a vertically positioned LVDT. An Omega-type load cell, model LCM203-50kN, is used for readings of the applied vertical load. The top loading cap is made of light Nylon 66-type material to minimize the initial seating load acting on the soil specimen. Pore-air pressure is controlled and supplied to the soil pores through the top loading cap via a coarse porous stone, which enables implementation of the axis-translation technique using the $s = u_a$ (i.e., $u_w = 0$) testing approach.

Step 7 – The core assembly of the biaxial system (Step 6) is ultimately placed inside a Wykeham Farrance-type pressure cell, in which specimens can be subjected to cell pressures of up to 1.7 MPa. The axial stress is applied by means of a type MTS Universal Machine with 1.0-MN capacity, which can be used for plane strain testing under either stress- or strain-controlled loading schemes.

In this work, a typical soil specimen was prepared via uniaxial consolidation of a slurry mixture, made of 75 % silty sand and 25 % kaolinite, into an acrylic biaxial mold. The slurry is prepared with water content about twice its liquid limit of

25.3 %. The slurry is then consolidated to dimensions of 80 mm × 80 mm × 135 mm. Load increments of 12.5, 25, 50, 100, 200, and 400 kPa are applied, resulting in a saturated unit weight of 20.1 kN/m^3. The sample is removed from the consolidation mold, from which two specimens are trimmed for plane strain testing in the biaxial apparatus. All specimens are trimmed to final dimensions of 90 mm × 60 mm × 30 mm (Fig. 11.7). The soil classifies as silty sand (SM) as per the USCS.

A series of eight suction-controlled tests were then performed on an equal number of identically prepared specimens of SM soil. Specimens were tested under initial net confining pressure, $p = (\sigma_3 - u_a) = 50$, 75, or 100 kPa, and constant matric suction, $s = 50$, 75, or 100 kPa, as depicted schematically in Fig. 11.8. The procedure is similar to conventional CD testing with axial strain control (Vardoulakis and Goldsheider 1981; Dresher et al. 1990; Alshibli et al. 2004). Upon completion of pore-fluid equalization stage, the soil specimen was sheared at a constant vertical strain rate of 0.004 mm/min (4.44 × 10^{-5} % vertical deformation per minute), which is considered to be low enough to prevent sudden increases in pore pressure during suction-controlled shearing in this particular type of intermediate materials (Fredlund and Rahardjo 1993).

Figures 11.9 and 11.10 show the stress-strain response of SM soil from plane strain tests performed under constant matric suctions, $s = 75$ kPa and 100 kPa, respectively. MIT notation for net mean stress, $p = (\sigma_1 + \sigma_3)/2 - u_a$, and deviator stress, $q = (\sigma_1 - \sigma_3)/2$, was adopted. As it is expected, soil strength is greatly influenced by the level of initial confinement, with highest strength obtained under $p = 100$ kPa and $s = 100$ kPa. In every case, peak strength is followed by softening behavior until an apparent residual state is finally attained. The initial

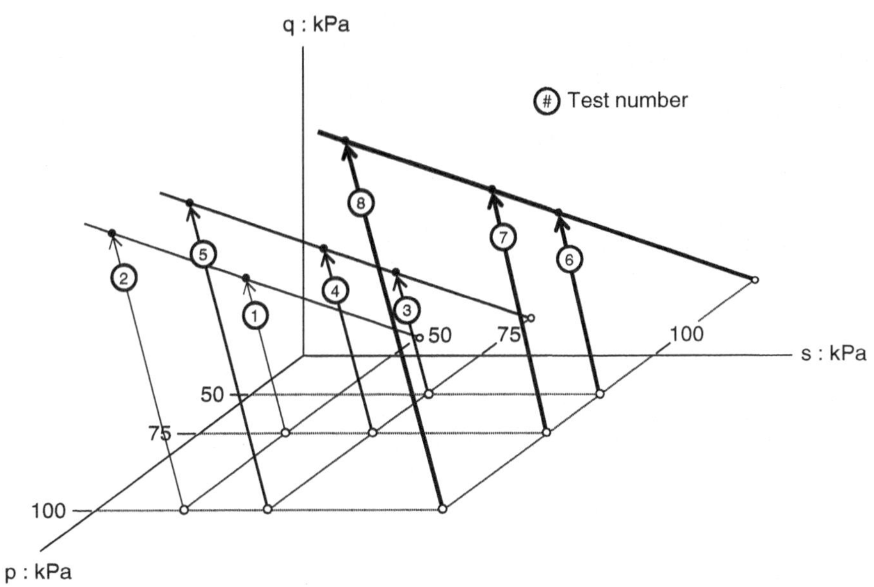

Fig. 11.8 Suction-controlled plane strain test paths induced on SM soil specimens

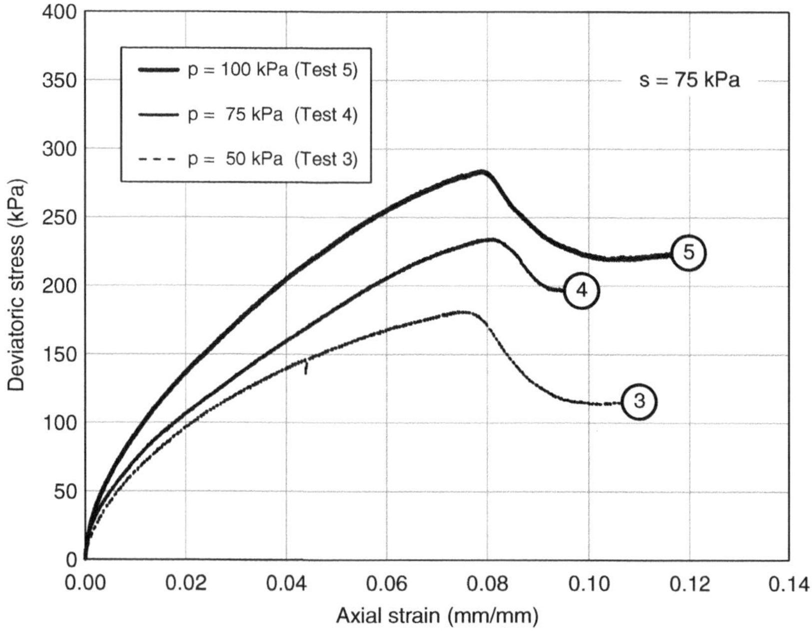

Fig. 11.9 Response of SM soil from s-controlled plane strain shearing, s = 75 kPa

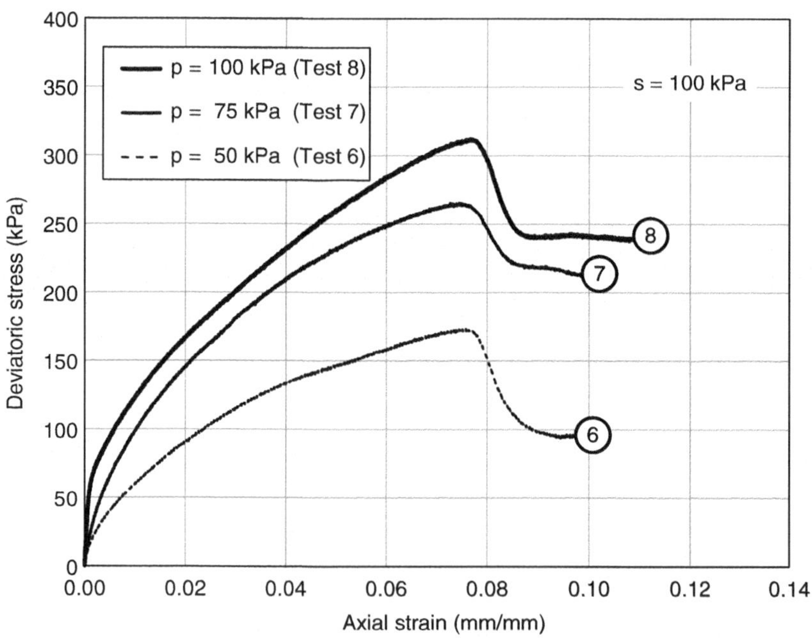

Fig. 11.10 Response of SM soil from s-controlled plane strain shearing, $s = 100$ kPa

void ratio of all identically prepared specimens was approximately $e = 0.65$. These results hint to a peak strength framework under plane strain conditions conceptually similar to the one postulated for peak shear strength of unsaturated soils under axisymmetric stress states (Fredlund et al. 1978), as follows:

$$\tau_{ff} = c' + (\sigma - u_a)\tan\phi' + (u_a - u_w)\tan\phi^b \tag{11.1}$$

where ϕ' = friction angle with respect to net mean stress, $(\sigma - u_a)$ and ϕ^b = parameter representing the increase in strength with suction. Figure 11.11 shows typical features of a failure surface induced by suction-controlled plane strain shearing on a compacted specimen of SM soil. The specimen corresponds to the test conditions, $p = 75$ kPa and $s = 50$ kPa. The deformed, ink-mesh-tracked geometry of the failed specimen shows a relatively uniform deformation in the minor principal axis X_3 (Fig. 11.6) throughout its entire height, as measured by the three DVRTs (Fig. 11.7 – step 5). A fully developed failure surface, making a 62° angle with the horizontal, was readily identified. Similar features, with a failure surface making a 65° angle with the horizontal, were observed for a soil specimen failed under matric suction, $s = 100$ kPa, further substantiating the results shown in Figs. 11.9 and 11.10.

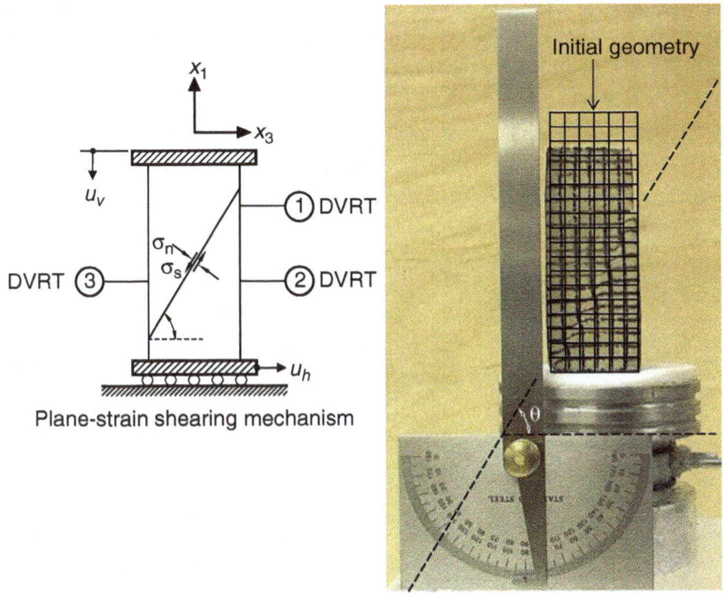

Fig. 11.11 Typical failure induced by s-controlled plane strain shearing of SM soil

11.4 Suction-Controlled Triaxial Testing

Triaxial testing continues to be the most universally used technique to characterize the shear strength and volume-change behavior of saturated and unsaturated geomaterials under axisymmetric stress sates. The implementation of the axis-translation technique to impose/control matric suction during triaxial testing of unsaturated soils is limited by the air-entry value of the ceramic disk(s) the specimen is normally resting on. This generally narrows down the suitability of this technique up to a maximum nominal value of suction of 1500 kPa. Critical geotechnical infrastructure, however, including earth slopes in tropical regions and nuclear waste disposal clay liners, involves the use of naturally occurring or compacted geomaterials subjected to a much wider range of moderate-to-high suction states throughout any given year: Fig. 11.12. Vapor pressure techniques have proved to be suitable for controlling the relative humidity in the pore-air phase of the test soil, thus allowing for the imposition of drying, wetting, and stress paths in evaluating strength, deformation, and flow characteristics of unsaturated soils under relatively high suction values (Blatz et al. 2008).

This paper introduces a servo/suction-controlled double-walled triaxial system that is suitable for unsaturated soil testing via either axis-translation or vapor transfer technique (Lu and Likos 2004). The experimental effort focused on suction-controlled testing of a relatively dense soil with a natural tendency to experience dilatancy and post-peak softening as it approaches critical state. Statically compacted specimens of silty sand were tested in the low-to-medium suction range (50–750 kPa), via the axis-translation technique, as well as the higher suction range (20–300 MPa), via the vapor transfer technique. Previous calibration investigation yielded suitable shearing rates of 0.0086 %/min (axial deformation per unit time) for the lower matric suction range and 0.0009 %/min for the higher total suction range (Patil et al. 2014).

Main features of the bottom assembly of the suction-controlled double-walled triaxial cell are illustrated in Fig. 11.13, including all of the following key items: 1, bottom pedestal with three 15-bar ceramics; 2, top cap with three porous stones;

Fig. 11.12 (**a**) Slopes. (**b**) Nuclear waste depositories. Geotechnical infrastructure subjected to moderate-to-high suction states (Photos courtesy of Prof. Jean-Louis Briaud and Prof. Lyesse Laloui)

Fig. 11.13 Bottom assembly of suction-controlled double-walled triaxial cell

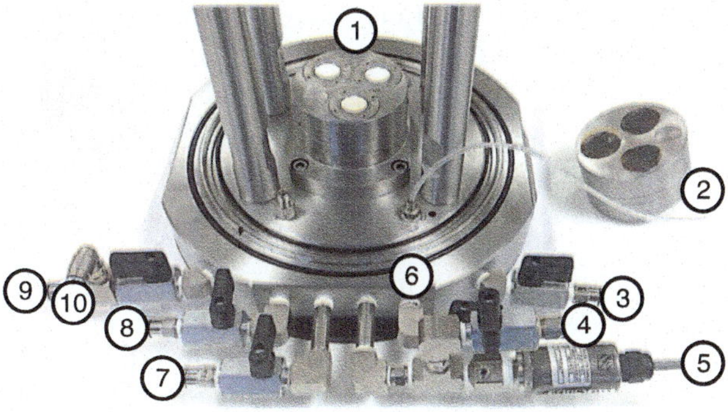

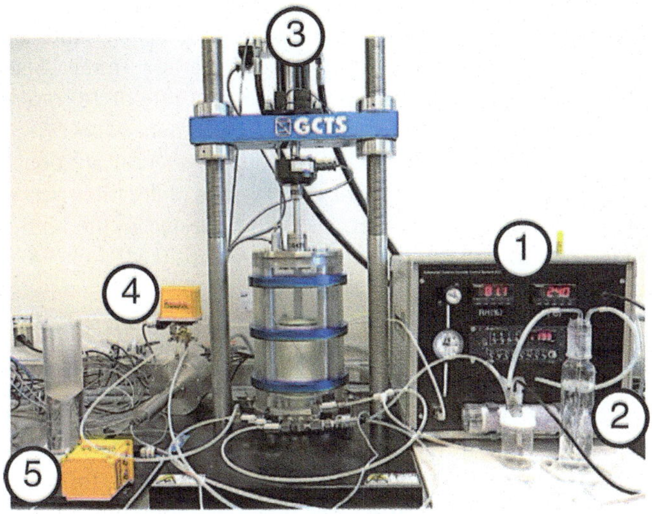

Fig. 11.14 Triaxial cell with automated relative humidity (Auto-RH) control system

3, inner-cell water inlet; 4, pore-water pressure inlet; 5, pressure transducer; 6, pore-air pressure inlet; 7, flushing inlet; 8, flushing outlet; 9, outer-cell water outlet; and 10, volume-change outlet (connected to volume-change device).

In the present work, as previously mentioned, the triaxial shear apparatus was upgraded, in order to extend its suction-controlled testing capabilities to higher total suction range, by combining its core system with a fully automated relative humidity (Auto-RH) control system: Fig. 11.14. A similar system has been used by Likos and Lu (2003) for assessing soil-water retention properties of clays. The fully integrated system allows for accurate assessments of shear strength and volume-change behavior of unsaturated soils, with direct control of the relative humidity in the soil pores prior to and during suction-controlled triaxial shearing.

Main features of the fully integrated triaxial system are illustrated in Fig. 11.14, including 1, auto-RH system; 2, gas bubbler, desiccant, and temperature probe; 3, core triaxial system; 4, automated volume-change device; and 5, automated flushing device.

Humidity in the pore-air of the soil can be ramped along paths of increasing or decreasing relative humidity, ranging from ~2 % RH to ~95 % RH, which corresponds to total suction states of ~500,000 kPa to ~10,000 kPa, and is typically accomplished in step increments of ~10 % RH. The "forced-flow" nature of the mixed-flow system significantly reduces the required pore-fluid equalization time (Likos and Lu 2006). The thermodynamic relationship between relative humidity of pore-water vapor (RH) and total suction (MPa) can be readily established via Kelvin's Law.

The soil used in this work classifies as silty sand (SM) according to the USCS, with 55 % sand, 37 % silt, and 8 % clay; specific gravity, $G_s = 2.67$; standard Proctor maximum dry unit weight, $\gamma_{d\text{-max}} = 1.87\text{g/cm}^3$; and optimum moisture content, OMC = 12.2 %. Specimens were statically compacted into a 70-mm diameter, 130-mm height, split mold via a triaxial loading frame. Specimens were prepared in nine lifts, under a constant displacement rate of 1.0 mm/min, to target void ratio, $e = 0.46$–0.49, with an average water content of 14.2 %, which corresponds to a degree of saturation of 81 % and initial suction of about 10 kPa. (The soil-water retention curve was obtained via pressure plate, filter paper, and relative humidity-based techniques.) The specimens to be tested in the high suction range were preconditioned to target total suction values of 20 MPa or 300 MPa, prior to suction-controlled triaxial testing.

Figure 11.15 shows the stress-strain and volume-change responses of compacted SM soil from suction-controlled axisymmetric shearing under constant matric suction, $s = 50$ kPa, and initial net mean stress, $p = 100$, 200, or 300 kPa. Ceramics with 1-bar air-entry values were used to induce a matric suction of 50 kPa. Additional tests were performed on identically prepared specimens using 3-bar ceramics for matric suction of 250-kPa and 15-bar ceramics for matric suctions of 500 kPa and 750 kPa. Results clearly show post-peak softening behavior, regardless of initial net mean stress, as the test soil approaches critical state condition under a constant matric suction. (Results from triaxial testing at higher suctions showed that the amount of post-peak softening increased with increasing matric suction.)

Previous consolidated-drained (CD) testing on fully saturated specimens of SM soil yielded purely compressive behavior, in terms of volumetric strain response. Results from suction-controlled tests, however, show a clearly manifest change from compressive to dilatant behavior, regardless of initial net mean stress, as the test soil approaches critical state condition under a constant matric suction even as low as 50 kPa. The amount of dilatancy, as expected, decreases with increasing net mean stress.

Figures 11.16 and 11.17 show the best-fit critical state lines (CSL) obtained from triaxial testing in the low-to-medium matric suction range (50–750 kPa) via

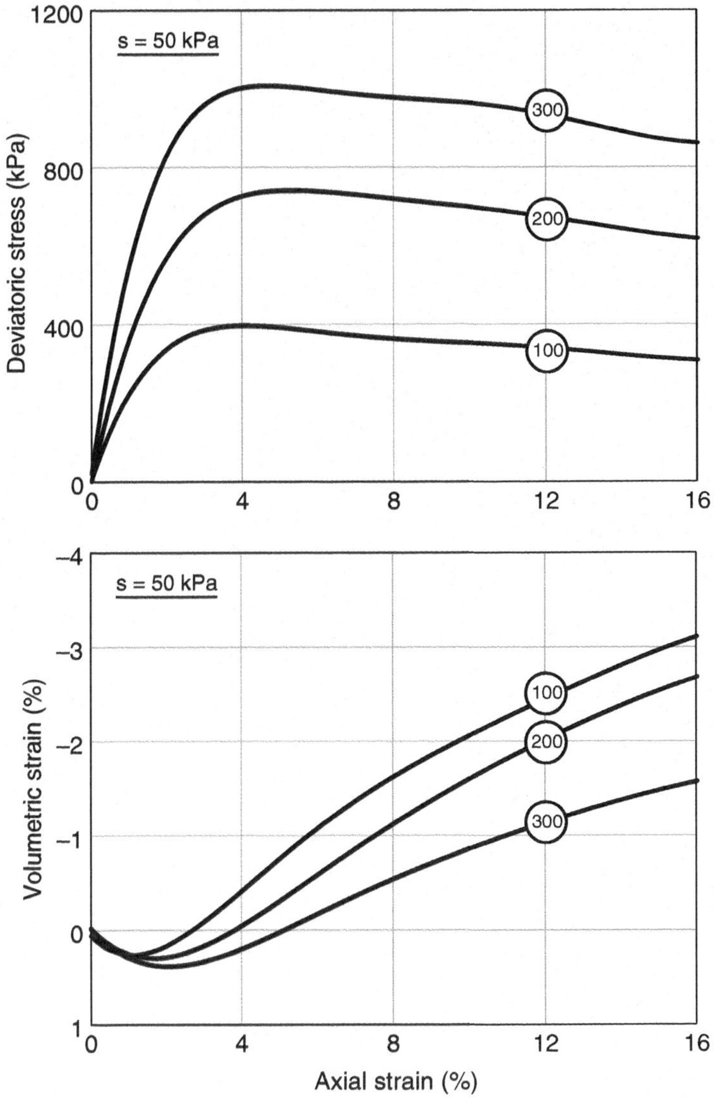

Fig. 11.15 Response of SM soil from s-controlled axisymmetric shearing, $s = 50$ kPa

axis-translation technique, as well as higher suction range (20–300 MPa) via vapor transfer technique, respectively. Suction is observed to exert a significant influence on the final position of the CSL. The slope of all critical state lines, however, remains reasonably constant, in agreement with the constitutive framework postulated by Alonso et al. (1990).

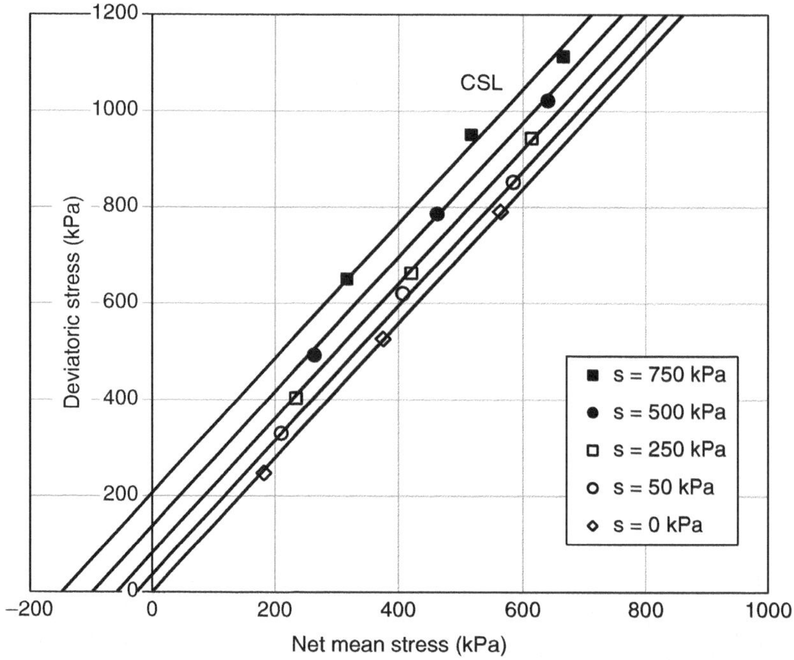

Fig. 11.16 CSLs of SM soil from s-controlled triaxial testing via axis-translation

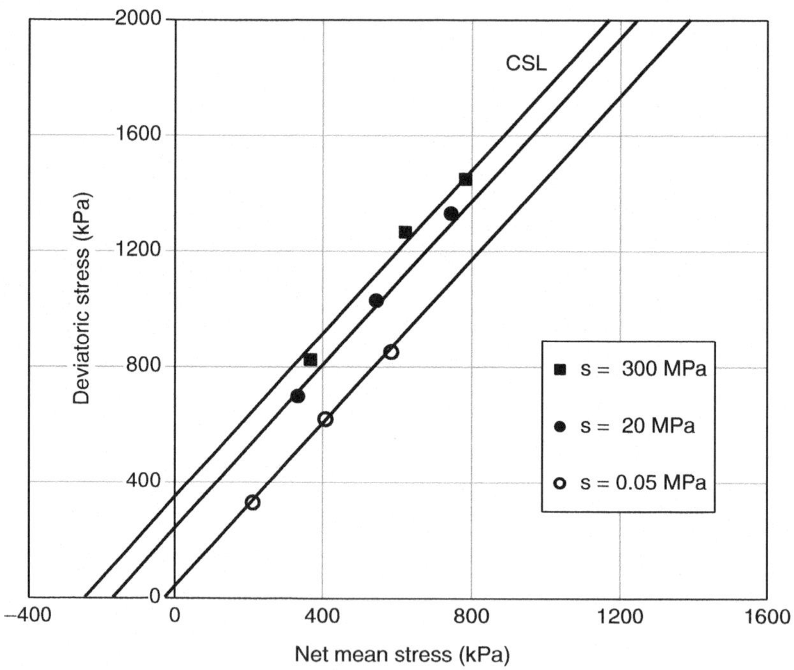

Fig. 11.17 CSLs of SM soil from s-controlled triaxial testing via relative humidity

11.5 Suction-Controlled Ring Shear Testing

A vast majority of geotechnical infrastructure made of compacted soil, or resting on unsaturated ground, involves a wide range of deformations. Calculation of foundation settlement, for instance, requires a good estimation of soil stiffness at relatively small strains. Analysis of earth slopes, embankments, and soil bearing capacity, on the other hand, requires good estimations of shear strength from peak to residual. Assessment of residual shear strength parameters at relatively low net normal stresses and suction states, as reflected by the range of experimental variables considered in this work, is of critical importance in slope stability analyses involving potentially shallow failures triggered by rainfall at relatively high degrees of saturation. To date, however, there is limited experimental evidence of unsaturated soil behavior at large deformations as the soil is subjected to controlled suction states. Research efforts have been deterred in the past by the lack of suitable testing tools and techniques. It is in this context that a suction-controlled ring shear (RS) apparatus plays a fundamental role in a thorough characterization of this type of materials.

This paper introduces a servo/suction-controlled RS apparatus that is suitable for testing unsaturated soils under large deformations via axis-translation technique. The design of its core system is based on the original Bromhead apparatus (Bromhead 1979). A detailed description of its full development, including main components and thorough performance verification against the original Bromhead device, is presented by Hoyos et al. (2011, 2014). The step-by-step assembling process of the RS system can be summarized as follows:

1. All actuators and the DA/PC system are switched on to allow the instruments to come into equilibrium and minimize the influence of temperature offsets.
2. A small piece of wet filter paper is placed over the top of 5-bar ceramics, prior to soil compaction, to ensure phase continuity between pore-water in the soil and water in the saturated 5-bar ceramics: Fig. 11.18a).
3. A 15-mm (0.59-in) thick specimen is statically compacted directly into a bottom annular platen: Fig. 11.18b. The specimen is then transferred to the RS frame and the platen tightly fixed onto the bottom base plate: Fig. 11.18c.
4. The vertical load shaft is brought up via a servo controller and the upper annular platen affixed to the top of the piston shaft: Fig. 11.18c. A vertical sitting load of 25 N is applied in order to bring the upper platen in full contact with the soil.
5. All drainage and flushing lines are filled with de-aired water and flushed several times to avoid any trapped air in the whole system.
6. The main RS cell is set into place and the top cover plate affixed to the cell wall: Fig. 11.18d. A pore-air pressure u_a line, from a PCP-15U panel, is connected to the top cover plate via a quick connector.

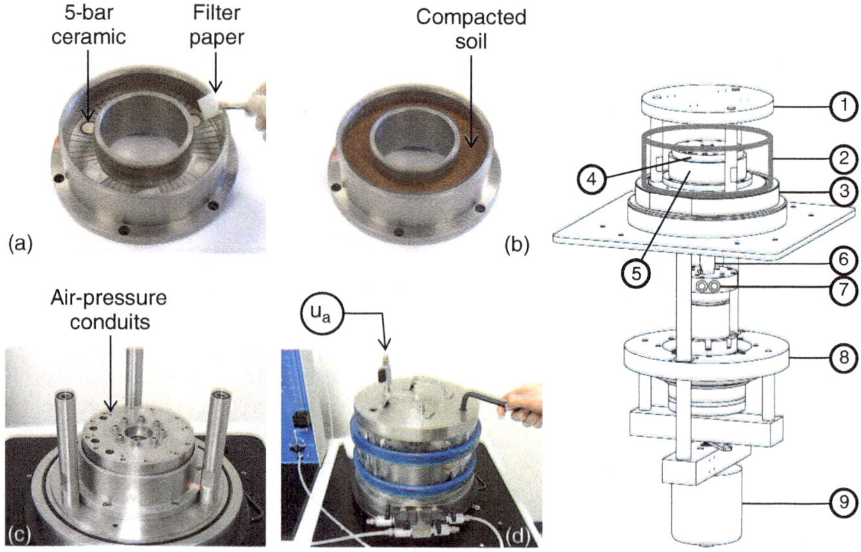

Fig. 11.18 RS assembly: (**a**) lower platen, (**b**) test specimen, (**c**) top platen, (**d**) main cell

7. Readings of the load-torque transducers are reset while one LVDT and angular deformation sensors are re-zeroed prior to RS testing.
8. The specimen is then subjected either to a suction-controlled single-stage or multistage RS test using the $s = u_a$ testing approach (i.e., $u_w = 0$).
9. When the test is finished, all pressures are gradually reduced back to atmospheric pressure and the failure surface examined via microscopic digital imaging. Key items of the isometric view in Fig. 11.18 are described by Hoyos et al. (2011).

The test soil used in this investigation classifies as silty clay (CL) according to the USCS, with 18 % sand, 50 % silt, and 32 % clay, $\gamma_{d\text{-max}} = 1.77$ g/cm^3, OMC = 17 %, LL = 37 %, PI = 20 %, and specific gravity, $G_s = 2.72$. During compaction, the upper platen is used to compress one single lift of a loose soil-water mix to target dry unit weight of 95 % of $\gamma_{d\text{-max}}$ under a constant displacement rate of 1.25 mm/min, which corresponds to initial degree of saturation of 88 % and matric suction of 100 kPa. (The soil-water retention curve was obtained via pressure plate and filter paper techniques.)

In this work, all suction-controlled RS tests were conducted at an equivalent horizontal shear displacement rate of 0.025 mm/min (equivalent to a rotational speed of 0.023°/min), which is slightly lower than that recommended for suction-controlled RS testing on silty or clayey soils (Vaunat et al. 2006; Infante Sedano et al. 2007). All RS tests were performed by following a multistage scheme in

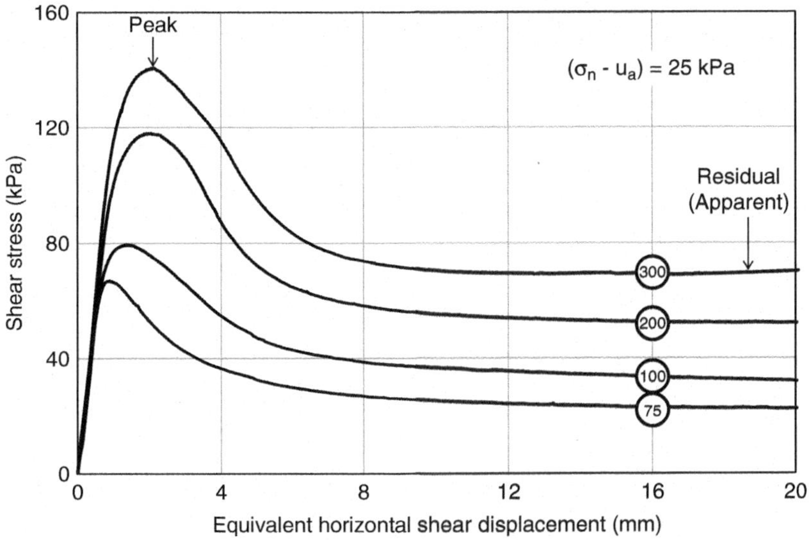

Fig. 11.19 Peak and residual response of CL soil from s-controlled ring shear testing

which residual strength assessments were made at three or more net normal stresses, $(\sigma_n–u_a) = 25$, 50, 75, 100, and 200 kPa, under matric suction, $s = (u_a–u_w) = 25$, 50, 75, 100, 200, or 300 kPa. Shearing was stopped when it was apparent a residual state had been reached.

Figure 11.19 shows the shear stress vs. equivalent horizontal shear displacement response of compacted CL soil, during first shearing stage, at matric suctions, $s = 75$, 100, 200, and 300 kPa, and net normal stress, $(\sigma_n–u_a) = 25$ kPa. Specimens exhibit a steady increase in shear stress up to a peak value, followed by a gradual decrease until an apparent residual state is attained. The peak is clearly more pronounced at higher suctions. It can also be readily noticed that suction has a significant effect on residual strength, with a considerable increase for $s = 300$ kPa.

Figure 11.20 shows the effect of matric suction on both the peak and the residual failure envelopes obtained from suction-controlled RS testing of compacted CL soil.

Results appear to confirm the direct correspondence observed between the nonlinear nature of peak failure envelopes (threshold value of matric suction beyond which they become nonlinear) and the air-entry value of the test soil obtained from SWCC (about 100 kPa for the CL soil). The residual failure envelope, however, remains reasonably linear for the range of suction values induced via axis-translation in the present work.

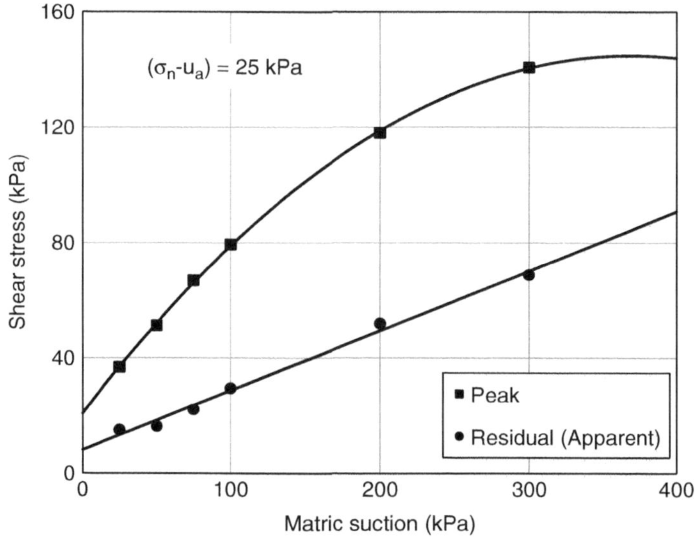

Fig. 11.20 Effect of matric suction on peak and residual failure envelopes from CL soil

11.6 Concluding Remarks

Civil and geotechnical engineers around the world, vividly aware of the thousands of tragic losses from natural disasters every year, are becoming increasingly aware that unsaturated soil research may provide more thorough and reliable analytical tools and design guidelines for crucial geotechnical infrastructure made of compacted soil or resting on unsaturated ground. The devices and protocols described in this paper offer the unlimited potential to advance our understanding of unsaturated soil behavior over a whole range of suction-controlled paths and modes of deformation, facilitating the assessment of more reliable material properties. However, the most exciting prospect of all is that future test series will reveal something completely unanticipated, opening new vistas on the fascinating nature of unsaturated soil behavior.

Acknowledgments The research work summarized in this keynote paper has been supported by the US National Science Foundation (NSF) and COLCIENCIAS, Bogotá, Colombia. This support is gratefully acknowledged. Any findings, conclusions, or recommendations expressed in this material are those of the authors and do not necessarily reflect the views of NSF or COLCIENCIAS. The authors would also like to acknowledge the contributions from current and former graduate students, Ujwalkumar Patil, Alejandro Pino, Eduardo Suescún, Jairo Yepes, William Douglas, Andrés Cruz, Diego Pérez-Ruiz, and Claudia Velosa, and research collaborators, Arcesio Lizcano and Manuel Padilla.

References

Alonso EE, Gens A, Josa A (1990) A constitutive model for partially saturated soils. Geotechnique 40(3):405–430
Alshibli KA, Godbold DL, Hoffman K (2004) The Louisiana plane strain apparatus for soil testing. Geotech Test J ASTM 27(4):337–346
Blatz JA, Cui Y-J, Oldecop L (2008) Vapour equilibrium and osmotic technique for suction control. Geotech Geol Eng 26:661–673
Borden RH, Shao L, Gupta A (1996) Dynamic properties of Piedmont residual soils. J Geotech Eng ASCE 122(10):813–821
Bromhead EN (1979) A simple ring shear apparatus. Ground Eng 12(5):40–44
Cho GC, Santamarina JC (2001) Unsaturated particulate materials: particle-level studies. J Geotech Geoenviron Eng ASCE 127(1):84–96
Crus JA, Hoyos LR, Lizcano A (2014) Response of intermediate unsaturated soils under suction-controlled plane strain conditions: preliminary investigation. In: Proceedings of the sixth international conference on unsaturated soils, vol 1, pp 93–98
Drescher A, Vardoulakis I, Han C (1990) A biaxial apparatus for testing soils. Geotech Test J ASTM 13:226–234
Fredlund DG, Rahardjo H (1993) Soil mechanics for unsaturated soils. Wiley, New York, 517 pp
Fredlund DG, Morgenstern NR, Widger RA (1978) The shear strength of unsaturated soils. Can Geotech J 15(3):313–321
Hoyos LR, Velosa CL, Puppala AJ (2014) Residual shear strength of unsaturated soils via suction-controlled ring shear testing. Eng Geol Elsevier 172:1–11
Hoyos LR, Velosa CL, Puppala AJ (2011) A servo/suction-controlled ring shear apparatus for unsaturated soils: Development, performance, and preliminary results. Geotech Test J ASTM 34(5):413–423
Infante Sedano JA, Vanapalli SK, Garga VK (2007) Modified ring shear apparatus for unsaturated soil testing. Geotech Test J ASTM 30(1):1–9
Khosravi A, McCartney JS (2012) Impact of hydraulic hysteresis on the small-strain shear modulus of unsaturated soils. J Geotech Geoenviron Eng ASCE 138(11):1326–1333
Likos WJ, Lu N (2006) Pore scale analysis of bulk volume change from crystalline swelling in Na^+ - and Ca^{2+}-smectite. Clay Clay Miner 54(4):516–529
Likos WJ, Lu N (2003) Automated humidity system for measuring total suction characteristics of clay. Geotech Test J ASTM 26(2):178–189
Lu N, Likos WJ (2004) Unsaturated soil mechanics. Wiley, Hoboken, 556 pp
Mancuso C, Vassallo R, d'Onofrio A (2002) Small strain behavior of a silty sand in controlled-suction resonant column-torsional shear tests. Can Geotech J 39:22–31
Ng CWW, Yung SY (2008) Determination of the anisotropic shear stiffness of an unsaturated decomposed soil. Geotechnique 58(1):23–35
Patil UD, Puppala AJ, Hoyos LR (2014) Assessment of suitable loading rate for suction-controlled triaxial testing on compacted silty sand via axis-translation technique. In: Proceedings of geo-congress 2014, Geo-Institute of ASCE, pp 1307–1316
Prakash S (1981) Soil dynamics. McGraw-Hill Book Company, New York, 426 pp
Sawangsuriya A, Edil TB, Bosscher PJ (2008) Modulus-suction-moisture relationship for compacted soils. Can Geotech J 45:973–983
Vardoulakis I, Goldscheider M (1981) Biaxial apparatus for testing shear bands in soils. In: Proceedings of 10th international conference on soil mechanics and foundation engineering, Stockolm, vol 4, pp 819–824
Vaunat J, Amador C, Romero E, Djeren-Maigre I (2006) Residual strength of a low plasticity clay at high suctions. In: Proceedings of fourth international conference on unsaturated soils, Carefree, Arizona, vol 1, 1279–1289

Chapter 12
Underground Space for Sustainable Urban Development: Experiences from Urban Underground Metro Constructions in India

T.G. Sitharam and S.D. Anitha Kumari

Abstract For thousands of years, underground has provided humans refuge, useful resources, physical support for surface structures, and a place for spiritual or artistic expression. More recently, many urban services have been placed underground. Over this time, humans have rarely considered how underground space can contribute to or be engineered to maximize its contribution to the sustainability of society. Tunneling works and other forms of underground construction have been taken up recently in India as part of urban mass transit operations. The experiences from the various underground metro constructions in India are consolidated along with the technologies used for the tunneling procedure. For this, case studies of the four underground metro constructions in India are presented. The different methods of tunneling and its procedure with their suitability in different situations are also looked into. From the stability point of view, tunneling can be a significant cause of settlement resulting in huge damages to surface structures in densely populated cities. This paper also focuses on the development of computational procedures including DEM and FEM to model the soil stresses and deformations that develop as a consequence of underground construction and underground space creation. Methods are being developed to access the risk of settlement-induced damage of these openings.

Keywords Sustainability • Tunnels • Tunneling methodology • Case studies • Numerical modeling • Seismic response

T.G. Sitharam (✉)
Department of Civil Engineering, IISc, Bengaluru 560012, India
e-mail: sitharam@civil.iisc.ernet.in

S.D. Anitha Kumari
Faculty of Engineering and Technology, M S Ramaiah University of Applied Sciences, Peenya, Bengaluru 560058, India

G.L. Sivakumar Babu et al. (eds.), *Sustainability Issues in Civil Engineering*,
Springer Transactions in Civil and Environmental Engineering,
DOI 10.1007/978-981-10-1930-2_12

12.1 Introduction

Thirty-two percent of Indian population is presently residing in its urban areas. This population is set to go higher and higher as rapid urbanization takes place and is expected to reach around 590 million by 2030. This has led to the fact that the urban infrastructure should be developed by keeping in mind the future growth of the existing cities. For any country to keep its pace of growth, adequate transport facilities are a must. The explosive growth of personal vehicles in the roads due to lack of efficient public transport systems has forced the governments to think about the introduction of alternate fast, affordable, safe, sustainable, and comfortable public transport modes. In this context, mass rapid transit system, MRTS, has significance as it reduces energy consumption and pollution, thus making it eco-friendly, provides more space occupancy, and above all is a cheap mode of transport. This has led to the introduction of metro rails, which are considered one of the most important modes of urban transport, in various cities in India. Even though the challenges associated with the metro rail at various stages starting from its planning to inception are considerably high, the long-term benefits associated with those rapid transit systems have forced the governments to implement them in ever-growing cities of India. Currently, the metro rails are operational in five cities, viz., Kolkata, New Delhi, Bangalore, Gurgaon, and Mumbai. Hyderabad Metro is also under construction, and it is in its testing stage, which is mainly over ground. Of these, Delhi Metro has the maximum operational length of 190.03 km with 48.06 km running underground. The timely execution and the huge operational success of Delhi Metro are a big boost to the metro systems in India. The expansion of the operational metro systems is under way. Four metro systems are under construction in Chennai, Hyderabad, Jaipur and Kochi. Around 11 cities have been identified for future metro systems in India.

Due to the fast growth of the cities, the amount of surface space available for the introduction of these metro systems at grade is less. Hence in congested urban cities, where space is scarce, the construction is either overground and/or underground. The underground systems help to reduce the cost and project completion time when the acquisition of land is nearly impossible or moving surface utilities which are expensive. Moreover, the harmful surface emissions from the vehicles can be avoided by collecting these within the underground and cleaning it thus making it eco-friendly and sustainable. In this paper, the urban underground metro construction in India with case studies of Delhi, Bangalore, Mumbai, and Kolkata is presented along with the different methods of tunneling adopted. As it is difficult to estimate the underground conditions and the behavior of these structures through experiments, numerical modeling has gained significance in analyzing the behavior of these structures. Hence, this paper also focuses on the development of computational procedures using DEM and FEM to model the soil stress deformations due to underground construction. The behavior of twin tunnels during seismic loading is also numerically modeled using FEM.

12.2 Methodology of Tunneling/Tunnel Construction

There are different tunnel construction methods adopted depending on the geological and hydrological conditions, cost and time considerations, limits of surface settlements/disturbance, and many other factors. Some of the common construction methods are:

(Ref: http://miningandblasting.wordpress.com/2011/07/27/tunnel-construction-methods-and-their-comparison/)

(a) Cut and cover tunneling
(b) Drill and blast technique
(c) Bored tunneling by TBM

12.2.1 Cut and Cover Tunneling

Cut and cover tunneling is generally preferred for shallow tunnels. This type of tunneling is done when excavation from the surface is possible economically. In this method, the tunnel is built inside the excavation, and once the structure is completed, the excavation is covered with backfill material. There are two methods adopted in cut and cover tunneling: (i) bottom-up construction and (ii) top-down construction.

In bottom-up construction, the tunnel is completed before it is covered up and the surface reinstated as shown in Fig. 12.1. The advantage of bottom-up construction is that the inside of the excavation is easily accessible for machinery and materials, and drainage systems can be installed outside the structure. However, it has the disadvantage that the surface cannot be reinstated to the final stage until the construction is completed. Moreover, temporary relocation of utilities is also required at times.

In top-down construction, the tunnel walls are constructed first and the final structural walls act as the excavation support. In case water is present, dewatering should be done. The first stage will be to excavate till the bottom of top slab of tunnel. Immediately after this, the roof is constructed and tied to the support of excavation walls. Once this is done, the top is backfilled and the ground is reinstated. The advantage here is that the obstruction on the surface can be minimized and can be opened to public use much before the complete construction. Figure 12.2 shows the various stages in top-down construction. This also has the advantage of reducing the total time taken for construction by overlapping various activities. However some disadvantages are also associated with this method. It provides limited space for the construction of the interior slabs. The accessibility to the tunnel will be limited unlike the bottom-up construction.

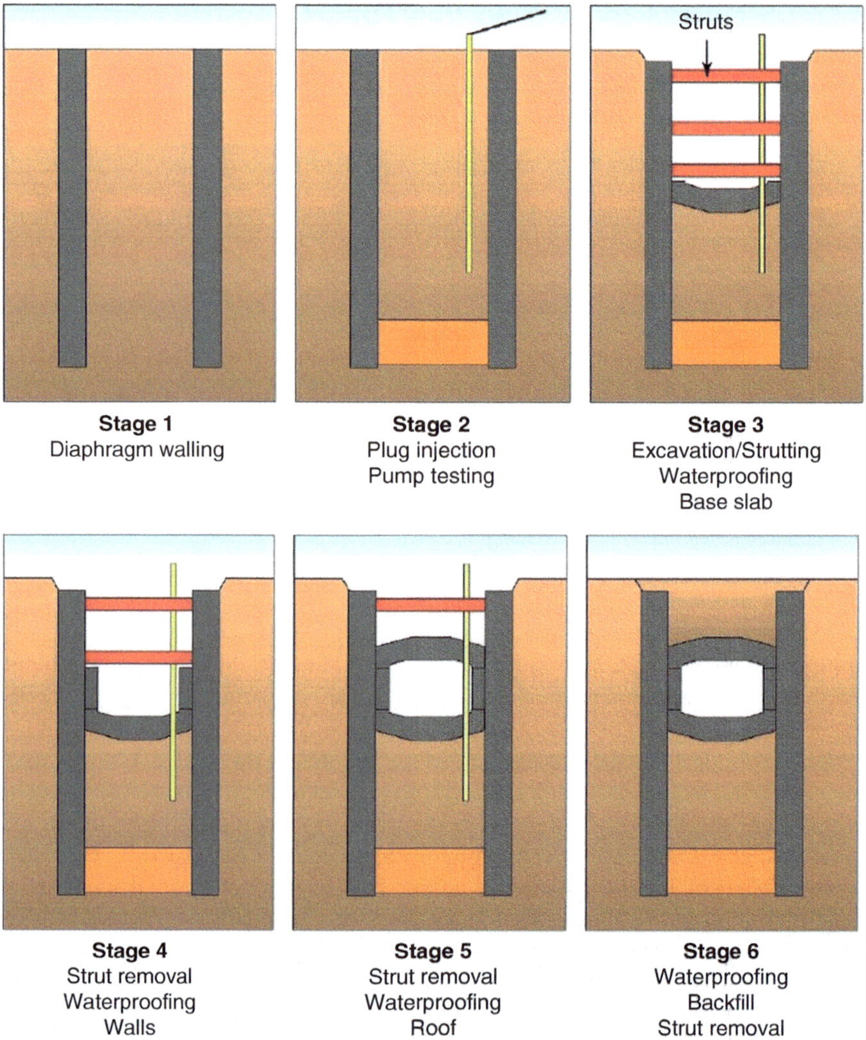

Fig. 12.1 Conventional bottom-up sequence of construction (Ref: http://www.scribd.com/doc/27950632/Tunnel-Construction)

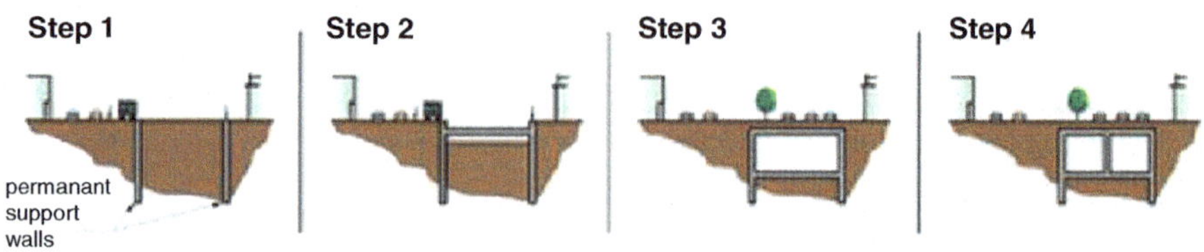

Fig. 12.2 Stages of top-down construction of tunnels (Ref: http://www.fhwa.dot.gov/bridge/tunnel/pubs/nhi09010/05.cfm)

12.2.2 Drill and Blast Technique

In this method, explosives and detonators are placed on a large number of blast holes drilled at designated tunnel surfaces. The detonation of explosives results in the collapse of the rocks. The debris of each blasting is carried to the surface and the new tunnel surface is reinforced before further blasting. Generally drill and blast technique is preferred in areas where shorter tunnels in hard rocks are required. In order to obtain an approximately circular cross section, it is necessary to have controlled blasting by knowing the amount of explosives in properly positioned and drilled holes. In order to direct the energy from the explosives for a proper alignment, various factors like the geological condition of the rock bed, the spacing, size, depth, and angle of the drill hole have to be precisely determined. The construction sequence followed in drill and blast technique is shown in Fig. 12.3. Some of the major disadvantages associated with drill and blast technique are the safety risks like production of poisonous gases and high volumes of dust and noise pollution. Moreover, due to continuous blasting, there is possibility of weakening of rocks around the tunnel. This requires the provision of additional support in the form of rock bolting and shotcreting.

12.2.3 Bored Tunneling by Tunnel Boring Machine (TBM)

TBM is one of the most widely used methods of tunneling for excavating long tunnels. The cutter head mounted on the face of the TBM consists of a number of disk cutters. These cutters while rotated under high pressure chip the rock mass, thus helping to excavate them. In addition to excavation, TBM can also be used to remove the excavated material and install the reinforced concrete lining. The various stages in the tunnel excavation using TBM are shown in Fig. 12.4. While planning the tunnel excavation, the tunnel is divided into various sections. For each section, excavation is done for a launching shaft and retrieval shaft. The TBM is assembled in the launching shaft and the tunnel is excavated. When the excavation reaches its destination, the TBM is dismantled and retrieved through the retrieval shaft. In this case precast lining is provided as permanent tunnel walls immediately after excavation. There are three major types of TBM in use: (a) hard-rock TBM,

Fig. 12.3 Construction sequence of drill and blast technique (Ref: http://www.mtr-kwuntonglineextension.hk/en/construction/construction-methods.html)

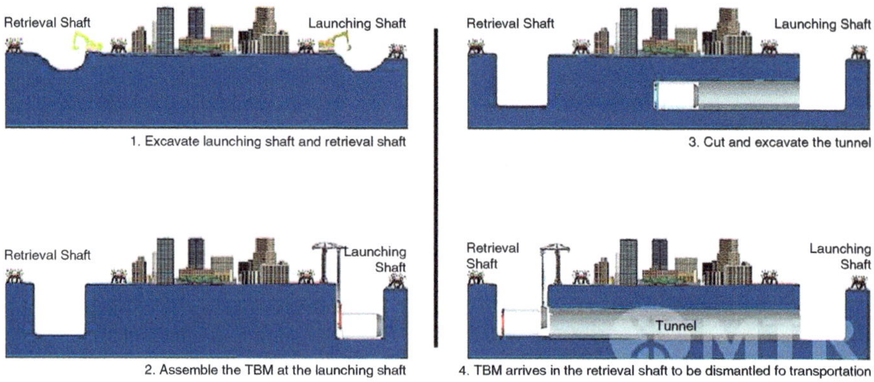

Fig. 12.4 Various stages of excavation using TBM (Ref: http://www.expressraillink.hk/en/construction/construction-methods.html)

(b) soft-ground TBM, and (c) mixed/hybrid TBM. The most important advantage with TBM is that it is suitable for almost all types of soils when the length of the tunnel is high at large depths. However, the handling of this sophisticated equipment requires highly skilled labor, and in case of any corrections required, this method is very expensive.

The above review of various tunneling methods indicates that each has its own advantages and disadvantages. However, the selection of a particular method of tunneling depends on the local geological condition, the length of the tunnel, the depth at which the tunnel is required, the presence of adjoining structures, etc. In the following sections, four different case studies of underground tunneling from different parts of India are presented. The details of each of the metro along with a brief description of the geological conditions of the area and the difficulties associated with the underground metro construction and its impact on the urban infrastructure are reviewed.

12.3 Case Studies

12.3.1 Bangalore Metro

The Bangalore Metro is divided into north-south and east-west corridors. Of the total 42.3 km of Phase 1, 8.8 km consists of twin tunnels with seven underground stations. There are two underground sections in each of these corridors spanning for a length of 4.0 km in north-south and 4.8 km in east-west. This underground section links the city center to the outer elevated sections of the network. The underground section consists of twin tunnels of 5.5 m diameter each with a distance of 5 m in between. The tunnels are at a depth of 12 m from the ground surface. Figure 12.5

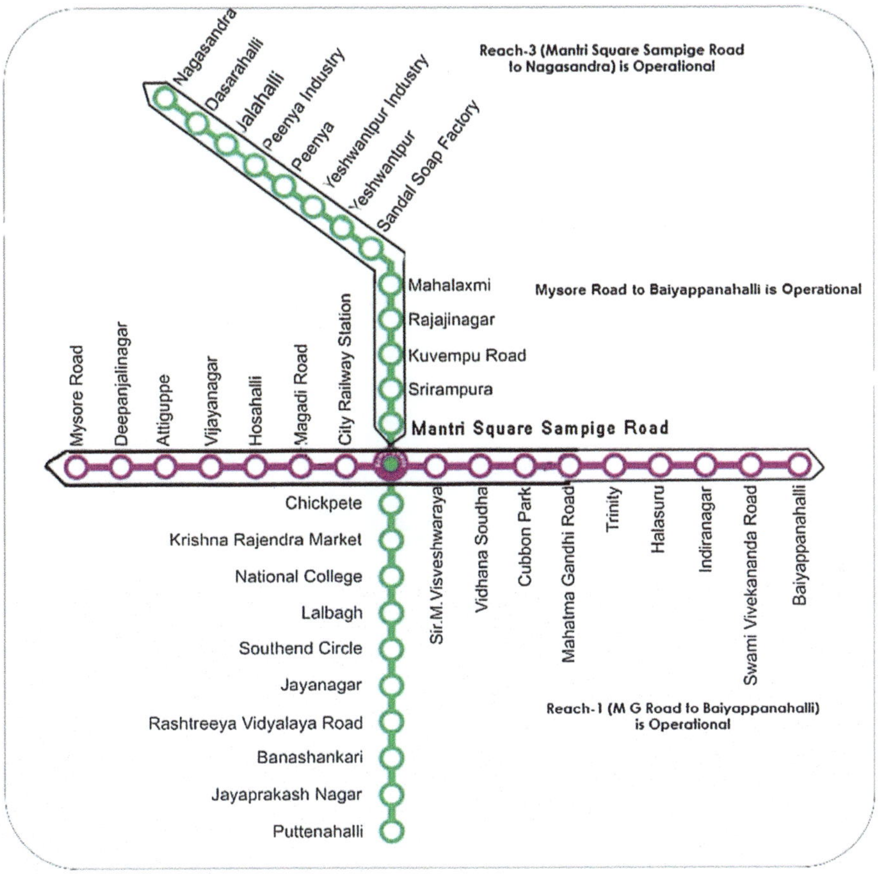

Fig. 12.5 Phase 1 alignment of Bangalore Metro (Ref: http://www.metrotraintimings.com/Bangalore/RouteMap.htm)

shows the Phase 1 alignment of Bangalore Metro. Currently, Reach-1 and Reach 3, 3A, are operational. Figure 12.6 shows the underground twin tunnel construction in Bangalore.

12.3.2 Details About Underground Tunnel/Tunneling Procedure

The geology of the metro project area covering different parts of Bangalore district consists of Precambrian granite and gneiss of which migmatite and gneiss are dominant. However, along the north-west direction across the western part of the district, zones of granite and granodiorite of 20 km width is also present. Small

Fig. 12.6 Underground twin tunnel TBM construction near Kalasipalyam, Bangalore. (**a**) North-south alignment, (**b**) east-west alignment

areas of charnockite are seen in the southwestern part and through the central part of the north-south direction small bodies of amphibolites, and schist is also observed. To know about the groundwater levels, data from exploratory wells through surveying and in situ measurements are done. Figure 12.7a and b illustrates the topography, groundwater level, depth to the bed rock, and the tunnel alignment for the north-south and east-west alignments, respectively. The depth to hard rock varies across the tunnel section in both the alignments which makes it extremely challenging as the tunnel will be passing through both the hard rock as well as fissured zones of saprolite. The north-south alignment in Fig. 12.7a indicates that large stretches of tunnel pass through the fissured zone as the thickness of this saprolite zone is more, thus making the hard bed rock deeper.

As pointed out earlier, the tunneling procedure and the tunnel types depend on the geological conditions of the site. The ground cover in the case of tunnels varies from 60 to 17 m and the underground alignment mostly follows the roads. The geology in the underground section is very challenging with three different layers of soil consisting of a very recently deposited fill as the top layer, which is followed by a layer of silts and sands with varying amounts of clay and gravel. Beneath this, fragmented weak to hard granitic gneiss type of rock is present. In addition to this varying geology, the water table varies from 2 m to 10 m belowground level which is significantly affected by the rainfall. The presence of major structures along the eastern alignment added with shallow depth of tunnels and mixed face conditions

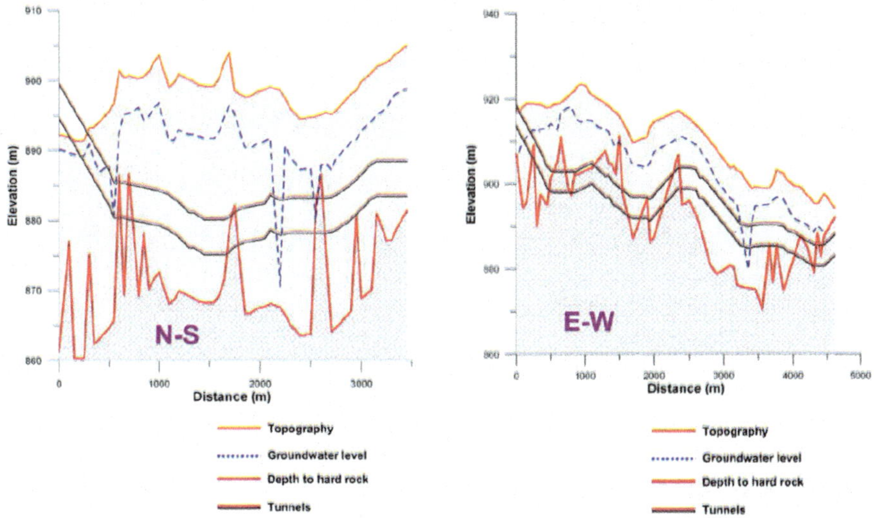

Fig. 12.7 Topography, groundwater level, depth to bed rock, and tunnel position for Bangalore Metro (Ref: http://bmrc.co.in/pdf/news/IISc-Report.pdf)

called for protection measures against subsidence over this section. Based on this, slurry TBMs are selected for tunneling so that the tunnel face will be supported during excavation, thereby minimizing the subsidence during the operation.

12.3.3 Delhi Metro

Delhi Metro is one of the most successfully constructed/operated metro systems in India which consists of four different lines covering a total length of 189.63 kms in Phase 1 and Phase 2 (Fig. 12.8). There are 143 stations, of which 38 underground stations are present in the 48 km of underground metro stretch, five stations are at-grade, and the remaining stations are elevated. Around 45 km of the planned Phase 3 is also underground. In the underground stretch, the stations are constructed by cut and cover method of tunneling. Depending on the geological condition, tunnels are excavated either by cut and cover method, New Austrian Tunneling Method (NATM), earth pressure balance, or slurry-type shield machine. For Phase 2, for the construction of 5.8 m tunnel diameter, DMRC used 14 tunnel boring machines. In this case the excavation diameter was 5.11 m with 14 m to 60 m overburden (http://www.delhimetrorail.com/projectsupdate/delhimasseia.pdf).

The average elevation of the city of Delhi is around 198 to 200 m above MSL. The level of ground water table varies from 20 m to 30 m at different parts of the city. It is observed that the groundwater level is falling at a rate of 3 m per year. Geological study of the subsurface along the Delhi Metro alignment indicates that the soil consists of fine-grained material like clay or silt with different amounts

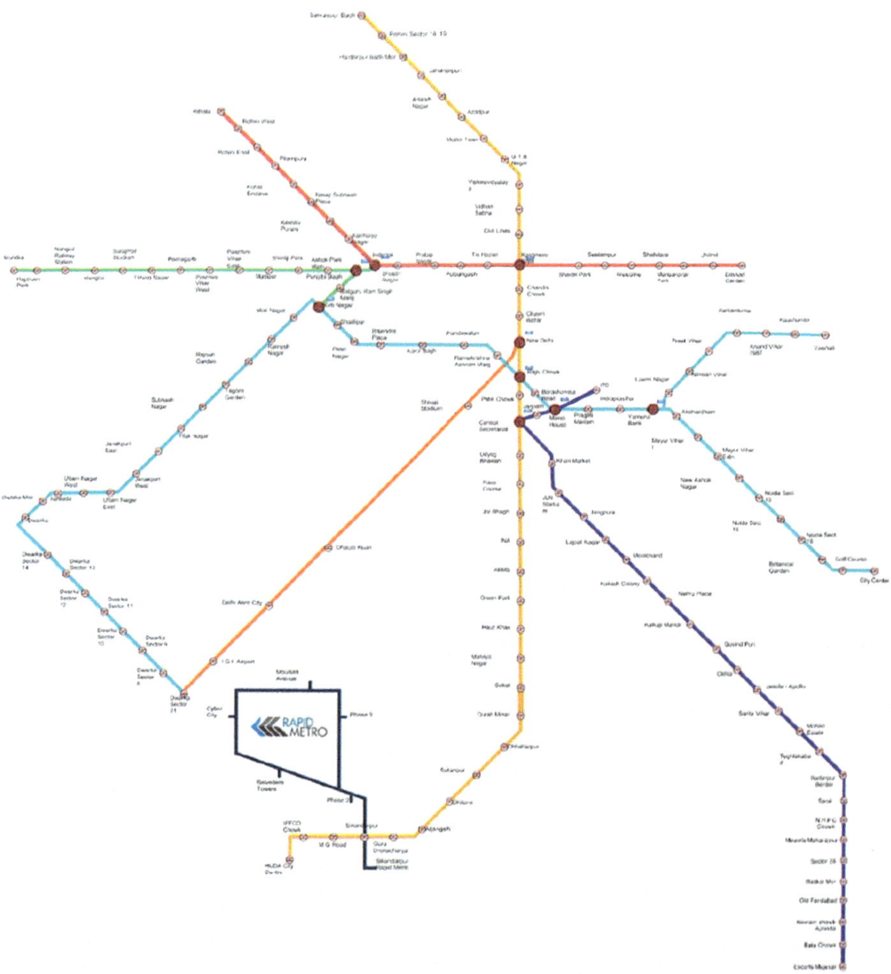

Fig. 12.8 Phase 1 and Phase 2 of the Delhi Metro map (Ref: http://www.delhimetrorail.com/route-map.aspx)

of fine sand content (http://www.delhimetrorail.com/eia%20report/Environment_Impact_Assessment_SECTOR%20_9_TO_IGI-AIRPORT_CORRIDOR.pdf). As there were huge variations in the groundwater level, the overburden was between 3.5 and 22 m. Based on the soil investigation carried out for the underground construction, cut and cover method and tunnel boring machines (TBMs) capable of excavating soft soils are adapted. Here, a minimum soil cover of 6 m is proposed for tunnels.

Phase 1 of Delhi Metro employed 3 TBMs, whereas during Phase 2, 14 TBMs were used to construct around 35 km of underground stretch. However, Phase 3 which is planned to pass under strategic locations including railway tracks,

Fig. 12.9 Tunnel boring machine which is assembled for a length of 96 m begins the tunneling work (Ref: http://www.delhimetrorail.com/whatnew_details.aspx?id=c4kJd1nWTgMlld)

high-rise building, flyovers, ring roads, and below-existing underground metro stretches will be using 34 TBMs to dig these underground strips. The TBM used for the Delhi Metro is specially designed to suit the needs of the project location. This is a 6.61 m diameter earth pressure balancing (EPB) shield. Around 3400 m can be bored by the combined effort of each of the TBM. Figure 12.9 shows an assembled underground TBM which is 96 m long used for tunneling work between Juma Masjid and Delhi Gate Metro stations.

12.3.4 Chennai Metro

Chennai Metro project is planned to create two corridors with a total length of 45.1 kms for the Phase 1 implementation. Corridor 1 is designed to have 23.1 kms of which 14.3 kms is underground and the remaining elevated. The total length of corridor 2 is 22 kms and a length of 9.7 kms is underground in this stretch. Figure 12.10 shows Phase 1 of Chennai Metro. The total length of the underground stations is 24 kms with 19 underground stations. The project involves the construction of 5.5 km of twin tunnels (Fig. 12.11) in one package and 3.1 km of twin tunnels in another package. The tunneling from Nehru Park to Egmore of Chennai Metro Rail which forms a distance of 948 m has been completed using TBM. The speed of operation of TBM depends on the soil conditions and the speed adopted is around 6–10 m/day as the soil is loose. Currently, the TBMs are able to make significant breakthrough for two different sections at Central Metro and Kilpauk Underground station. In the Central Metro underground tunneling, TBM retrieval method was the first of its kind in India. Here, just before the breakthrough, the shaft was filled with bentonite and water till the level of the track slab. The twin tube

Fig. 12.10 Chennai Metro Rail project Phase 1 (Ref: http://chennaimetrorail.gov.in/)

tunneling is performed using five TBMs. The four TBMs in use are EPB shield with 6.6 m diameter. Figure 12.12 shows the tunnel boring machine.

For the project, geotechnical investigations are done using data from 75 boreholes. The data indicates that the project area is composed of granites, gneisses, and sedimentary formations like shale and sandstone with or without fossiliferous beds of various ages. In some places, doleritic dikes are also found which are covered with younger alluvium of thickness varying from 3.0 to 30.0 m. Thus, the ground

Fig. 12.11 Twin tunneling work at Washermanpet, Chennai Metro (Ref: http://www.masterbuilder.co.in/the-tunneling-heroes-of-indian-metros/)

Fig. 12.12 Tunnel boring machine (TBM) (Ref:http://www.thehindu.com/news/cities/chennai/life-of-a-metro-the-first-look-underground/article3502044.ece)

formed by granite, sand, silt, and clay with 300 mm diameter boulders is excavated using EPB. This EPB has a mixed ground cutterhead and specialized small grippers located around the machine's shield. These grippers will hold the cutterhead firmly in harder ground and provide necessary forces to pull the cutterhead back from the tunnel face. In addition to this, for a distance of 3.9 kms stretching from Saidapet ramp area to Gemini station, two Herrenknecht-shielded TBM will be used for boring twin tunnels. The cutterhead diameter of this TBM is 6.63 m and the length is around 80 m.

12.3.5 Kolkata Metro

Kolkata Metro is the first metro system constructed in India. Kolkata is situated on the banks of Hooghly River, and the geology is formed by the Ganga-Brahmaputra river system. Initially, the development of Kolkata city was in the north-south direction for a length of 50 km along the east bank of Hooghly River. This city rests over large thickness of fluvial-marine sediments which forms the pericratonic tertiary basin. This basin has three structural units, viz., westernmost shelf or platform, the Central hinge or shelf/slope break, and deep basinal part in the east and southeast (Diptendra and Chattopadhyay 2009). The water table level in this area is very shallow and may touch the ground level during monsoons. The subsoil exploration done during the construction of north-south line indicated erratic deposition of materials, different thickness of layers, and the presence of bedding planes. Due to this along the metro alignments, soil exploration up to a depth of greater than 30 m from ground level is done.

Three north-south corridors of total length 97.5 km was identified as early as 1971 for Kolkata Metro. This is the first underground railway built in the history of India. Trains first started running in October 1984, and by September 1995, the full stretch that had been initially planned was completed and operational. Currently, the network has one operational line (Line 1) and one under construction (Line 2), with four further lines in various stages of planning. Being the country's first metro, the 17 km stretch of the underground tunnel in the north-south metro took 23 years for the construction. The entire underground part is constructed using cut and cover methodology. Currently the east-west corridor is being built which acts as a second metro for the city. Figure 12.13 shows the route map of the existing metro and the planned east-west corridor of metro. This corridor consists of a total of 14.67 km with 8.9 km underground and 5.77 km elevated. Around 520 m length of this underground tunnel is passing 30 m below Hooghly River as an underwater metro line (Fig. 12.14). Earth pressure balance TBM which is capable of boring up to 35 m per day is used for the construction of the tunnels in this corridor. However, till the first breakthrough, the TBM is used to bore up to 15 m per day. During the initial tunneling process, the presence of stiff clay made the process extremely difficult. All the stations are constructed using cut and cover technology. This TBM consists of a cutterhead which is a rotating cutting wheel, a main bearing, and a thrust system. The diameter of this machine is 6.5 m and it is 120 m long. This TBM can also be used to install the concrete channels which are prefabricated. TBM is the most preferred method of underground tunneling as it causes minimum disturbance to the existing structures and traffic.

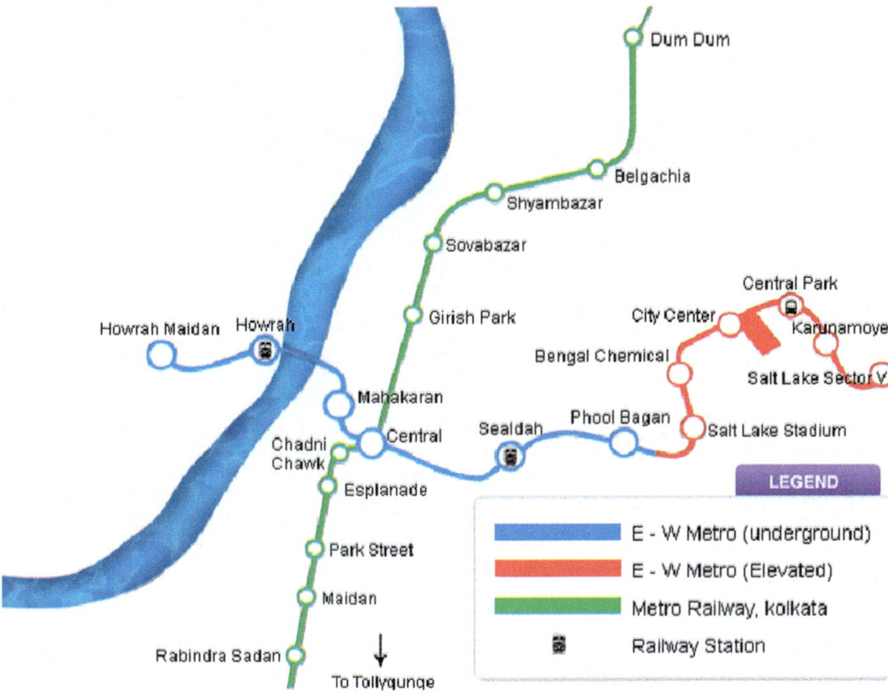

Fig. 12.13 Kolkata Metro

12.4 Numerical Modeling

Having reviewed the history of various metro systems in operation and under construction in India, it is necessary to understand the behavior of the underground soil/rock during and after the construction processes. However, experimental/field monitoring of the processes is time consuming, difficult, and impractical in certain cases which underline the necessity for numerical modeling of the process. Some of works reported on the numerical modeling of underground sections include Coulthard (1999), Ng et al. (2004), Anitha et al. (2012), Anitha and Sitharam (2014), etc. In the following sections, a comprehensive note of the numerical simulations of the tunnel stability using two different numerical methods, viz., discrete element method and finite element method, is presented.

12.4.1 Discrete Element Method

The inherent discrete nature of the soil/rock particles results in transferring of forces through the contacts between the particles. DEM is an explicit finite difference

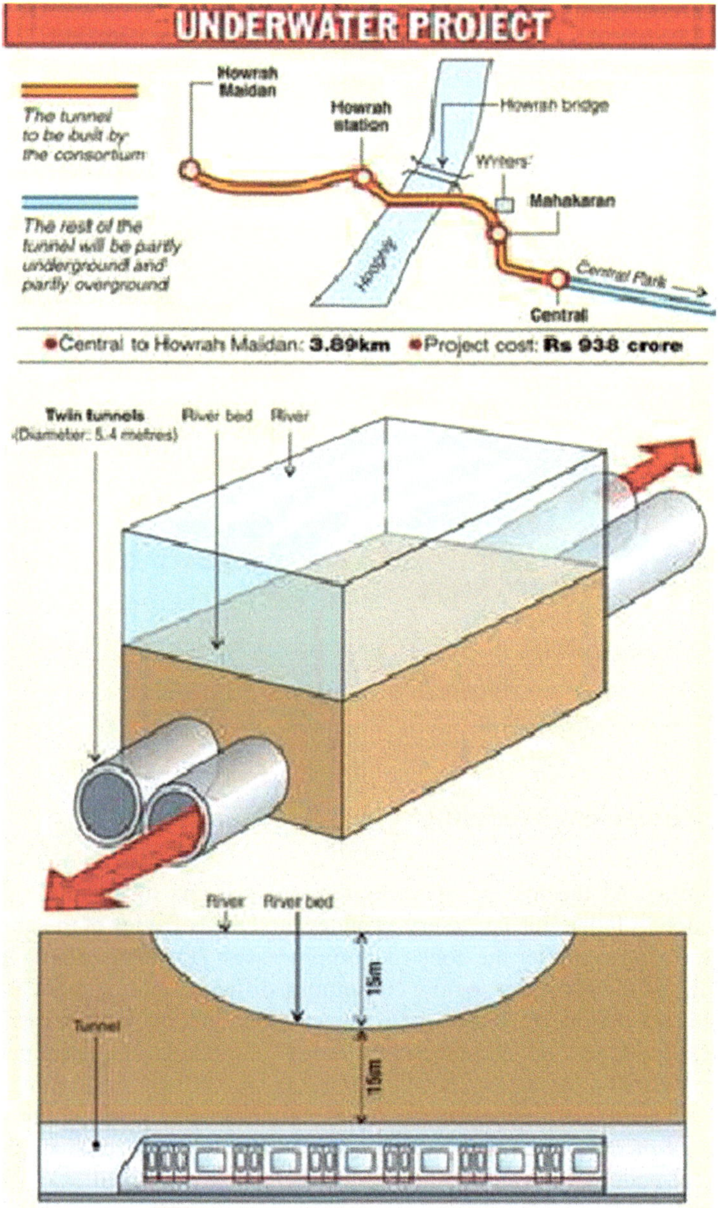

Fig. 12.14 Underwater metro line (Ref: http://www.kmrc.in/route_map.php; http://www.skyscrapercity.com/showthread.php?t=1603028&page=8)

method wherein the interaction of the particles is monitored contact by contact and the motion of the particles modeled particle by particle (Cundall 1971). In discrete element method, the assembly is modeled as a conglomeration of particles which are discrete in nature and interact only through the contacts. In order to model the weak rocks, which are bound by cementing materials, contact bonds are employed. This idea is based on the bonded particle model suggested by Potyondy and Cundall (2004) wherein the rock mass is represented as a dense packing of circular or spherical particles which are bonded together at their contact points. These contact bonds are breakable in nature, and the macro behavior of the rock mass as a whole depends on the strength of these bonds. The excavation of the underground structure in spite of its depth disturbs the soil and rock masses. This results in the settlement of the ground surface which affects the safety of the existing buildings. Due to several practical reasons and site conditions, tunnels are excavated on soft to hard soils/weak rocks at shallow depths. In order to understand the stability of tunnels constructed in a soft ground, simulations are done on an unlined tunnel and tunnel with lining.

12.4.2 Modeling the Assembly Using DEM

A weathered rock assembly having a dimension of 25 m × 10 m × 25 m shown in Fig. 12.15 is modeled for the study. PFC^{3D} (2008) is used for the numerical simulation of the model. The model constitutes 50,000 spherical particles. Table 12.1 shows the properties used for the simulation of the assembly. Once the particles are allowed to settle under gravity, a confining stress of 0.8 MPa which corresponds to the overburden pressure is applied to the sample. To introduce the bonding between the particles forming the rock mass, contact bonds are installed. In this assembly, at a depth of 18 m from the ground surface, a tunnel of radius 3 m is

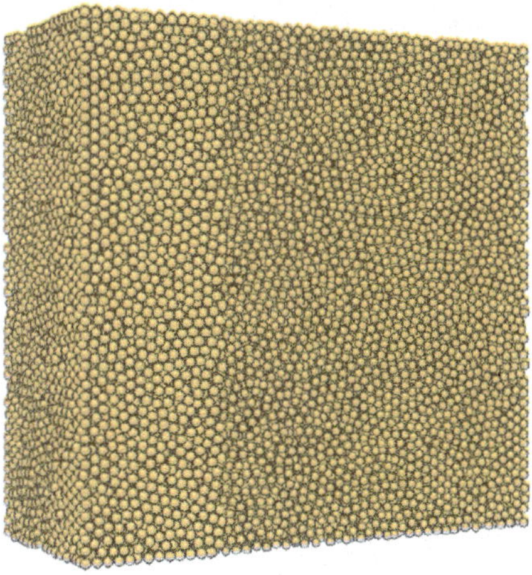

Fig. 12.15 Sample (25 m × 10 m × 25 m)

Table 12.1 Properties of particles used for the simulation of the assembly

Property	Value
Normal stiffness of particles	100 MN/m
Shear stiffness of particles	100 MN/m
Wall stiffness	1000 MN/m
Density of particles	2600 kg/m^3
Normal contact bond strength	20 kN
Shear contact bond strength	20 kN
Porosity	0.4
Number of particles	50,000
Interparticle friction	0.45
Particle size	0.075–0.1 m

excavated. The effect of excavation of the tunnel on the contact force and stress variation is studied for a tunnel unlined/without any support and for a tunnel which is lined.

12.4.3 Finite Element Method

Finite element method is widely used in rock mechanics applications and can be applied to a large spectrum of soil/rock mechanics problems. In finite element method, the assembly is modeled as a continuum consisting of elements which are connected at discrete points called nodes. In two-dimensional analysis, the assembly is discretized into triangle or quadrilateral shapes, and all the forces are transmitted through nodes. The analysis of the problem is basically done in terms of these nodal forces and nodal displacements. The FEM analysis is performed using the commercial software PLAXIS.

12.4.4 Modeling the Assembly Using FEM

A 25 m wide and 32 m deep assembly is modeled. At a depth of 18 m from the ground surface, a 3 m radius tunnel is excavated. After defining the geometry, it is divided into 15 noded triangular elements as shown in Fig. 12.16, and the properties used for the model material and the lining are shown in Table 12.2. Mohr-Coulomb elastoplastic constitutive model is adopted for this study (Plaxis User Manual 2009). Since the ground surface is horizontal, K_0 procedure is followed and gravity loading is applied to the assembly. Following this, loading phase is initiated.

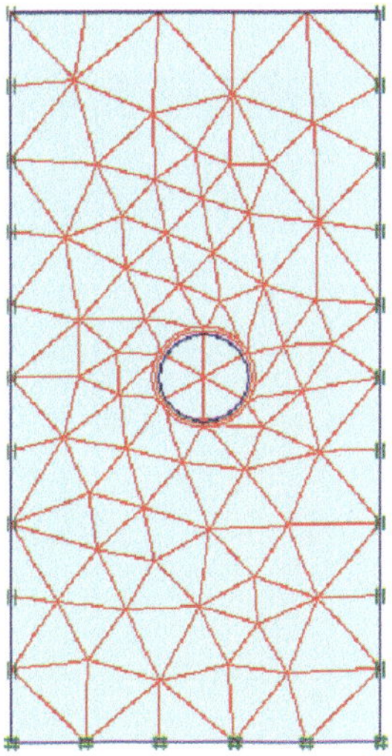

Fig. 12.16 Discretized model

Table 12.2 Properties used for the weak rock material

Model	Mohr-Coulomb
Behavior	Drained
Unit weight	26.5 kN/m^3
Young's modulus	5e6 kN/m^2
Poisson's ratio	0.2
Cohesion	20 kN/m^2
Angle of internal friction	24
Behavior of lining	Elastic
Lining stiffness , EA	1.4e7 kN/m
Lining flexural rigidity, EI	1.43e5 kNm
Lining thickness	0.35 m
Weight/m length	8.4 kN/m/m

12.4.5 Distribution of Contact Forces/Stresses for Unlined and Lined Tunnels

Figure 12.17a and b shows the stress variation and contact force variation around the unlined tunnel modeled using FEM and DEM, respectively. The vertical stress near the crown of the tunnel is very low for the unlined tunnel. However, the tunnel failure is progressing uniformly in all directions around the tunnel as observed from the DEM analysis. The thickness of the lines in Fig. 12.17b indicates the magnitude of the contact force. It is observed that there is significant reduction in the contact force in the area surrounding the tunnel leading to the ultimate failure of the unlined tunnel system. The collapse of the tunnel is progressing from all sides, more being

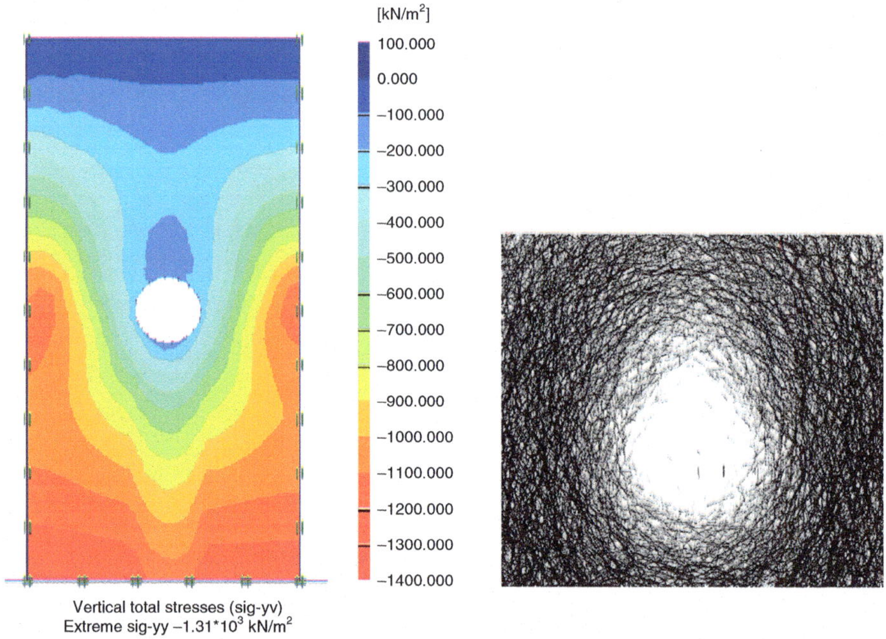

(a) Vertical stress - FEM **(b) Contact forces – DEM**

Fig. 12.17 Stress and force distribution – FEM and DEM analysis, unlined tunnel. (**a**) Vertical stress – FEM. (**b**) Contact forces – DEM

from the top. The analysis results indicate that the unlined tunnel system has completely collapsed immediately. The FEM analysis results also show trends similar to the discrete element simulations; the deformations are very high near the crown of the tunnel indicating a complete collapse of the system. The stress contours along the depth of the considered section points to the stress distribution due to the excavation.

The variation of the stress and contact force for the lined system is shown in Fig. 12.18a and b. For the lined tunnel, the distribution of stresses around the tunnel is uniform which results from the confinement provided by the lining thereby restricting the movement of particles. The stress contours and the contact force distribution around the tunnel opening indicate how the system has stabilized through the introduction of lining even though there is a reduction in the overall contact force and some collapse of the assembly. The arching action provided by the ground due to the confinement introduced by the lining increases the overall stability of the system. Thus the studies helped to identify the requirement of a lining and how the stress or displacement varies over the tunnel crown. Another point worth mentioning is the suitability of both the methods for analyzing the underground structures. However, depending on the surface conditions and the

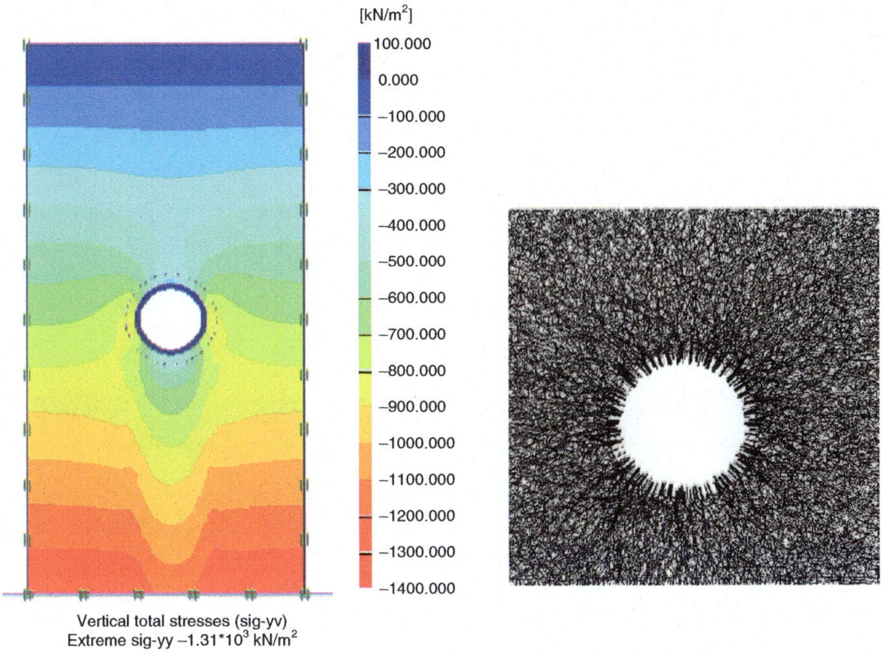

(a) Vertical stress - FEM **(b) Contact forces – DEM**

Fig. 12.18 Stress and force distribution – FEM and DEM analysis, lined tunnel. (**a**) Vertical stress – FEM. (**b**) Contact forces – DEM

presence of more fissures or discontinuum, suitable method has to be selected for the precise modeling of the stress and displacement around the underground openings.

12.4.6 FEM Simulations of Seismic Response of Twin Tunnels

As seen in various case studies, twin tunnels are being constructed in different underground metros. As these tunnels are passing through heavily crowded and congested urban city centers, it is very important to investigate the seismic response of such structures. Even though it was thought that the underground structures are immune to earthquake structures, recent earthquakes have devastated underground structures thereby emphasizing the requirement of seismic analysis of underground tunnels. Hence in the following section, the behavior of twin tunnels under seismic loading is presented.

Tunnels of 4 m radius which are aligned horizontally are simulated using FEM (Fig. 12.19). The crown of the tunnel is at a depth of 8 m from the ground surface,

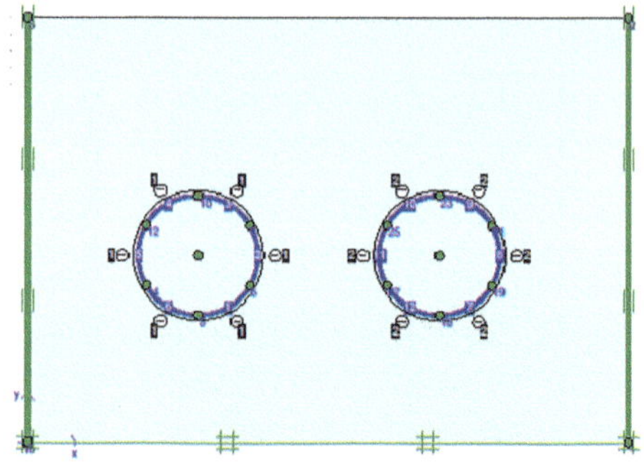

Fig. 12.19 Horizontally aligned circular twin tunnels

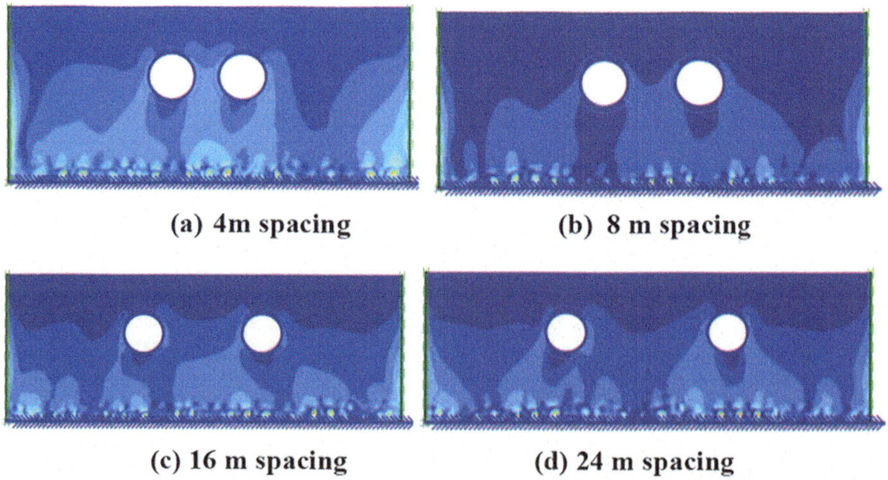

Fig. 12.20 Variation of vertical stresses at different tunnel spacing. (**a**) 4 m spacing. (**b**) 8 m spacing. (**c**) 16 m spacing. (**d**) 24 m spacing

whereas the bottom boundary is at a clear distance of 14 m from the base of the tunnel. A clear distance of 24 m is provided on the sides to avoid the boundary effects. The earthquake effect is studied by applying a loading obtained from the strong motion data of USGS (2011) which is having a peak horizontal acceleration value of 0.2 g. This acceleration is applied to the bottom of the model, and the effect of this loading is studied by varying the spacing between the tunnels. Figure 12.20a–d shows the vertical stress distribution around the tunnels when the tunnel spacing is varied from 4 m to 24 m. When the spacing between the tunnel increases, the stress overlap decreases leading to a more stable configuration. The stress distribution around the tunnels becomes more uniform as the spacing between them increases. However for closely spaced tunnels, nonuniform stress distribution is observed which can be attributed to the stress interference from both the tunnels. It is also observed that when the tunnels are spaced closely, both the tunnels and the rock mass in between them acted as a single body, whereas as spacing increases, they act as independent entities. It was also found that, on an average, the horizontal displacement was around 10 to 15 times that of the vertical displacement.

12.5 Conclusions

The paper consolidates the various details and experiences from the completed and ongoing metro constructions in India with emphasis on the underground engineering for sustainable development. Underground tunnels for the metros are constructed in subsurface whose properties vary from soft clay to hard rocks. The underground part of the Kolkata Metro Rail which is the oldest metro system in India was completely done using cut and cover methodology. The challenges faced during the construction and implementation of this metro paved the way for the successful planning, construction, and implementation of the other metros. Currently the most advanced tunneling technologies including the earth pressure balance tunneling boring machine are being used for the ongoing metro projects. This is evident from the fact that the required 4 years to construct 700 m underground tunnel using cut and cover technology took just less than 2 months using TBM. For the first time in India, underwater metro line is also being constructed across river Hooghly as part of Kolkata Metro system. The detailed investigation of the geology of underground along with subsurface explorations has helped to identify the appropriate tunneling methodology and equipments for the excavation and installation of these structures. These advanced techniques help to build these underground structures with minimum disturbance to the traffic and surface constructions and also minimum environmental impact in a congested urban environment.

Since the behavior of subsurface soil under various loading conditions is not predictable due to the nonhomogeneous strata, the changes occurring during the construction as well as execution of the underground projects are very important. To address this, numerical simulations are done using different methods to understand the variations in stresses and displacements around the tunnels. DEM and FEM were used to study the effect of loading on lined and unlined tunnels. The results indicate that in the case of unlined tunnels, the system collapsed completely due to the low bonding capacity of the soil on which the tunnel is excavated. However in the case of lined tunnels, the lining provided additional support to the structure, and a uniform distribution of the stresses around the tunnel is observed along with the arching action. The formation of the arch action along with the loosening of the ground was captured well by the DEM compared to FEM as it considers the discontinuous nature of the soil strata.

Even though it was considered that underground structures are safe against earthquakes, recent reports show that these are also vulnerable to the impacts of earthquake loading. Hence in this study, the effect of seismic loading on the behavior of underground twin tunnels placed at different spacing is reported. The studies indicate that the spacing between the tunnels affects the stress and displacement distributions. The interference effects due to the adjacent tunnel will be nullified if the distance between the tunnels is more than three times the diameter of the tunnel and the distributions will be uniform.

References

Anitha Kumari SD, Sitharam TG (2014) Tunnels in weak ground, discrete element simulations. ISRM J 1(1):15–22

Anitha Kumari SD Vipin KS, Sitharam TG (2012) Seismic response of twin tunnels in weathered rocks. Geotechnical special publication GSP-225, ASCE, pp 3268–3274

Coulthard MA (1999) Applications of numerical modeling in underground mining and construction. Geotech Geol Eng 17:373–385

Cundall PA (1971) A computer model for simulating progressive large-scale movements in blocky rock mechanics. In: Proceedings Symposium International Society Rock Mecha. Nancy, France, 2: 129–136

Das D, Chattopadhyay BC (2009) Characterization of soil over Kolkata municipal area. IGC 2009, Guntur, INDIA, pp 11–14

Itasca Consulting Group Inc (2008) PFC^{3D} (Particle Flow code in 3 Dimensions) Manual, Version 4.0 Minneapolis

Ng CW, Lee KM, Tang DKW (2004) Three-dimensional numerical investigations of new Austrian tunnelling method (NATM) twin tunnel interactions. Can Geotech J 41(3):523–539

PLAXIS User Manual Version 9.0

Potyondy DO, Cundall PA (2004) A bonded-particle model for rock. Int J Rock Mech Min Sci 41:1329–1364

USGS (2011) http://nsmp.wr.usgs.gov/data_sets/upland.html. Last Accessed 25 June 2011

http://bmrc.co.in/pdf/news/IISc-Report.pdf

http://chennaimetrorail.gov.in/newsletter/February-2014.pdf

http://chennaimetrorail.gov.in/pdf/project_brief_updated_aug08.pdf

http://en.wikipedia.org/wiki/Drilling_and_blasting

http://miningandblasting.wordpress.com/2011/07/27/tunnel-construction-methods-and-their-comparison/

http://www.delhimetrorail.com/eia%20report/Environment_Impact_Assessment_SECTOR%20_9_TO_IGI-AIRPORT_CORRIDOR.pdf

http://www.delhimetrorail.com/projectsupdate/delhimasseia.pdf

http://www.delhimetrorail.com/route-map.aspx

http://www.delhimetrorail.com/whatnew_details.aspx?id=c4kJd1nWTgMlld

http://www.expressraillink.hk/en/construction/progress-update.html

http://www.fhwa.dot.gov/bridge/tunnel/pubs/nhi09010/05.cfm

http://www.gautrain.co.za/construction/2008/10/tunnel-excavation-methods/

http://www.kmrc.in/

http://www.masterbuilder.co.in/the-tunneling-heroes-of-indian-metros/

http://www.metrotraintimings.com/Bangalore/RouteMap.htm

http://www.nbmcw.com/articles/metro-and-tunneling/29091-tunnelling-by-epb-tunnel-boring-machine-in-dmrc.html

http://www.p3planningengineer.com/productivity/tunneling/tunneling.htm

http://www.scribd.com/doc/27950632/Tunnel-Construction

http://www.skyscrapercity.com/showthread.php?t=1603028&page=8

http://www.thehindu.com/news/cities/chennai/life-of-a-metro-the-first-look-underground/article3502044.ece

Chapter 13
Sand–Tire Chip Mixtures for Sustainable Geoengineering Applications

S. Bali Reddy and A. Murali Krishna

Abstract The applications of scrap tire-derived recycled materials in civil engineering applications have been increasing largely because of their potential economic and environmental benefits. This paper first evaluates sand–tire chip (STC) mixture properties and then discusses about the application of STC mixtures in geoengineering applications through laboratory model studies. Locally available sand and tire chips of 20 mm long with 10 mm square cross section are adopted for preparing the STC mixtures. Tire chips are mixed with sand in various percentages, ranging from 10 to 70 % with an increment of 10 %. Index and mechanical properties of sand–tire chip mixtures are determined for different proportions. Based on large direct shear tests, it was found that internal friction angle values are increased with TC contents up to 30 %. The study indicates that the optimum percentage of tire chips of the selected size is in the range of 30–40 % by weight, which is the equivalent of 50–60 % by volume. Model tests on retaining wall models by using various STC mixtures ranging from 10 to 50 % were discussed. The paper concluded that the STC mixture at the optimum ratio results in lightweight material with 20 % less unit weight with better strength parameters and compressibility behavior, which can effectively be used for geoengineering applications.

Keywords Scraped tires • Sand–tire chip (STC) mixtures • Sustainability • Retaining walls • Displacements • Earth pressure

13.1 Introduction

Scrap tires are undesired urban waste material and the volume is increasing every year worldwide. Scrap tires have characteristics that make them not easy to dispose (Alqaissi 2012). The reuse of scraped tires is an essential step in creating a sustainable future. Research into the reuse of waste tire has often led to fill material

S.B. Reddy (✉) • A.M. Krishna
Indian Institute of Technology Guwahati, Guwahati, India
e-mail: sodom@iitg.ernet.in; amurali@iitg.ernet.in

G.L. Sivakumar Babu et al. (eds.), *Sustainability Issues in Civil Engineering*, Springer Transactions in Civil and Environmental Engineering,
DOI 10.1007/978-981-10-1930-2_13

that provides greater economy than those traditionally practiced. Scrap tire-derived materials in geotechnical engineering application can be in three forms like tire crumbs (length < 10 mm), tire chips (length 10 mm to 50 mm), and tire shred (length > 50 mm) as per ASTM D6270-08.

Scraped tires can be used either as a whole tire or in shredded form in civil engineering applications because it possesses light weight. The average value of specific gravity for shredded tire chips was reported by researchers (Ahmed 1993; Humphrey et al. 1993) to be 1.02–1.24, which is less than half of that of conventional sand. Waste tires used for reinforcing soft soil in road construction (Bosscher et al. 1997) and drainage characteristic of fill material significantly influence the stability of retaining structures under saturated conditions. Fill material with low permeability under a saturated condition induces slope failure due to generation of excessive pore pressure. Compacted tire shred has hydraulic conductivity values equivalent to that of typical coarse gravel which range from 2.0 cm/s to 0.75 cm/s (Ahmed 1993). The shear strength of tire-derived aggregate (TDA) determined using triaxial shear has been reported by Bressette (1984), Ahmed (1993), Benda (1995), and Zornberg et al. (2004) and using direct shear by Humphrey et al. (1993), Cosgrove (1995), and Gebhardt (1997). Different researchers (Foose et al. 1996; Ghazav and Sakhi, 2005) used different sand–tire chip mixture properties by using different methods. This paper evaluates the properties of STC mixtures and model test results for geoengineering applications.

13.2 Materials Used

13.2.1 Sand

Locally available cohesionless dry sand is used in this study. Figure 13.1 shows the particle size distribution of the sand used in this study. The sand achieved a maximum dry unit weight of 16.0 kN/m^3 in a vibration test and the minimum dry unit weight observed in the loosest state was 13.6 kN/m^3. Large-sized (shear box of 300 × 300 × 300 mm) direct shear tests were conducted for sand with different normal stresses to find the shear strength parameters. Figure 13.2 shows the direct shear results of sand at various normal stresses. The properties of sand are shown in Table 13.1.

13.2.2 Tire Chips

Scrap tire-derived recycled materials in the form of tire chips are adopted for the study. The recycled tire chips of 20 mm long with 10 × 10 mm square cross section, cut from scrap tires, are used (after Reddy et al. 2015). The index and engineering

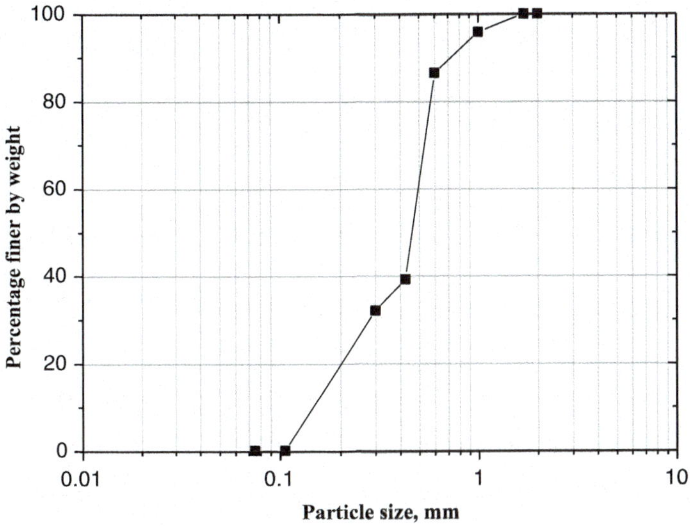

Fig. 13.1 Grain size distribution of the sand

Fig. 13.2 Direct shear test results of the sand (After Reddy and Krishna 2015)

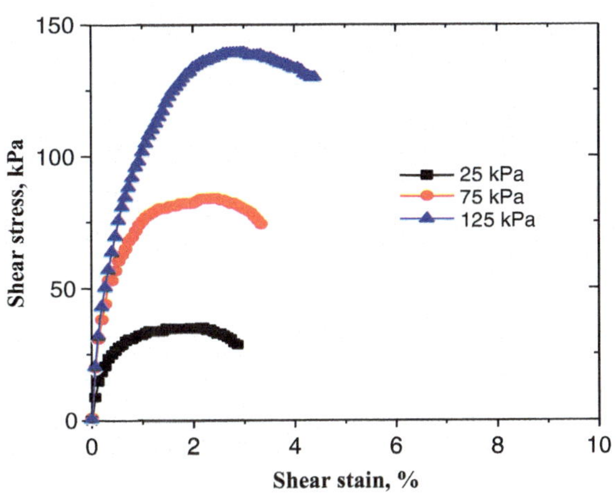

Table 13.1 Properties of sand and tire chips (Modified after Reddy et al. 2015 with permission from ASCE)

Property	Sand	Tire chips
Specific gravity	2.62	1.08
Maximum void ratio (e_{max})	0.94	0.97
Minimum void ratio (e_{min})	0.64	0.65
Maximum dry unit weight (kN/m^3)	16.00	6.45
Minimum dry unit weight (kN/m^3)	13.60	5.39
Angle of internal friction (°)	48	28
USCS classification symbol	SP	–

properties, such as specific gravity of tire chips, unit weights, angle of internal friction, and maximum and minimum void ratios of tire chips, are determined and reported in Table 13.1. Tire chip properties obtained are in the range of results reported by Humphrey and Manion (1992) and Balunaini et al. (2009).

13.3 Sand–Tire Chip (STC) Mixture Characterization

STC mixtures were prepared manually by adding TC to sand in some selected percentages by mass. Percentages of TC considered in the mixtures were 10, 20, 30, 40, 50, and 70 % (after Reddy et al. 2015 with permission from ASCE). The calculated amounts of sand and *TC* were placed together and mixed thoroughly to form STC mixture (Fig. 13.3). A sufficient amount of STC mixture is collected from the sample prepared and is used for the testing, and the corresponding weights of sand and *TC* are recorded to find the achieved percentage of *TC* in the mixture (Eq. 13.1). The targeted (%targeted) and achieved (%achieved) percentages of tire chips are shown in Table 13.2. This achieved percentage was within (±1):

$$TC(\%) = \frac{W_{tc}}{W_{tc} + W_s} \tag{13.1}$$

The dry unit weights of the mixtures were determined by using a cylindrical mold typically used for relative density tests. The dimensions of the mold used for the compaction were 28 cm diameter and 24 cm height. The mixtures are prepared as per calculated quantities by mixing manually the tire chips and sand by weight. For finding the maximum unit weight, the mold was filled in three layers with the mix giving each layer 56 blows using a modified Proctor compaction hammer of weight 4.53 kg. The compacted dry unit weights of each mixture were determined. For

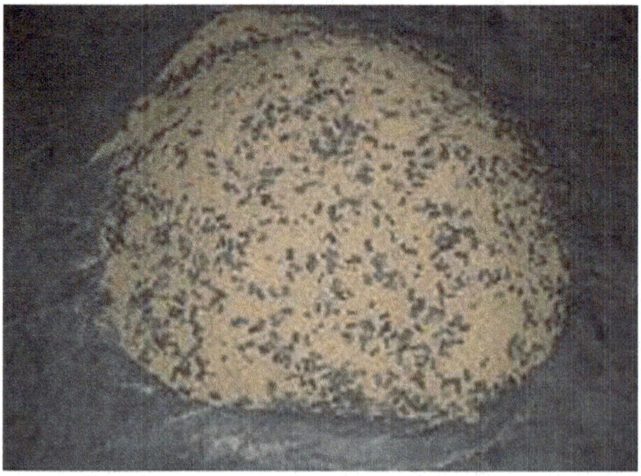

Fig. 13.3 Sand–tire chip (STC) mixture

Table 13.2 Target and achieved percentage of Tire chips in the STC mixture

$TC_{targeted}$ (% by weight)	$TC_{achieved}$ (% by weight)	
	Compacted state	Loose state
10	9.72	9.45
20	19.60	19.46
30	29.35	29.29
40	39.30	36.95
50	49.60	49.88
70	71.17	69.00
100	100	100

STC mixtures beyond 40 %, compaction problems existed due to higher tire chip contents by volume than sand, which requires special attentive compaction for obtaining uniform compacted samples (after Reddy et al. 2015 with permission from ASCE).

Specific gravity (G) values of mixtures are measured using pycnometer and also calculated using Eq. 13.2. Using the dry unit weights and specific gravity values of STC mixtures, the void ratios of the STC mixtures were calculated in the compacted and loose states (Eqs. 13.3, 13.4, 13.5, 13.6, and 13.7):

$$G_{m(\mathrm{cal})} = \frac{W_m}{\frac{W_s}{G_s} + \frac{W_{tc}}{G_{tc}}} \tag{13.2}$$

$$e_{\mathrm{total}} = \frac{V_{\mathrm{voids}}}{V_{\mathrm{solids}}} \tag{13.3}$$

The volume of the voids, V_{voids}, is formulated as

$$V_{\mathrm{voids}} = V_m - (V_{\mathrm{sand}} + V_{\mathrm{tire}}) \tag{13.4}$$

The volume of sand and tire chips in the mixtures, V_{sand} and V_{tire}, was evaluated using Eqs. 13.5 and 13.6:

$$V_{\mathrm{sand}} = \frac{W_s}{G_s} \tag{13.5}$$

$$V_{\mathrm{tire}} = \frac{W_{tc}}{G_{tc}} \tag{13.6}$$

$$e_{\mathrm{total}} = \frac{V_m - \left(\frac{W_s}{G_s} + \frac{W_{tc}}{G_{tc}}\right)}{\frac{W_s}{G_s} + \frac{W_t}{G_{tc}}} \tag{13.7}$$

A large direct shear test setup, having a shear box of 30 x 30 x 30 cm (Fig. 13.4), was used to determine the shear strength properties of STC mixtures. In the test program, the tire chip contents, ranging from 10, 20, 30, 40, 50, and 70 %, by total weight of STC mixture, were considered. When preparing the specimens, the

Fig. 13.4 Large-sized direct shear box with sand–tire chip mixture (After Reddy et al. 2015 with permission from ASCE)

amount of STC mixture in each proportion was determined using maximum unit weight values obtained from unit weight tests and the volume of the shear box. The required quantity of STC mixture was prepared by mixing the sand and tire chips manually to maintain the uniform mixture. The mixed material was steadily poured into the shear box in three layers. Each layer was compacted by using a rectangular wooden plate and imparting uniform compacting energy that was adopted in the unit weight tests. A similar compaction procedure was adopted by Ghazavi (2004) to make compacted sand–tire chip mixture samples. Direct shear tests were carried at three normal stresses of 25, 75, and 125 kPa. The shearing rate was kept at 1.25 mm/min for all the tests. Shearing force and shear displacements were measured during testing. When calculating the shear stresses on the specimen, area corrections were applied (after Reddy et al. 2015 with permission from ASCE).

13.4 Results and Discussion

The results of dry unit weight values of all STC mixtures are presented in Fig. 13.5. From Fig. 13.5, it can be observed that the compacted, loose dry unit weight, and specific gravity values of the mixtures of sand and tire chips decrease linearly with increasing percentages of tire chip contents in the sand–tire chip mixture, which was due to lower specific gravity of the scrap tire. By using 30 % tire chips in the STC mixture, there was approximately a 20 % reduction in dry unit weight. Figure 13.6 presents the measured and calculated specific gravity values of the different STC mixtures, which are well matched. For any targeted STC mixture, the specific gravity can be calculated using specific gravity values of sand and tire chips alone, along with the proposed weights for that mixture.

Using the dry unit weight and specific gravity of STC mixtures, the void ratio of the STC mixtures is calculated in compacted and loose state, which are plotted in

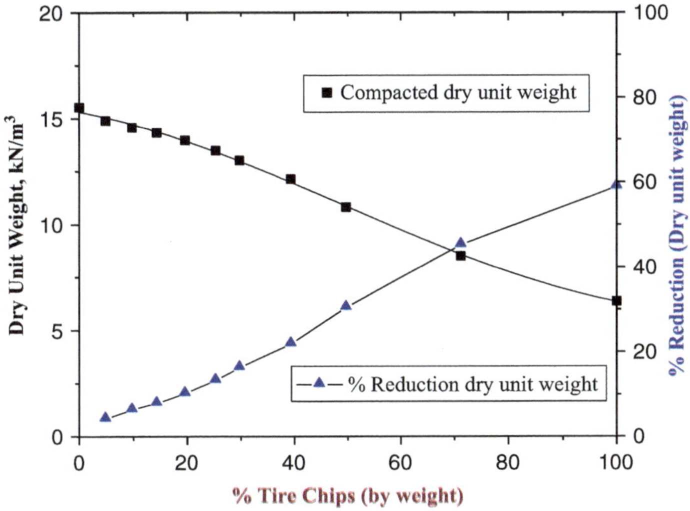

Fig. 13.5 Dry unit weight of STC mixtures

Fig. 13.6 Specific gravity values of STC mixtures (After Reddy et al. 2015 with permission from ASCE)

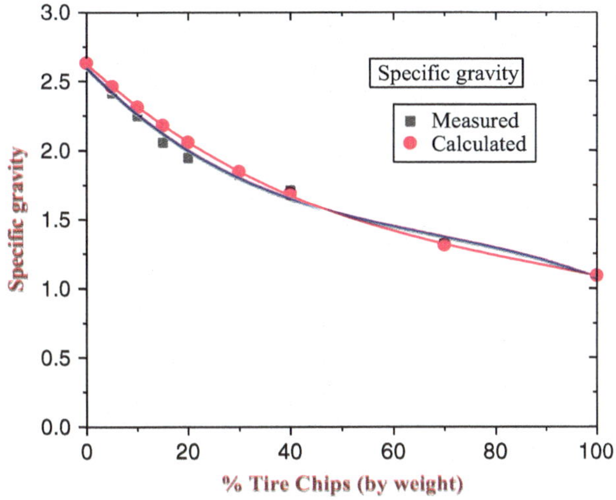

Fig. 13.7. As seen in Fig. 13.7, it can be observed that the void ratios at compacted and loose state of the mixture followed a decreasing trend up to 40 % tire chip content, followed by an increasing trend beyond 40 %. The sand particles filled the voids between the tire chips, and the minimum–maximum void ratio of the mixture decreased. This would continue until all the voids between the tire chips are completely filled with sand particles. When all the voids between the tire chips are filled with sand particles, any further addition of tire chips would increase both the overall volume of the mixture and consequently its minimum–maximum void

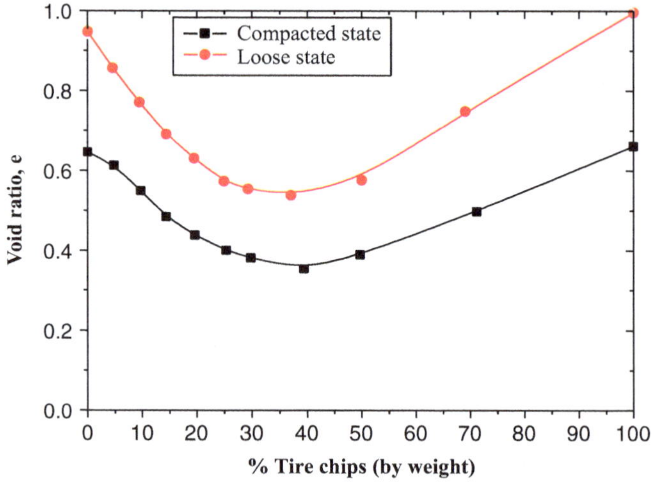

Fig. 13.7 Total void ratios of sand–tire chip mixtures (After Reddy et al. 2015 with permission from ASCE)

ratio (after Reddy et al. 2015 with permission from ASCE). A similar trend was also reported by Lade et al. (1998) and Balunaini et al. (2009).

STC mixture weight ratios were converted to volume ratios using the weights of sand and tire chip contents in each mixture, along with specific gravity of the individual materials, as shown in Eq. 13.8. The weight and volume ratios of tire chip contents are shown in graphical form in Fig. 13.8:

$$\%V_{\text{tirechips}} = \left(\frac{V_{\text{tire}}}{V_{\text{sand}} + V_{\text{tire}}}\right) \quad (13.8)$$

Large direct shear tests are performed on sand alone and STC mixtures at three normal stresses 25 kPa, 75 kPa, and 125 kPa. The variation of shear stress–shear strain curves for different sand–tire chip mixtures at normal stress 125 kPa is presented in Fig. 13.9. The figure indicates that the peak shear stress values were increased, with an increase in *TC* content up to 30 %. Thereafter, an increase of tire chip content in the STC mixture reduced peak shear stress. The increased shear strength behavior (up to STC 30 %) is attributed to the reinforcement effect of tire chips in the STC matrix, which was also reported by Zornberg et al. (2004). Further, initial stiffness of the mixtures was decreased, with an increase in *TC* content, and the amount of tire chips had a significant influence on the shear behavior of the mixtures. However, the strain corresponding to peak shear stress increased, with the increase in the percentage of tire chips as compared to sand alone, indicating the behavior changing from brittle to ductile with the addition of tire chips. This ductile behavior facilitated the use of STC mixtures in resilient structures (after Reddy et al. 2015).

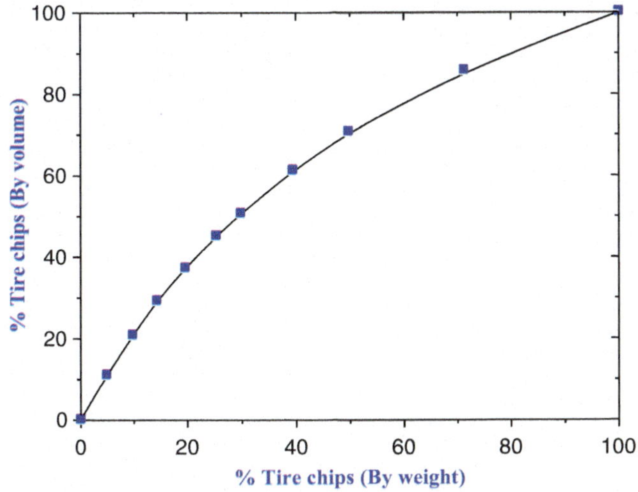

Fig. 13.8 Weight ratios and volume ratios of tire chips (After Reddy et al. 2015 with permission from ASCE)

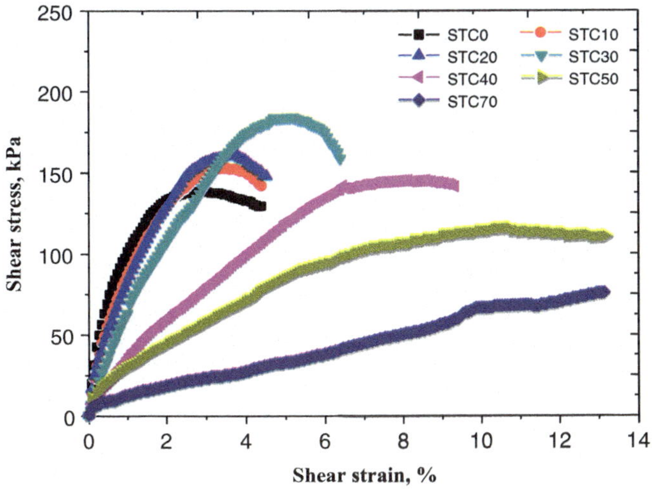

Fig. 13.9 Stress versus strain at normal stress 125 kPa

Figure 13.10 shows shear envelopes (shear stress versus normal stress) of STC mixture specimens. Tire Chips (*TC*) up to 50 %, the peak shear stress and for the remaining (70 % and pure *TC*), shear stresses at a strain of 10 % were considered for shear envelopes. For different normal stresses, the shear resistance of STC mixture is higher than that of pure sand, but the increasing trend maintains up to 30 % of *TC*, beyond which the shear resistance decreases (after Reddy et al. 2015).

Based on the peak shear strength values, c-φ and φ fitting parameters are determined and presented in Fig. 13.11. It is observed that by adding *TC* of up to

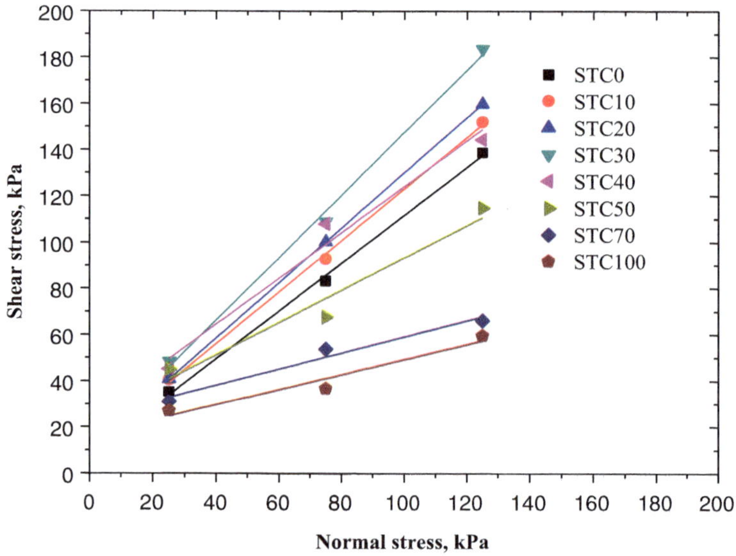

Fig. 13.10 Variation of shear stress with normal stress (After Reddy et al. 2015 with permission from ASCE)

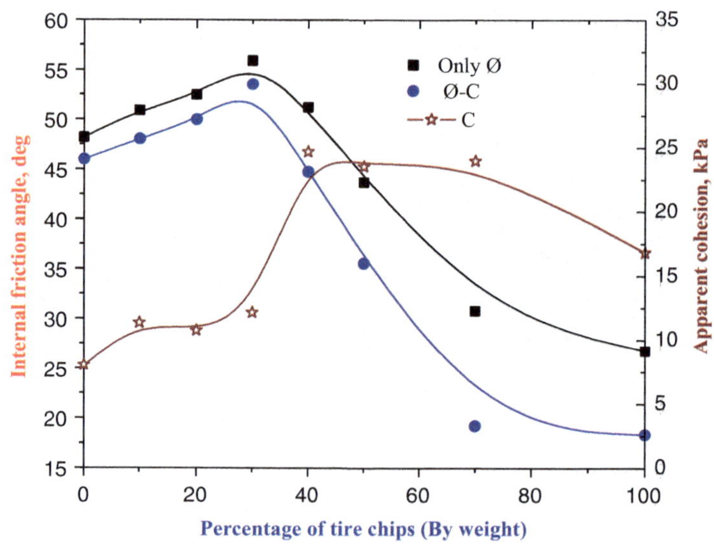

Fig. 13.11 Angle of internal friction and apparent cohesion values of STC (After Reddy et al. 2015 with permission from ASCE)

30 % by weight will increase internal friction angle values and then decreases slightly with further addition of %*TC*. The increase can be explained by the fact that sand particles filled up the voids in the mixture and thus enhancing the friction angle. This led to the increase of friction angle, while adding tire chip particles more than 30 % in mixtures would lead to the decrease of friction angle; more percentage of tire chips creates more voids in the mixture. The apparent cohesion intercept was also increased with increasing percentage *TC* up to 40 %.

13.5 Geoengineering Applications

13.5.1 Retaining Walls

The use of STC mixtures as backfill material in retaining structures was investigated through laboratory-scale model studies by Reddy and Krishna (2015) and Dammala et al. (2015). Retaining wall models of 600 mm height were constructed in a model container of 1200 mm × 600 mm in plan and 1000 mm height (Fig. 13.12). The model container is made of Perspex sheets of 10 mm thickness and braced by a steel frame made of steel angle sections that also facilitate for easy lifting and handling. The model wall of 600 mm high and 580 mm length was made with eight hollow rectangular steel sections, each of 25 mm wide and 75 mm height and 580 mm length, which were joined together using steel rods of 12 mm diameter. These steel rods were further connected to a bottom plywood base forming a rigid connection. Backfill of width 800 mm was adopted that was filled using sand or

Fig. 13.12 Model container with model wall (After Reddy and Krishna 2015)

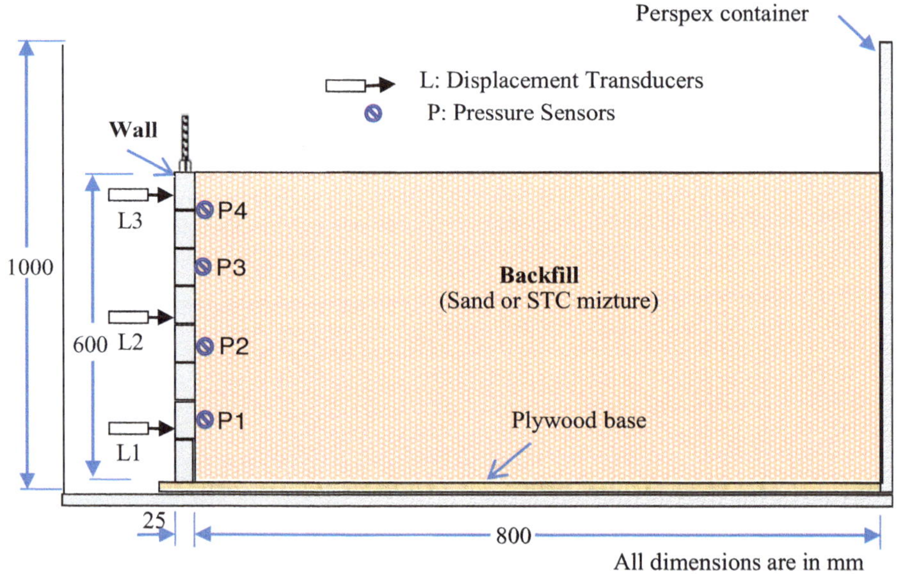

Fig. 13.13 Schematic diagram of model wall configuration

STC mixtures in different model tests. STC mixtures were prepared by manual mixing, maintaining the selected TC percentage levels. After providing temporary support to the wall facing, backfilling was done stagewise by free-falling technique (placing the calculated amount of backfill material based on layer thickness and target density) and compacting manually to achieve the target density. Figure 13.13 shows a schematic of model wall configuration with the locations of the various instrumentations. To monitor the lateral deformation of the wall, linear variable differential transformers (LVDTs), *L*1, *L*2, and *L*3, were positioned on the front face of the wall at 186 mm, 430 mm, and 580 mm from the base of the wall, respectively. Four pressure sensors (each of 50 kPa capacity), *P*1, *P*2, *P*3, and *P*4, are placed inside the wall, in contact with facing at 487 mm, 337 mm, 187 mm, and 37 mm elevations from the top to measure horizontal soil pressures against the facing.

The model wall was built with pure sand (control case) and various STC mixtures (10, 20, 30, 40, and 50 %) as backfill materials. Horizontal wall displacements along the height of the wall were monitored using three LVDTs positioned as shown in Fig. 13.13. The horizontal wall displacements measured after the support removal (without surcharge loading) are shown in Fig. 13.14. Figure 13.14 shows the variation of horizontal displacements along the height of the wall for all the model tests. For STC0 backfill tests, the horizontal displacements for the repeated tests are reasonably matched to each other confirming the repeatability of the tests. Maximum wall displacement is observed to be about 1.75 mm in this case. Among different tests, STC30 backfill showed the lowest wall top displacement as 0.65 mm. Percentage reduction of displacements is calculated by using top

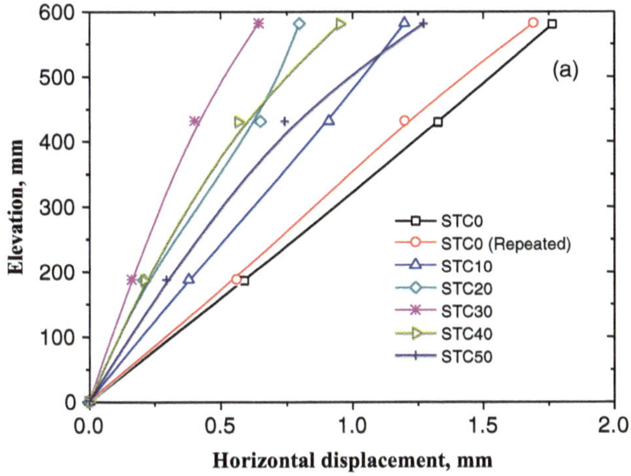

Fig. 13.14 Displacement profiles of model wall with STC mixtures (After Reddy and Krishna 2015 with permission of Springer)

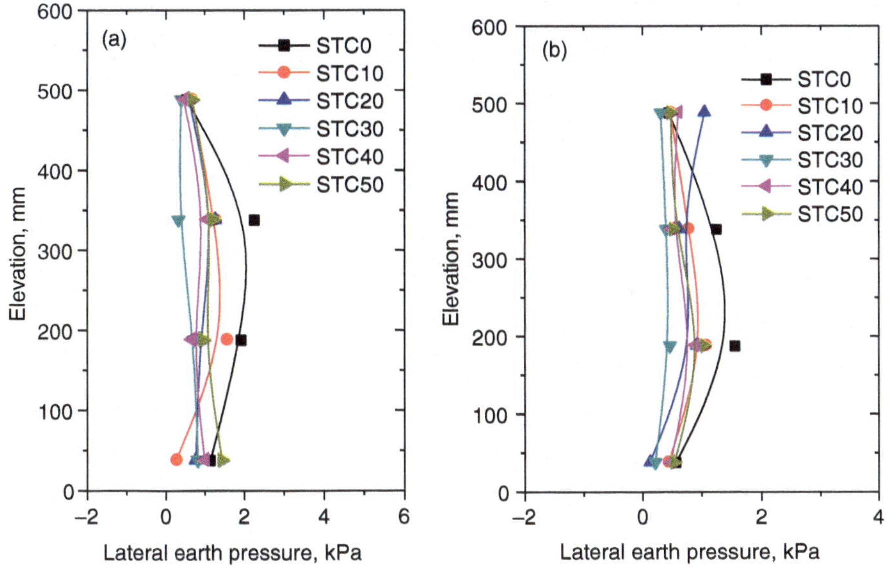

Fig. 13.15 Measured lateral earth pressures: (**a**) At-rest condition. (**b**) After support removal (After Reddy and Krishna 2015 with permission of Springer)

displacements of control case and STC mixture values. It revealed a 60–65 % reduction in top displacements as compared to control case (pure sand).

Lateral earth pressures are monitored at two conditions (at rest and after support removal). Four earth pressure sensors were placed at different elevations on the retaining wall for a purpose (Fig. 13.13). Figure 13.15 shows the earth pressure response under both the condition for all the model walls with different STC mixtures. Figure 13.15a shows the lateral earth pressures along the height of the wall for at-rest condition while Fig. 13.15b for support removal condition. The figures indicate that lateral pressures fall in the range of 0–2 kPa under both the conditions. Though there is no consistent trend in relative variations of pressures for different model walls, it can be observed that the pressures after support removal condition are low compared to the at-rest condition. Further it can also be seen that the earth pressures were affected by the STC mixture giving the lowest earth pressure for STC30 model.

13.5.2 Foundations/Embankments

Alqaissi (2012) studied the influence of tire chip on the behavior of strip footing on sandy soil by performing a series of laboratory model tests. The tests were conducted inside the model box made of steel plate of 3 mm thickness, with 280 mm internal length, 250 mm width, and 250 mm depth. The front side of the box (250 mm × 250 mm) was made of hard glass of 10 mm thickness to view the deformation mechanism during the test. The model strip footing was made of steel with dimensions of 30 × 30 × 270 mm long shown in Fig. 13.16. The model filled with sample by dropping the sand or mixture from the cone of 2.5 mm aperture size, which is moved horizontally by hand at a height of 500 mm that gives a relative density of about 64 % at which the is considered to be in medium condition then the dial gauges are installed at their position carefully to record the soil and footing settlements. They conducted model tests with tire inclusion of 10, 20, and 30 %, for chips having a size of 5–9 mm and 10–20 mm. As seen in Fig. 13.17, the settlement of strip footing on sand–tire chip mixture was about

Fig. 13.16 Model container for typical foundation (After Alqaissi 2012)

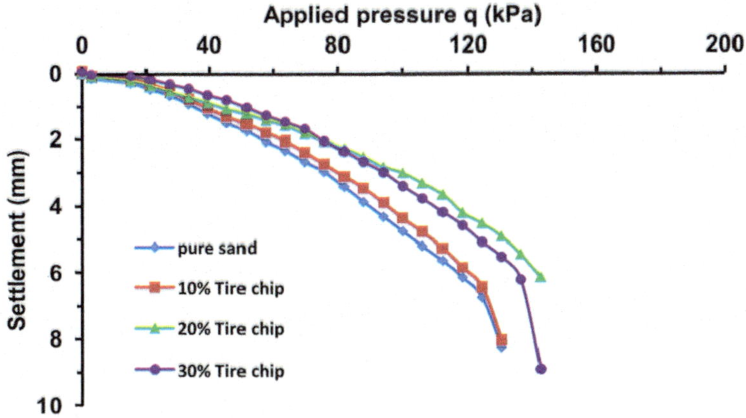

Fig. 13.17 Effect of tire chip contents on settlement (After Alqaissi 2012)

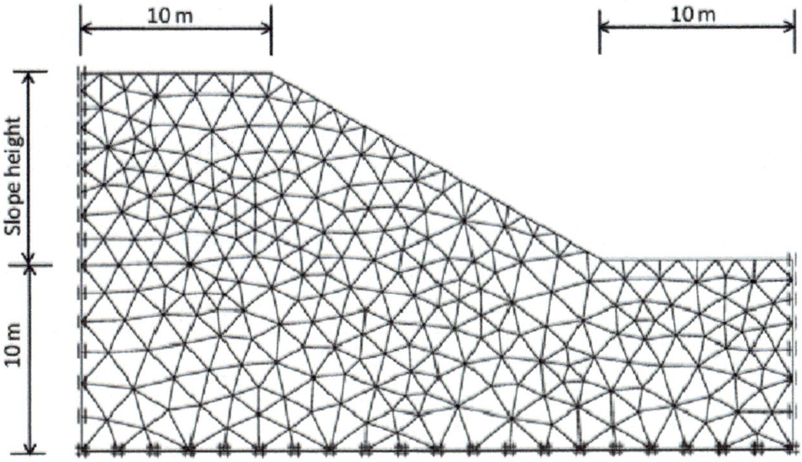

Fig. 13.18 Finite element mesh used in PLAXIS for the stability analysis of 10 m height embankment with slope angle of 30° (After Shahin et al. 2011)

22 % less than in the case of pure sand, and the ultimate bearing capacity of the mixture noticeably increased up to two times its value in the case of pure sand up to 20 % tire chips.

Shahin et al. (2011) studied slope stability analysis using bauxite residue sand–tire mixtures with variation of embankment heights with slope angle of 30° using 2D plane strain conditions (Fig. 13.18). Tire grain (4 mm) range varied from 10 to 40 % by weight. They observed that in all cases (Fig. 13.19), the factor of safety decreases with the increase of embankment height, as expected, and for all embankment heights, significant improvement in slope stability occurs with increasing tire grain content (especially at 30 %).

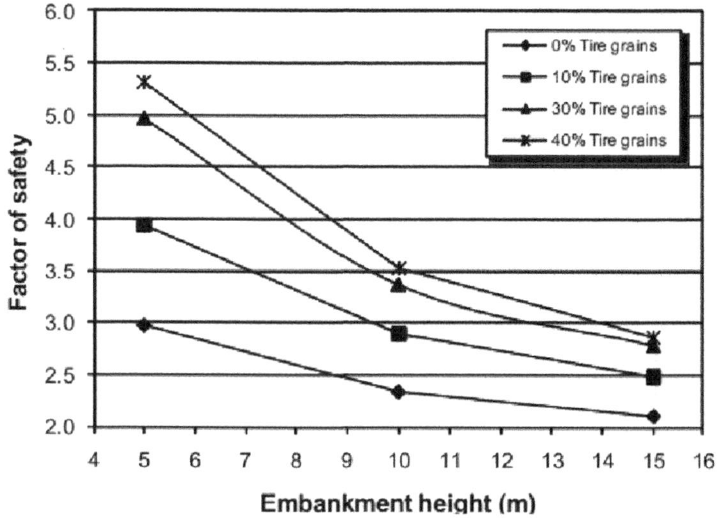

Fig. 13.19 Effect of tire grain contents on factor of safety with embankment height (After Shahin et al. 2011)

13.6 Sustainability Aspects of Sand–Tire Chip Mixtures

Scrap tires are undesired urban waste and the volume is increasing every year. Tires have characteristics that make them not easy to dispose and potentially combustible. Furthermore, sites available for disposal of tires are getting fewer. Therefore, it is crucial to find ways for their alternative utilization by means of recycling/reuse. The reuse of waste scrap tire materials is an essential step in creating a sustainable future in civil engineering applications. Mixing of tire chips (without metal reinforcement) with soil forms an insulating layer that inhibits ignition in the fill:

- Lightweight embankment fill to increase slope stability or reduce settlement.
- Scrap tire is a low-cost material.
- Free draining.
- Vibration damping.
- Optimum STC mixture work as lightweight backfill material for retaining walls.
- Optimum STC mixture work as ground improvement for foundations, pavements, and embankments.
- Optimum STC mixture is effectively working, being economic friendly, compared to advanced ground improvement methods.
- Optimum STC mixture is effectively working in an environmentally friendly nature.

13.7 Conclusions

This chapter presented the evaluation of index and mechanical properties of the STC mixtures through laboratory investigations and its optimum mixture proportion. Model studies reported in the literature on the use of STC mixtures for sustainable geoengineering applications were also discussed. The following are the conclusions drawn from the work discussed in this chapter:

1. Dry unit weights and specific gravity values are decreased with increasing percentage of tire chips in sand. About twenty percent (20 %) reduction in dry unit weight, as compared to sand alone, has been observed in 30 % STC mixture.
2. The addition of percentage TC to sand significantly increased peak shear strength. The apparent cohesion intercept was also increased with increasing percentage TC.
3. Adding TC of up to 30 % will increase internal friction angle and then decreases slightly with further addition of %TC.
4. From the results it can be stated that as the percentage TC more than 30 % (by weight) is added to the sand, the peak strain of the specimen increases but the stiffness and strength decrease.
5. The optimum STC mixture indicates that this material behaves almost as stiffly as a conventional sand backfill. The deformations on the surface of the backfill are small, and the reinforcement provides a small reduction of the lateral earth pressure.
6. The results presented from model tests on retaining walls showed significant reduction in wall deformations and earth pressures for the model walls with STC mixtures as backfill.
7. Settlement of strip footing on sand–tire chip mixture was about 22 % less than in the case of pure sand.
8. Ultimate bearing capacity of the mixture noticeably increased up to two times its value in the case of pure sand up to percentage 20 % tire chips.

Overall, the chapter highlights the advantage of using scrap tire-derived materials (in the form of tire chips) along with conventional geomaterials in various geoengineering structures. This also serves as the sustainable application in geoengineering. Further studies on different geo-structures using different forms of tire materials and base geo materials may be essential to evolve practical design guidelines.

Notations

$TC(\%)$	Percentage of tire chips
W_{tc}	Weight of tire chips, kg
W_s	Weight of sand, kg

$G_{m(cal)}$	Calculated specific gravity of mixture from density test
W_m	Weight of sand–tire chip mixture, kg
W_{tc}	Weight of tire chips, kg
G_s	Specific gravity of sand
G_{tc}	Specific gravity of tire chips
e_{total}	Total void ratio
V_m	Volume of mold
V_{tire}	Volume of tire chips in the mixture in m^3
V_{sand}	Volume of sand in the mixture in m^3

References

Ahmed I (1993) Laboratory study on properties of rubber-soils. Ph.D. thesis, School of Civil Engineering, Purdue University, West Lafayette

Alqaissi (2012) Behavior of Strip footings resting on sand mixed with tire-chips. J Eng Dev 16 (4). ISSN 1813–7822

ASTM (2008) Standard practice for use of scrap tires in civil engineering applications, ASTM Standard D6270-08. American Society for Testing and Materials, West Conshohocken

Balunaini U, Yoon S, Prezzi M, Salgado R (2009) Tire shred backfill in mechanically stabilized earth wall applications. FHWA/IN/JTRP-2008/17. West Lafayette, Indiana

Benda CC (1995) Engineering properties of scrap tires used in geotechnical applications. Report no. 95-1. Vermont Agency of Transportation, Montpelier

Bosscher PJ, Edil TB, Kuraoka S (1997) Design of highway embankments using tire chips. J Geotech Geoenviron Eng 123(4):295–304

Bressette T (1984) Used tire material as an alternative permeable aggregate. Report no. FHWA/CA/TL-84/07. Office of Transportation Laboratory, California Department of Transportation, Sacramento

Cosgrove TA (1995) Interface strength between tire chips and geomembrane for use as a drainage layer in a landfill cover. In: Proceedings of geosynthetics'95, vol 3. Industrial Fabrics Association, St. Paul, MN, pp 1157–1168

Dammala P, Sodom B, Adapa M (2015) Experimental investigation of applicability of sand tire chip mixtures as retaining wall backfill. IFCEE:1420–1429. doi:10.1061/9780784479087.128

Foose GJ, Benson CH, Bosscher PJ (1996) Sand reinforced with shredded waste tires. J Geotech Eng 122(9):760–767

Gebhardt MA (1997) Shear strength of shredded tires as applied to the design and construction of a shredded tire stream crossings. MS thesis, Iowa State University

Ghazavi M (2004) Shear strength characteristics of sand mixed with granular rubber, Geotechnical and Geological Engineering 22: 401–416

Ghazavi M, Sakhi MA (2005) Influence of optimized tire shreds on shear strength parameters of sand. Int J Geomech 5(1):58–65

Humphrey DN, Manion WP (1992) Properties of tire chips for lightweight fill. In: Proceedings of the conference on grouting, soil improvement, and geosynthetics, vol 2. ASCE, New Orleans, Louisiana, pp 1344–1355

Humphrey DN et al (1993) Shear strength and compressibility of the tyre chips for use as retaining wall backfill. Transportation Research Record No. 1422, Lightweight Artificial and Waste Materials for Embankments over Soft Soils, Transportation Research Board, Washington, DC, pp 29–35

IS: 2720 (part 4) (1985) Methods of test for soils: grain size analysis. Bureau of Indian Standards, New Delhi

Lade PV, Liggio CD Jr, Yamamuro JA (1998) Effects of non-plastic fines on minimum maximum void ratios of sand. Geotech Test J TGJODJ 21(4):336–347

Reddy SB, Krishna AM (2015) recycled tire chips mixed with sand as lightweight backfill material in retaining wall applications: an experimental investigation. Int J Geosynth Ground Eng. doi:10.1007/s40891-015-0036-0

Reddy SB, Kumar DP, Krishna AM (2015) Evaluation of optimum mixing ratio of sand tire chips mixture for geo-engineering applications. J Mater Civ Eng. doi:10.1061/(ASCE)MT.1943-5533.0001335)

Shahin MA, Mardesic T, Nikraz HR (2011) Geotechnical characteristics of bauxite residue sand mixed with crumbed rubber from recycled car tires. J GeoEng 6(1):63–72

Zornberg JG, Viratjandr C, Cabral AR (2004) Behavior of tire shred–sand mixtures. Can Geotech J 41(2):227–241

Chapter 14
Evaluation of Bioreactor Landfill as Sustainable Land Disposal Method

P. Lakshmikanthan, L.G. Santhosh, and G.L. Sivakumar Babu

Abstract Sustainable municipal solid waste management has become a challenge to the engineers in the present world. Good pretreatment methods coupled with landfilling are looked as sustainable means of disposal of municipal solid waste. Bioreactor landfill is one such sustainable option. In the present study two landfill simulators (dry and leachate recirculation) were used to investigate the effect of leachate recirculation on the stabilisation process of mechanically biologically treated (MBT) municipal solid waste (MSW). The simulator with a leachate recirculation had a higher degree of waste stabilisation towards the end of the experiment due to higher moisture content and micro-organisms. The results observed at the end of 380 days prevail that the process combination of above operational parameters adopted in bioreactor was a more efficient approach for stabilisation of MSW. After 1 year of operation, the residues of the simulators were analysed, and it was found that the settlement and gas production were greater in leachate recirculation simulator than the dry simulator. The carbon content reduced in the bioreactor simulator by more than 60 % compared to the dry simulator. It was also observed that the biodegradation time for MBT-MSW was reduced in bioreactor simulator compared to maximum values presented in the literature.

Keywords Municipal solid waste • Bioreactor landfill • Sustainability • Stabilisation • Landfill gas

P. Lakshmikanthan (✉) • L.G. Santhosh
Centre for Sustainable Technologies, Indian Institute of Science, 560012 Bengaluru, India
e-mail: lakshmikanthancp@gmail.com; lgsanthu2006@gmail.com

G.L. Sivakumar Babu
Department of Civil Engineering, Indian Institute of Science Bangalore, Bengaluru 560012, Karnataka, India
e-mail: gls@civil.iisc.ernet.in

G.L. Sivakumar Babu et al. (eds.), *Sustainability Issues in Civil Engineering*,
Springer Transactions in Civil and Environmental Engineering,
DOI 10.1007/978-981-10-1930-2_14

14.1 Introduction

Use of bioreactor landfills for landfilling is relatively new to Asian countries. Modern technology and research in the area of landfill engineering have paved way to engineered landfills and bioreactor landfills. In the engineered landfills which are often termed as conventional landfills and have inerts as main constituents, the stabilisation times are longer and bioreactors have emerged as alternatives. Bioreactor landfill is defined as a sanitary landfill operated for the purpose of transforming and stabilising the readily and moderately decomposable organic waste constituents within 5–10 years following closure and needs to have control and understanding of microbiological processes (Pacey et al. 1999). This process is in contrast to a traditional landfill that stores layers of compacted garbage in a dry condition to minimise the leachate which then biodegrades within its boundaries. There are three types of bioreactor operational processes which can be used: aerobic, anaerobic and hybrid or facultative. In an aerobic bioreactor, air (oxygen) is added to increase aerobic activity and accelerate waste stabilisation of the landfill. In an anaerobic bioreactor landfill, moisture is added in the form of leachate to obtain optimum moisture levels of at least 40 %t by weight to accelerate biodegradation of the waste. Biodegradation occurs in the absence of oxygen (anaerobically) and produces gas. The application of leachate storage and recirculation technique in landfill has proven its beneficial effect on waste biodegradation in landfill (Chiemchaisri et al. 2002). A hybrid or facultative landfill accelerates waste degradation by employing a sequential aerobic-anaerobic treatment to decompose organics in the upper sections and collect gas from lower sections. There are four reasons indicated as justification for use of bioreactor technology: (a) to enhance potential for waste to energy conversion, (b) to store and/or treat leachate, (c) to recover air space as well as overall landfill space and (d) to ensure sustainability in terms of reduced greenhouse gas emissions. The fourth justification, i.e. to ensure sustainability, has considerable significance and results in reduced costs associated with avoided long-term monitoring and maintenance and delayed the location of a new landfill (Reinhart et al. 2002). In sustainable landfills, processes, control and use of products and residues are best handled with full knowledge of what is likely to happen within landfill, and there are minimal negative effects on the environment. This can be achieved after the stabilisation of waste within a landfill and the stabilised waste can be mined to release the space for refilling. Landfill mining in a landfill should be undertaken when the landfilled wastes are sufficiently stabilised. Stages of stabilisation depend to a large extent on parameters that control the chemical and biological processes (e.g. chemical and biological compositions, stress, moisture content, temperature and micro-organisms) occurring in the landfill waste.

In the present study, two landfill simulators (dry and bioreactor) were used to analyse conventional landfill and bioreactor landfill in terms of stabilisation and sustainability. The evaluation is carried out by analysing the geotechnical properties and the chemical properties of waste initially and after 1-year monitoring. Laboratory experiments were conducted on mechanically biologically treated (MBT) municipal solid waste (MSW) retrieved from Mavallipura landfill situated in the outskirts of Bangalore. The MBT waste is referred to as MSW in this paper.

14.2 Test Methods

Although there are several tests conducted on MSW landfills, the bioreactor landfill area needs further research and understanding of the physical, chemical and biological processes over a period of time. Specially designed cells were fabricated (Fig. 14.1) in order to monitor the MSW for dry and bioreactor condition. The MSW setups were monitored for 1 year. In the dry condition leachate was not recirculated. In the bioreactor setup, the leachate was recirculated at regular intervals. MSW was placed in both the setups that had a water content of 44 % and bulk density of 10.3 kN/m^3. The geotechnical and chemical characterisation of MSW were measured before placing the MSW in the cell and after 1 year of monitoring. Specific gravity, hydraulic conductivity, compressibility, shear strength and carbon content analysis tests were conducted.

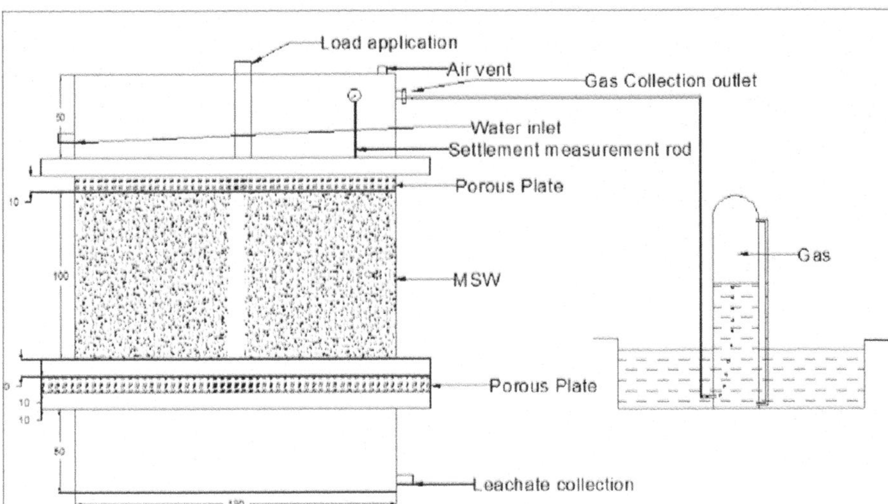

Fig. 14.1 Experimental setup for MSW gas collection

14.2.1 Physical Composition, Total Volatile Solids and Particle Size Distribution of MSW

The physical characterisation of MSW is done in order to study the effect of the physical composition on the stability characteristics of the MSW. The characterisation of MSW size <35 mm was used in this study. The waste is segregated by hand sorting into paper, plastics, inerts (rubber, leather), glass, stones and the organic fraction of the waste. Table 14.1 shows the MSW composition prior testing.

The organic content of the MSW was studied in the laboratory by the TVS method according to the APHA (1965) (American Public Health Association) standard methods. Organic content of the compost reject <35 mm particle size was calculated as the ratio of the weight loss to the initial specimen weight after heating to a temperature of 550 °C in a muffle furnace. Organic content of the waste was calculated as the ratio of the weight loss to the initial specimen weight after heating from a temperature of 105 °C to a temperature of 550 °C in a muffle furnace. The initial decomposable organic content of the waste was found to be 55 % and the inerts constituted to 45 %. The TVS of MSW decreased from 54 % to 38 and 29 in dry simulator and bioreactor simulator, respectively.

Sieve analysis of the MSW is conducted in the laboratory in accordance with the ASTM D422 standards. Sieve analysis was performed with opening sizes from 0.075 to 35 mm using the waste components after drying. Figure 14.2 presents the size distribution of MSW samples for the dry and bioreactor conditions. As can be seen, particle size of the MSW samples increases with age. In fresh waste samples, 50 % of the elements are smaller than 30 mm. The percentage passing increases from 90 to 98 for the bioreactor condition. This indicated the reduction in the MSW particle size in the bioreactor case. This is probably due to the effect of the biodegradation progress, which acts by disintegrating the MSW particles and making the MSW a finer material over time.

Table 14.1 Composition of MSW

Type of waste	Percentage (%)
Clothes	6.34
Plastics	28
Glass	1.28
Leather	0.8
Coconut	5.56
Metal	–
Stones	1.96
Rubber	0.88
Wood	0.16
Organic content	54.2

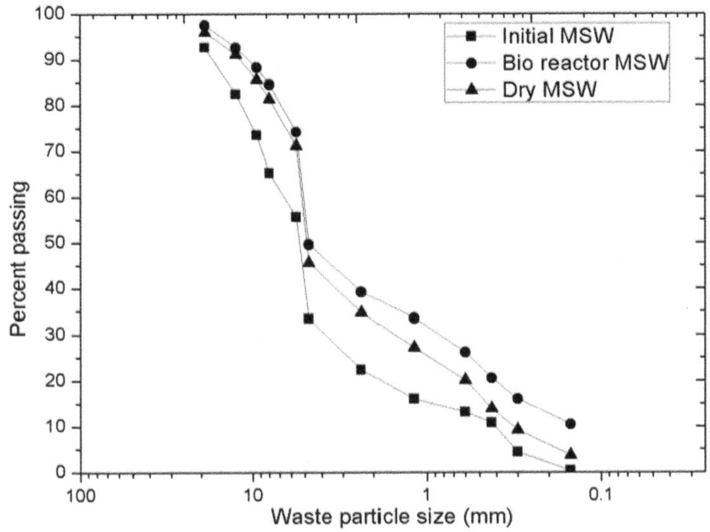

Fig. 14.2 Particle size distribution of MSW

14.2.2 *Gas Quantification and Settlement*

The gas production and settlement were monitored for two cases. A static load of 100kPa was applied on both the cases. The field capacity of the MSW was calculated as 45 %. The leachate that was recirculated was brought from the landfill site. Around 1.6 L of leachate was added initially in the bioreactor case, whereas no leachate was added in the dry case. The quantity of gas was measured by the downward displacement of water as shown in Fig. 14.1. The quality of gas was measured using a gas chromatograph. Elemental analysis was conducted in a CHN analyser in order to find out the elemental carbon, nitrogen and hydrogen contents in the MSW initially and after 1 year in both the considered cases.

14.2.3 *Geotechnical Characterisation*

14.2.3.1 Specific Gravity

The specific gravity tests were conducted in accordance with ASTM test method ASTM D 854–02 and ASTM D 5057–90. Pycnometer and density bottle methods were employed on particles passing through the 35 mm sieve in laboratory for finding the specific gravity of MSW. Five tests were conducted on representative MSW samples for the initial, dry and bioreactor cases.

14.2.3.2 Hydraulic Conductivity

Laboratory tests were conducted to determine hydraulic conductivity by constant head and falling head methods. The tests were performed according to ASTM D2434. The water content and the bulk density of MSW were 44 % and 10.3 kN/m^3. The waste was compacted into the mould of 50 mm internal height and 80 mm internal diameter. The permeability was calculated based on Darcy's law.

14.2.3.3 Compressibility

Compressibility testing was carried out in an oedometer in order to determine the compressibility characteristics of fresh MSW and one-year-old MSW. The size of the sample was 100 mm diameter and 25 mm height. The water content and the bulk density of MSW were 44 % and 10.3 kN/m^3. The maximum load applied was 920 kPa. The waste was compacted into the mould using a circular tamping plate and placed in between the porous stones.

14.2.3.4 Consolidated Undrained Shear Strength

The consolidated undrained triaxial tests were performed according to ASTM D4767. The MSW was compacted statically and then placed in the triaxial cell. The sample size was 50 mm diameter and 100 mm height. The water content and the bulk density of MSW were 44 % and 10.3 kN/m^3. The samples were initially subjected to saturation, consolidation and finally compression. The confining pressures applied were similar to the drained tests (50, 100, 150 kPa). The deviator stress-axial strain graphs were plotted.

14.2.3.5 Consolidated Drained Shear Strength

The consolidated drained tests were performed in accordance with the ASTM D3080. Laboratory triaxial testing was conducted on waste samples of water content and the bulk density of MSW were 44 % and 10.3 kN/m^3, with unit weight of 10 kN/m3. The sample size was 50 mm diameter and 100 mm height. Shearing was done at a strain rate of 0.5 mm/min. The confining pressures applied were 50, 100 and 150 kPa.

14.3 Results and Discussions

14.3.1 Gas Quantification and Settlement

The gas production in the bioreactor case was around 20 L compared to 5 L in the dry case (Fig. 14.3a). The gas production was less in the dry case due to the limited availability of the moisture and nutrients which was supplied through leachate circulation in the bioreactor case. The results of gas composition analysis found 37–39 % of methane in the bioreactor case and 30–32 % in the dry case. The gas production had decreased after 350 days in the bioreactor case. In bioreactor case, there was 30 % settlement compared to around 16 % settlement in dry case (Fig. 14.3b). This indicated a higher settlement in bioreactor case. CHN elemental analysis showed a 61 % reduction in carbon content in the bioreactor case compared to 14 % in the dry case as shown in Table 14.2. This may be attributed to the inadequacy of moisture and nutrients to the micro-organisms in the dry case. However, there was no significant decrease in the nitrogen and hydrogen content in both the cases.

14.3.2 Geotechnical Characterisation

14.3.2.1 Specific Gravity

Specific gravity of the initial MSW sample in the present study was found to be 1.33. The specific gravity for the dry case and bioreactor case was found to be 1.26

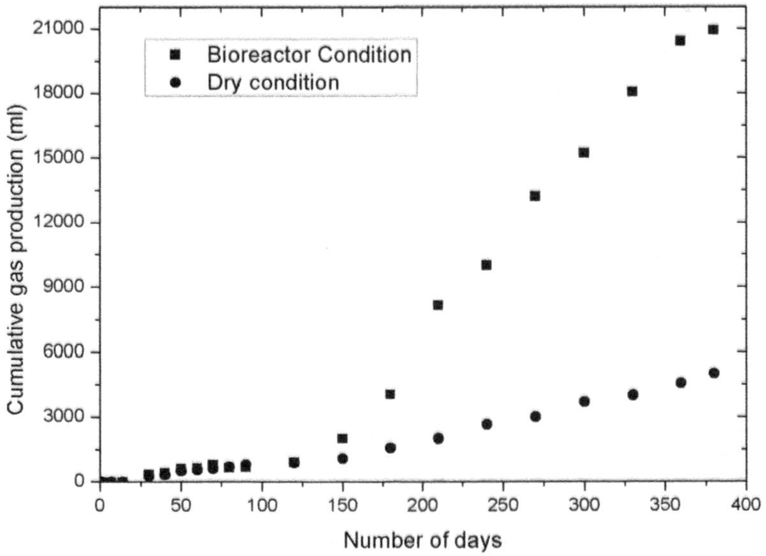

Fig. 14.3a Variation of landfill gas production with time for the considered cases

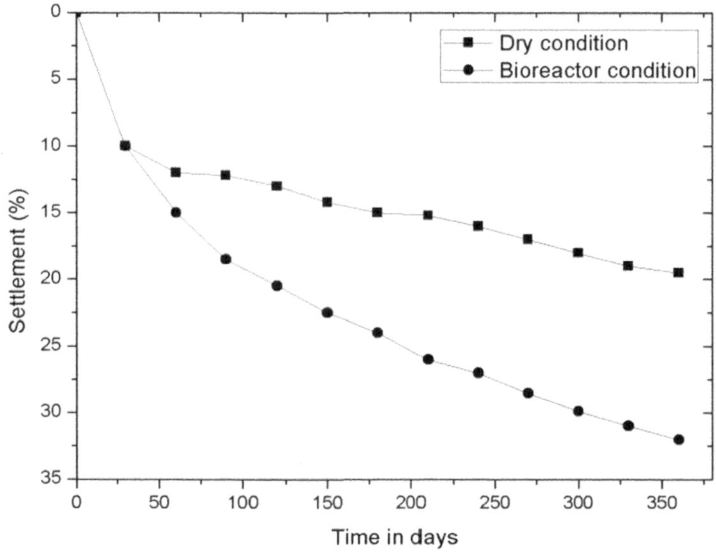

Fig. 14.3b Variation of MSW settlement with time for the considered cases

Table 14.2 Elemental analysis of MSW

Parameter	N %	C %	H %
Initial MSW	1.6389	19.279	2.5872
After 1 year			
Dry condition	1.4822	16.697	2.4686
Bioreactor condition	0.12771	7.4557	1.2817

and 1.20, respectively. Hettiarachchi (2005) reported values of 1.59 and 1.67 for mixed simulated waste. Gabr and Valero (1995) reported the specific gravity of the MSW from the tests done on the entire particle size distribution as 2.0 and for the finer fraction (<No. 200 sieve) as 2.4. The lower values of specific gravity can be attributed to the presence of decomposed organic matter.

14.3.2.2 Hydraulic Conductivity Tests

The hydraulic conductivity obtained from the tests was 6.4×10^{-4} cm/s and 4.35×10^{-3} cm/s for falling head and constant head method, respectively. The hydraulic conductivity decreased to 5×10^{-6}cm/s and 3.2×10^{-5}cm/s in the dry case and bioreactor case, respectively. Therefore, there is a decrease in the permeability of MSW with age. This is in agreement with Reddy et al. (2009b) who stated that the decrease in permeability in aged MSW is attributed to the increase in the smaller particles resulting from degradation.

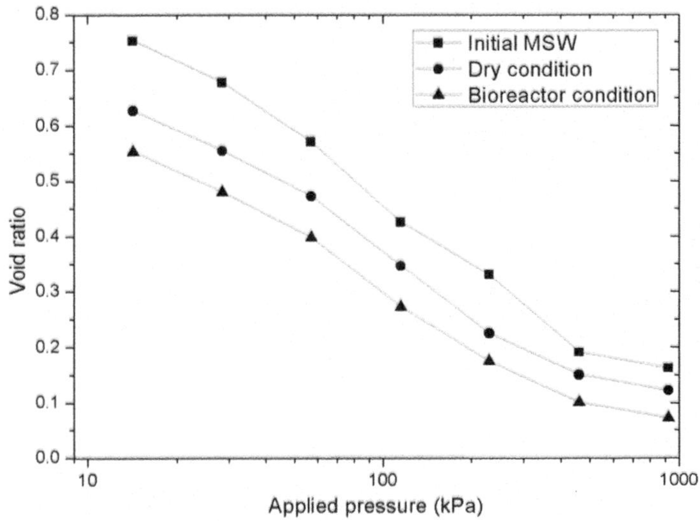

Fig. 14.4 *e-log (p)* of MSW samples for the considered cases

14.3.2.3 Compressibility Tests

The compression index values decreased with the age of waste (Fig. 14.4). The compression index values obtained were 0.347 for the initial MSW, 0.27 for the dry case and 0.264 in the bioreactor case. Hossain (2002) reported 0.16–0.25 as compression ratios for fresh MSW. The compression index values published by Reddy et al. (2009a) were in the range of 0.24–0.33. The void ratio also decreased from 0.79 to 0.70 and 0.63 in dry case and bioreactor case. The compression index values and void ratios decreased with the age of waste. This is in agreement with Hossain et al. (2006) concluded that with degradation the MSW structure will change and the MSW particles break down leading to a decrease in the MSW void ratio and thus a decrease in the MSW. However, there was no significant difference in the compression index values of dry and bioreactor cases.

14.3.2.4 Consolidated Drained Shear Strength

Deviator stress increased with axial strain without showing any peak in the stress-strain curve (Fig. 14.5a, b). MSW shear strength in triaxial compression has been defined in the literature as the mobilised shear stress at 5–25 % axial strain. Though the tests were conducted for strains up to 35 %, shear strength at 20 % strain was used for generating the Mohr-Coulomb shear strength envelopes. The friction angle (Φ) values were found to be 33° for the initial MSW samples. The Φ values decreased to 27° and 22° in dry and bioreactor case. The cohesion values did not

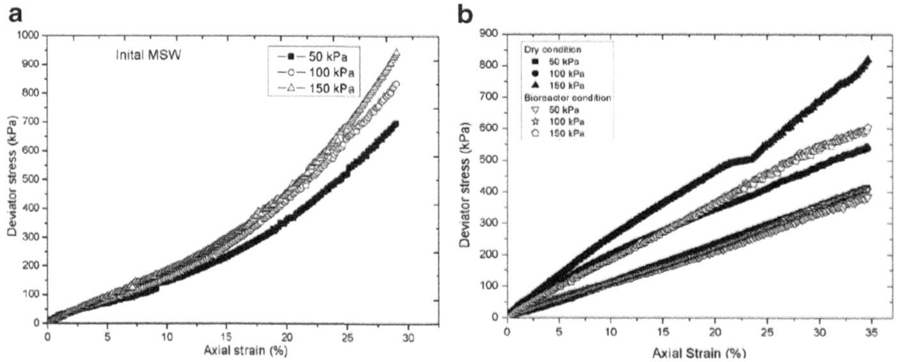

Fig. 14.5 (**a**). Deviator stress-strain curves of consolidated drained tests of initial MSW samples. (**b**). Deviator stress-strain curves of consolidated drained tests for dry and bioreactor case

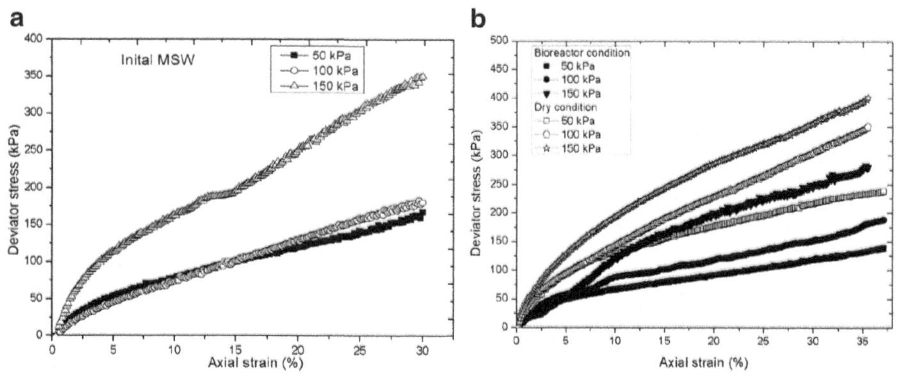

Fig. 14.6 (**a**). Deviator stress-strain curves of consolidated undrained tests of initial MSW samples. (**b**). Deviator stress-strain curves of consolidated undrained tests for dry and bioreactor case

vary considerably. There is a reduction in the Φ values with the increase in the age of waste.

14.3.2.5 Consolidated Undrained Shear Strength

The deviator stress increased with axial strain without showing any peak (Fig. 14.6a, b). The friction angle (Φ) values were found to be 32° for the initial MSW samples. The Φ values decreased to 25° and 28° in dry and bioreactor case. Though there is a reduction in the Φ values with the increase in the age of waste, the Φ value in bioreactor case was greater than in the dry case. The cohesion values did not vary considerably

14.4 Summary and Conclusions

This study summarises the evaluation of the dry type landfill and bioreactor landfill. The geotechnical and the chemical properties of initial MSW samples and the MSW residue after of 380 days of monitoring were studied and analysed. The following conclusions can be drawn from this study:

1. The landfill gas production was greater in the bioreactor case (21 L) than in the dry case (5 L). The methane content also was higher in the bioreactor case (37–39 %) than in the dry case (30–32 %).
2. The bioreactor was able to effectively reduce the carbon, nitrogen and hydrogen content in the MSW. There was a reduction of 60 % in carbon content, 90 % in nitrogen and 50 % in hydrogen compared to 13 % in carbon content, 10 % in nitrogen and 4.5 % in hydrogen. Stabilisation of waste is faster in the bioreactor case.
3. The MSW particle sizes decreased with increase in age. The aged MSW exhibited a reduced compression index, specific gravity and hydraulic conductivity values.
4. The friction angle values also decreased from 33° to 22°, whereas no significant change in the cohesion values was observed.

It can be concluded that the bioreactor case is a better sustainable option compared to the dry case. Since this study is restricted to laboratory experiments, large-scale experiments and field-scale studies are required to prove the bioreactor landfill as an effective sustainable technology.

References

American Public Health Association, Water Pollution Control Federation, & Water Environment Federation (1965) Standard methods for the examination of water and wastewater, vol 11. American Public Health Association, Washington, DC

Chiemchaisri C, Chiemchaisri W, Nonthapund U, Sittichoktum S (2002) Acceleration of solid waste biodegradation in tropical landfill using bioreactor landfill concept. In: Proceeding of the 5th Asian symposium on academic activities for waste management, Kuala Lumpur, Malaysia, 9–12 September 2002

Gabr MA, Valero SN (1995) Geotechnical properties of municipal solid waste. Geotech Test J ASTM 18(2):241–251

Hettiarachchi CH (2005) Mechanics of biocell landfill settlements. PhD dissertation, Department of Civil & Environmental Engineering, New Jersey Institute of Technology, Newark

Hossain MS (2002) Mechanics of compressibility and strength of solid waste in bioreactor landfills. PhD dissertation, Department of Civil Engineering, North Carolina State University, Raleigh

Hossain MS, Penmethsa KK, Hoyos L (2006) Permeability of municipal solid waste in bioreactor landfill with degradation. Geotech Geol Eng J 27:43–51

Pacey JF, Yazdani R, Reinhart D, Morck R, Augenstein D (1999) The bioreactor landfill: an innovation in solid waste management. SWANA, Silver Springs

Reddy KR, Hettiarachchi H, Gangathulasi J, Bogner JE (2009a) Geotechnical properties of fresh municipal solid waste at Orchard Hills landfill. Waste Manag 29:952–959
Reddy KR, Hettiarachchi H, ParakallA N, Gangathulasi J, Bogner J, Lagier T (2009b) Hydraulic conductivity of MSW in landfills. J Environ Eng 135(8):677–683
Reinhart DR, McCreanor PT, Townsend T (2002) The bioreactor landfill – its status and future. Waste Manag Res ISWA 30:172–186

Part III
Sustainable Environmental and Water Resources

Chapter 15
Assessment of Landfill Sustainability

G.L. Sivakumar Babu, P. Lakshmikanthan, and L.G. Santhosh

Abstract Assessing landfills in terms of sustainability is a difficult task and needs to be addressed comprehensively. In this paper, we present an initial operationalisation of this approach for a pragmatic legal landfill assessment and its application to assess two prototypic Swiss landfills as part of a pilot study. Thereby, the six generic criteria of SPA are specified using 18 functional key variables (FKVs), such as 'control of pollutant release' or 'resilience to intended human impacts'. The first results from the pilot study indicate that SPA and its generic criteria provide a purposeful guiding framework for achieving a systemic and comprehensive SD assessment that seems (i) to be feasible for practical application, (ii) sensitive for relevant SD issues and (iii) transparent for the addressees of the assessment results. The results show that thermal treatment of the unrecyclable part of the waste stream is the preferred option for waste management when compared to modern landfilling. Furthermore, Eco-indicator 99 method is used to investigate the human health, ecosystem quality and resource use impact categories.

Keywords Municipal solid waste • Landfills • Sustainability • LCA • LCCA

15.1 Introduction

Sustainability can be defined as a creation and maintenance of the conditions under which humans and nature can exist in productive harmony that permit fulfilling the social, economic and other requirements of present and future generations. 'Sustainable development' was defined in 1987, in the report of the World Commission

---

G.L. Sivakumar Babu (✉)
Department of Civil Engineering, Indian Institute of Science Bangalore, Bengaluru 560012, Karnataka, India
e-mail: gls@civil.iisc.ernet.in

P. Lakshmikanthan • L.G. Santhosh
Centre for Sustainable Technologies, Indian Institute of Science, 560012 Bengaluru, India
e-mail: lakshmikanthancp@gmail.com; lgsanthu2006@gmail.com

G.L. Sivakumar Babu et al. (eds.), *Sustainability Issues in Civil Engineering*, Springer Transactions in Civil and Environmental Engineering,
DOI 10.1007/978-981-10-1930-2_15

on Environment and Development, as 'development which meets the needs of the present without compromising the ability of future generations to meet their own needs'. According to the current understanding based on results of the Johannesburg Summit, the environment is seen as one of the three pillars of sustainable development and points out its essential interconnection with the other two pillars, namely, the economic and social pillar (Fig. 15.1). Sustainability is a complex, multidimensional phenomenon, the measurement of which requires a comprehensive set of indicators, showing the developments in its various dimensions. The assessment of sustainability is carried out through common linking indicators that indicate the social, economical and environmental effects. Indicators of sustainability are changes we can observe in the world which indicate progress towards increased sustainability.

In this study the concept of sustainability has been applied to landfills for which the social, economic and environmental indicators are developed and analysed. Landfill has been defined by ISWA (1992) as 'the engineered deposit of waste onto and into land in such a way that pollution or harm to the environment is prevented and, through restoration, land provided which may be used for another purpose'. Landfills will continue to represent an important end point of waste management. This is crucial for sustainable development as waste problems of the whole society are eventually shifted to specific areas in which they might only affect a limited number of people (contradicting intra-generative equity), and today's waste problems could be transferred to future generations because of landfills' potential long-term reactivity (contradicting inter-generative equity) (Lang et al. 2007). Therefore, the assessment of the sustainability of landfills is a difficult task, which requires comprehensive site-specific landfill data analysis.

Fig. 15.1 The three pillars of sustainability

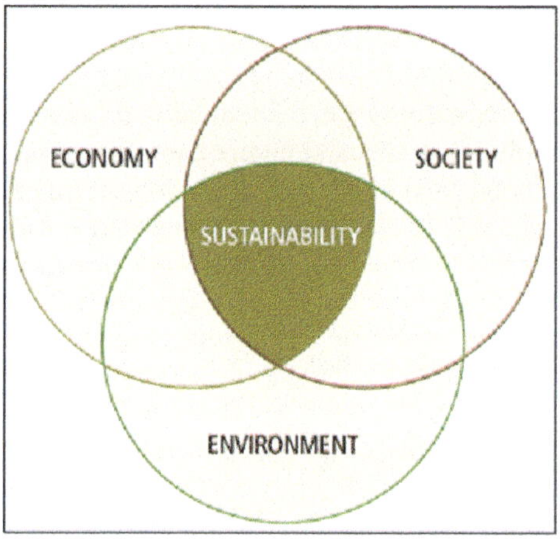

15.2 Description of Applied Scenarios

Bangalore City, one of the eight metros in India, produces about 4500 t per day (tpd) of municipal solid wastes (MSW). Mavallipura landfill site located in the outskirts of Bangalore has been selected for this study. It has been reported that till recently, about 60 % of the MSW collected was dumped at about 60 known and unknown (unrecorded) dumping sites around Bangalore. Further, among these more than 35 sites received a mixture of domestic and industrial waste (Lakshmikantha 2006). In this paper two cases are considered for which sustainability is assessed. The two scenarios considered for the study are given below.

15.2.1 Case 1: Open Dumps

The open dumps are places which do not have any liner systems installed, and the area is temporarily or permanently used as waste disposal sites. There are no initial costs incurred in this method, but the environmental consequences are very high as the leachate may pollute the soil and groundwater and the emissions could pollute the air. The boundary in this case is the area of the dump site and only the transportation charges apply. Compaction and levelling are seldom done at the site. Figure 15.2 shows the system boundary.

15.2.2 Case 2: Bioreactor Landfill System

A bioreactor landfill changes the aim of landfilling from the storage of waste to the treatment of waste. A bioreactor landfill is a system that enhances the degradation of refuse by microbial action. Microbial degradation may be promoted by adding certain elements (nutrients, oxygen or moisture) and controlling other elements (such as temperature or pH). The most widely used and understood method of creating a landfill bioreactor is the recirculation of leachate, as the factor that limits

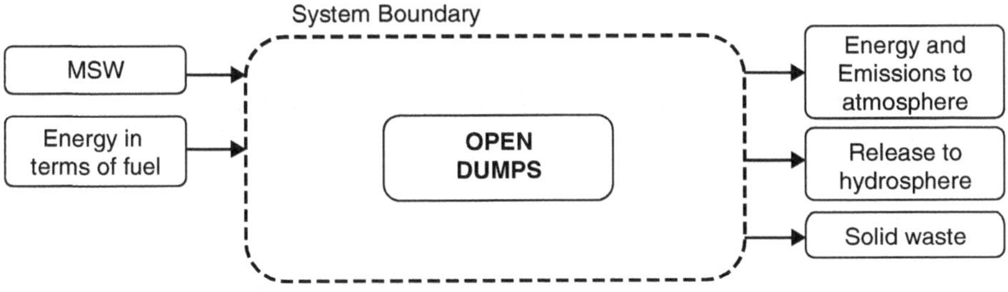

Fig. 15.2 System boundary of open dump

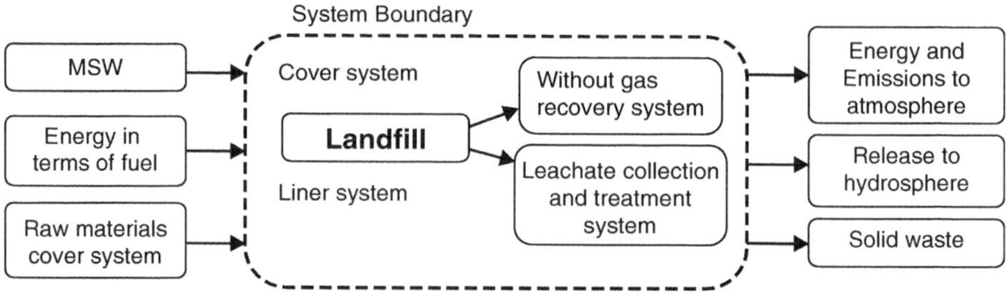

Fig. 15.3 System boundary of landfill without gas recovery system

microbial activity in a landfill is water. The recirculation of leachate increases the moisture content of the refuse in the landfill and, therefore, promotes waste degradation. There is a provision for leachate recirculation and landfill gas (LFG) collection (Fig. 15.3).

The best sustainable option in terms of minimal environmental consequences, costs and minimal social impacts is selected by comparing the impacts caused by each disposal method.

15.3 Methodology

Landfill sustainability is addressed through the indicators that represent the basic pillars of sustainability. Figure 15.4 shows the framework for assessing landfill sustainability. In order to quantify the sustainability indicators, an input and output analysis of the considered scenarios has been carried out.

15.3.1 Input Analysis

Energy inputs are those that are derived from non-renewable sources (diesel). The fuel that is required for transportation and management of waste, electricity needed for operation and maintenance, cover systems and liner systems, leachate collection and treatment system and gas collection and conversion systems are considered inputs to the system. The first scenario does not include all these things except the land. Energy consumed for the transportation of wastes to the landfill from the generation places is calculated by considering three mean distances 10, 20 and 30 kms from the disposal site. The density of the waste in the compacted trucks is considered as 425 kg/m ^{3}, and each compacted truck has a capacity of 6 tonnes of MSW. Assuming an efficiency of the trucks as 3 km per litre of diesel and the energy content of diesel as 36.7 MJ/L, the energy required for the transportation of MSW through the three mean distances is given in Table 15.1.

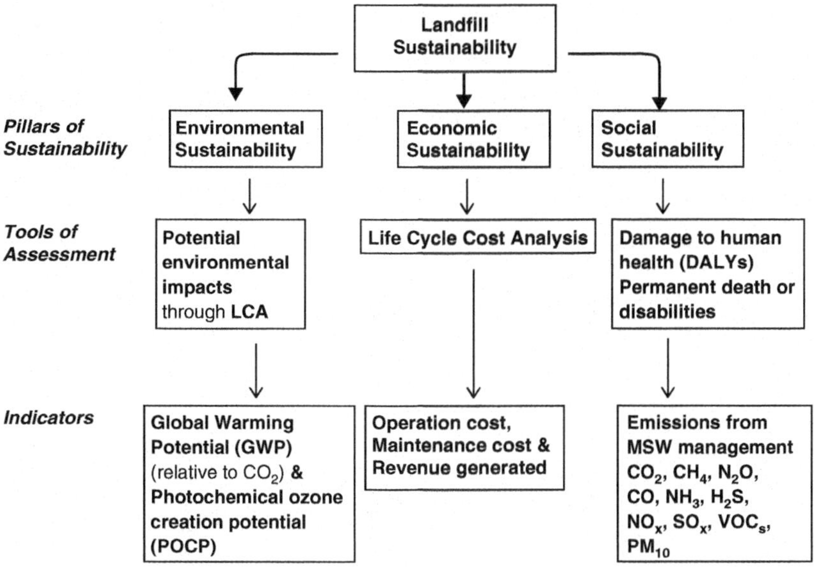

Fig. 15.4 Framework for assessing landfill sustainability

Table 15.1 Energy required for transportation of MSW for the three considered mean distances

Distance in km	Distance (to and from) in km	Energy required in MJ/tonne
10	20	42.8
20	40	85.6
30	60	122

Energy consumed for the management of MSW in the landfill site is calculated by assuming the capacity of the landfill as 2,090,000 t/year (based on 2011 population and generation rate of 0.6 kg/capita/day), four machines working in situ (two bulldozers and two roller compactors) and the diesel consumption of 15 l/h. Assuming the working hours per day as 8 h/day and 300 days/year, the energy consumed was calculated as 3 MJ/tonne of MSW. Table 15.2 summarises the total inputs to the disposal system.

15.3.2 Output Analysis

The outputs of the landfill systems are in the form of landfill gas that is generated by the decomposition of MSW, the leachate that is being generated and finally the leftover inert waste that can be used as compost. Also the emissions from the trucks and bulldozers that are used for transportation and management of MWS are considered as outputs from the system. The quantity of landfill gas that would be

Table 15.2 Inputs to the landfill system

Parameters	Values
Quantity of MSW	13.8×10^6 tonnes of MSW (for 25 years, design is done according to CPHEEO manual 2000)
Volume of daily cover	0.1 % of volume of waste (1.909×10^6 tonnes of MSW) (for 25 years, design is done according to CPHEEO manual 2000)
Volume of cover system	0.08 % of volume of waste (1.5×10^6 tonnes) (for 25 years, design is done according to CPHEEO manual 2000)
Total average rainfall in Bangalore	931 mm/year (based on 100-year data, Indian Meteorological Department)
Energy in terms of fuel	125 MJ/tonne of MSW (3 l of diesel/tonne of MSW)

Table 15.3 Diesel consumption for transportation and management of MSW

	Transportation				Management	
Parameters	Emissions in g/L of diesel	Emissions in g per tonne of MSW			Emissions in g/L of diesel	Emissions in g per tonne of MSW
Distance(to and from) in km		20	40	60		
Diesel consumption (litres/tonne of MSW)		1.2	2.3	3.3		0.5
CO_2	2663[1]	3195.6	6124.9	8787.9	2663[1]	1331.5
CO	11.95	14.34	27.485	39.435	4.52	9.04
HC	1.75	2.1	4.025	5.775	0.68	1.36
NO_x	2.36	2.832	5.428	7.788	6.28	12.56
$PM_{2.5}$	0.62	0.744	1.426	2.046	0.58	1.16

1 = www.ec.gc.ca

generated after 15 years by assuming the values given in Table 15.3 (assuming only 40 % of the total waste generated is landfilled and has around 90 % of degradable organic content) was calculated as 4.49×10^{12} l from 13.8×10^6 t of MSW and 815,000 l of biogas per tonne of MSW using the Buswell and Mueller equation. According to this relation, the methane fraction from degradation of glucose is given by

$$C_6H_{12}O_6 \rightarrow 3\ CH_4 + 3CO_2$$

This equation is considered in order to calculate the maximum emissions from the waste. It was assumed that the landfill gas contained 50 % methane and 50 % carbon dioxide. For the landfill systems with gas recovery system, there is energy savings associated with the conversion of the landfill gas (LFG) to electrical energy.

Road transport emits mainly CO_2, NO_x, CO and NMVOCs; however, it is also a small source of N_2O, CH_4 and NH_3. Therefore, the only major direct greenhouse

Table 15.4 Total emissions from the considered landfill disposal system

Emissions from the system	Open dumps in g	Bioreactor landfill in g
CH_4	268,950	53,790
CO_2	289721.4	78555.9
CO	48.47	48.47
HC	7.135	7.135
NO_x	22.35	20.338
$PM_{2.5}$	3.206	3.206

gas emission is CO_2. Emissions of CO_2 are directly related to the amount of fuel used. The kilometre travelled-based CO, HC, NO_x and $PM_{2.5}$ emission factors of emission control technology of Euro 0 light-duty diesel trucks (LDDTs) are 11.95, 1.75, 2.36 and 0.62 g/km, respectively (Kebin et al. 2010). The kilometre travelled-based emission factors of CO, HC, NO_x and $PM_{2.5}$ of Euro I heavy-duty diesel trucks (HDDTs) are 4.52 ± 2.56, 0.68 ± 0.19, 6.32 ± 1.58 and 0.58 ± 0.34 g/km. The emissions calculated based on the above-mentioned values for the transportation and the management of MSW are given in Table 15.3.

The total outputs include methane, carbon dioxide and other gases (NO_x, $PM_{2.5}$, PM_{10} and SO_x) that are released from the landfill and are emitted by the vehicles. The landfill system with and without recovery of landfill gas is given in Table 15.4. The efficiency of the gas collection system is assumed as 80 %. The transportation distance considered here is the maximum distances of 60 km. Emissions for management of waste in the open dumps are considered nil as there are no management activities undertaken.

15.4 Sustainability Assessment

15.4.1 Environmental Sustainability

The environmental sustainability is characterised by the extent of damage done to the ecosystems. The extent of damage is assessed by some impacts and indicators. According to the life cycle characteristics of waste treatment/disposal, its environmental impacts are classified into five kinds: energy depletion potential (EDP), global warming potential (GWP), acidification potential (AP), eutrophication potential (EP) and photochemical oxidant potential (POP). However, in this paper only energy depletion potential (EDP), global warming potential (GWP) and eutrophication potential (EP) are considered. The characterisation factors of the greenhouse gases that are considered for calculation are given in Table 15.5. Transportation of MSW mainly contributes to the acidification and human toxicology impacts.

The impacts of the respective scenarios are calculated by multiplying the equivalency factors (given in Table. 15.5) to the respective quantities. The

Table 15.5 Characterisation factors based on equivalency factors from IPCC 2001 GWP for 20 years and Eco-indicator 95

IPCC 2001	
Resources	Characterisation factors
Global warming potential(GWP)	
CH4	62
CO_2	1
CO	1.57
Eco-indicator 95	
Acidification potential (AP)	
NOx	0.7
SOx	1
NH_3	1.88
Eutrophication potential(EP)	
NOx	0.13
NH_3	0.33
Photochemical ozone creation potential (POCP)	
CH_4	0.007
Benzene	0.189
Ethene	1
Hydrocarbons, unspecified	0.398

Table 15.6 Indicators of the considered cases on environment for disposal of per tonne of waste

Indicators	Case 1 (in g)	Case 2 (in g)
Global warming potential (GWP) (relative to CO_2)	16,674,976	3,335,056
Photochemical ozone creation potential (POCP)	1885.4	379.3

equivalency factors are multiplied by the quantity of the gases released. The total impacts in disposing 1 tonne of waste in the two cases are given in Table 15.6. The impacts presented in Table 15.7 are the sum of all the impacts from transportation and waste degradation.

Scenario 1 projects the maximum environmental consequences. The reason for this is the absence of the liner system, the gas recovery system and the leachate collection and treatment system. The GWP and photochemical ozone creation potential (POCP) are maximised in this case and therefore severely affect the environment. Therefore, this is the least considered option in terms of environmental consequences. Scenario 4 (bioreactor landfill) emerges to be the best option. The global warming potential and photochemical ozone creation potential (POCP) are the minimum. The advantages of bioreactor landfills described by Warith (2002) are (1) enhancement in the LFG generation rates, (2) reduced environmental impacts, (3) production of end product that does not need landfilling, (4) overall

Table 15.7 Cost details for a landfill

Sl. No.	Item	Cost Rs × 10^5
1	*Initial fixed cost*	
	Site selection and site characterisation cost	26.88
	Design and detailed engineering cost	17.50
	Site development cost	160.30
	Total	**204.68**
2	*Yearly running cost (active)*	
	Phase development cost	427.25
	Phase operation cost	164.75
	Phase closure cost	175.95
	Total	**737.95**
3	*Yearly running cost (post-closure)*	
	Post-closure care cost	37.00
	Total	**37.00**
	Total	**979.63**

Note: All the above-mentioned prices are of base year 1998 as given in the CPHEEO manual 2000

reduction of landfilling cost, (5) reduction of leachate treatment capital and operating cost and (6) reduction in post-closure care, maintenance and risk. Hence, the bioreactor landfill proves to be the environmentally sustainable option.

15.4.2 Economic Sustainability

Life cycle costing (LCC) is a tool or technique that enables comparative cost assessments to be made over a specified period of time, taking into account all relevant economic factors both in terms of initial capital costs and future operational and asset replacement cost. Life cycle cost analysis (LCCA) is a technique used to evaluate the economic consequences over a period of time of mutually exclusive project alternatives. LCCA was applied to equipments initially. The understanding and uses of this tool have improved immensely, and it is being applied to various fields, products and processes. In this study the costs considered are the direct costs (initial costs and operation and maintenance costs). The cost details given in the manual developed by the Central Public Health and Environmental Engineering Organisation (CPHEEO) have been used in this analysis. The costs are summarised in Table 15.7.

The above cost does not include gas recovery system. The first two scenarios do not include the gas recovery system, whereas scenarios 3 and 4 include the gas recovery system. Capital costs vary according to the type of plant used to process the methane. California's capital costs varied from \$606 per kW to \$6811 per kW in 2001 (California Energy Commission, Landfill Gas-to-Energy Potential in California, p. 13). It is assumed that the cost of 1 MW plant is Rs. 333×10^5.

The total gas generated is calculated by using the IPCC first-order decay method. Bangalore generates 4602 t/day of waste. Assuming the collection efficiency as 80 %, waste generation as 0.6 kg/capita/day (Chanakya et al. 2009) and with present population as 9,588,910 (Census 2011), the methane generated over a period of 25 years is calculated as 9.510^6 m^3/year. Using calorific value of methane (lowest) as 9000 kcal/m^3, energy generated in 1 year is computed as 358 TJ, corresponding power being 11 MW. Assuming that electricity is being sold at a price of Rs. 2 per kWh, the revenue generated due to this would be Rs. 1,98,800,000. The average life expectancy of a landfill could range from 30 to 50 years. Therefore, the cost analysis is done for the landfill systems for 50 years. Table 15.8 gives the cost and saving details of the four considered scenarios over a period of 50 years.

The cost details show that there is a considerable amount of saving and earnings in scenario 4 over a period of 50 years. The dump sites do not incur any cost except the transportation costs but may cause immense environmental consequences. Therefore, it is not considered in the cost comparison with the other systems. The waste in the bioreactor is stabilised at a faster rate than the open dumps. In case 4 the existing landfill site can be mined every 25–30 years and used again. This reduces the overall costs for scenario 4. The bioreactor landfill generates revenue in

Table 15.8 Total cost details of the considered scenarios over a period of 50 years

	Case 1	Case 2
Initial fixed cost (Rs × 10^5)		204.68
Yearly running cost (active) (Rs × 10^5)	–	737.95
Yearly running cost (post-closure) (Rs × 10^5)	–	37
Gas recovery system (Rs × 10^5)	–	333.33
Total		**1312.96**
Cost over a period of 50 years (Rs × 10^5)	–	1549.9 (operation and maintenance of the same site)
Total		**1549.9**
Total cost over a period of 50 years (Rs × 10^5)		**2862.86**
Revenue generated from electricity (Rs × 10^5)	–	1988
Cost savings over a period of 50 years (Rs × 10^5)	–	204.68
Usage of the same landfill site every 25 years for 50 years		3976
Revenue generated from electricity for 50 years		
Total		**6168.68**
Total savings over a period of 50 years	0	**+3305.82**

Note: All the above-mentioned prices are of base year 1998 as given in the CPHEEO manual 2000. The '–' sign indicates a loss and '+' sign indicates a gain/savings terms of power production and also creates *jobs*. Hence, the bioreactor landfill option is an economically viable alternative.

15.4.3 Social Sustainability

Social sustainability is assessed in terms of the damage that is caused on human health and is measured in terms of the disability-adjusted life year (DALY). The disability-adjusted life year is a measure of overall disease burden, expressed as the number of years lost due to ill health, disability or early death. Traditionally, health liabilities were expressed using the expected or average number of years of life lost (YLL). This measure does not take the impact of disability into account, which can be expressed by 'years lived with disability' (YLD). DALYs are calculated by taking the sum of these two components. In a formula:

$$\mathrm{DALY} = \mathrm{YLL} + \mathrm{YLD} \tag{15.1}$$

Carcinogenic substances cause a number of deaths each year. In the DALY health scale, death has a disability rating of one. If a type of cancer is (on average) fatal 10 years prior to the normal life expectancy, we would count ten lost life years for each case. This means that each case has a value of ten DALYs. During a summer smog period, many people have to be treated in hospital for a number of days. This type of treatment in a hospital has a rating of 0.392 on the DALY scale. If the hospital treatment lasts 0.01 years on average (3.65 days), each case would be weighted 0.004 DALYs. All damage factors are expressed per kg emission. The unit of damage is DALY. The characterisation factors for calculating the respiratory effects on humans caused by organic and inorganic substances as given by Eco-indicator 99 are given in Table 15.9. The DALYs for both the cases are shown in Fig. 15.5.

Table 15.9 Characterisation factors for calculating the respiratory effects on humans caused by organic and inorganic substances as given by Eco-indicator 99

Component	Substances	Damage factor	Normalised damage factor	Weighted damage factor
Air	Dust (PM_{10})	3.75E-04	2.44E-02	9.74E+00
Air	Dust ($PM_{2.5}$)	7.00E-04	4.55E-02	1.82E+01
Air	NO	1.37E-04	8.90E-03	3.56E+00
Air	NO_2	8.87E-05	5.76E-03	2.30E+00
Air	NO_x	8.87E-05	5.76E-03	2.30E+00
Air	NO_x (as NO_2)	8.87E-05	5.76E-03	2.30E+00
Air	CH_4	1.19E-08	1.44E-06	7.93E-04
Air	CO_2	2.00E-07	2.42E-05	1.33E-02

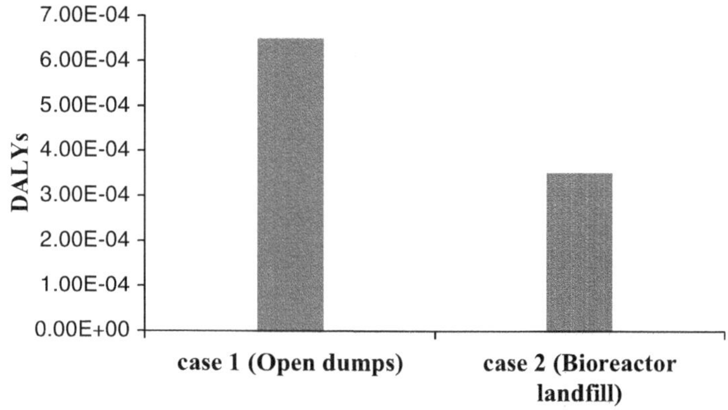

Fig. 15.5 Summary of human health damage per tonne of MSW

Analysis of the health damage per tonne of MSW processed shows that the bioreactor landfill is preferred over the open dumps due to the collection of respiratory compounds such as nitrogen oxides (NOx) and sulfur oxides (SOx) (Fig. 15.3). On the other hand, case 1 has the highest human health impacts. This is due to the fact that human health damages are sensitive to greenhouse gas emissions due to their climate change potential. In this scenario the main contributors to the impacts are methane, carbon dioxide and other respiratory compounds. It is evident from Fig. 15.5 that the maximum health damage from 1 tonne of waste is in case 1. The effective gas collection system reduces the health damage in case 2. Therefore, bioreactor landfill option is a better socially sustainable choice.

The life cycle analysis (LCA) tool has been adopted to assess the environmental sustainability in terms of the potential impacts to the environment. LCA serves as a decision-making tool in selection of the most sustainable, economic and environment-friendly land disposal options. According to the life cycle characteristics of waste treatment/disposal, its environmental impacts are classified into five kinds: energy depletion potential (EDP), global warming potential (GWP), acidification potential (AP), eutrophication potential (EP) and photochemical oxidant potential (POCP). However, in this study, global warming potential (GWP) and photochemical oxidant potential (POP) are considered as these indicate the maximum negative effect on the environment.

The LCC evaluation was taken as the major economic indicator to assess the economic sustainability of the system. LCC involves evaluation of all costs in a life cycle such as capital cost, operational and maintenance cost and environmental cost.

15.5 Results and Discussion

The sustainability of a landfill is assessed by developing indicators representing the environmental, economical and social sustainability. Life cycle analysis, life cycle cost analysis and disability-adjusted life year are the tools employed for developing the indicators. The environmental indicators (global warming potential (GWP) and photochemical oxidant potential (POCP)) were found to be minimal in bioreactor landfill. There were jobs and revenue generated in the bioreactor landfill option. The health impacts are also minimal in case 2. Therefore, it can be concluded that bioreactor landfill is an environmental, economical and socially sustainable means of land disposal option for waste.

References

Central Public Health, & Environmental Engineering Organisation (India) (2000) Manual on municipal solid waste management. Central Public Health and Environmental Engineering Organisation, Ministry of Urban Development, Govt. of India

Chanakya HN, Shwetmala K and Ramachandra TV (2010) Small-scale decentralized and sustainable municipal solid waste management potential for Bangalore anchored around total recycle and biomethanation plants. National Conference on Urban, Industrial and Hospital Waste Management 2010, Ahmadabad Management Association

ISWA (1992) 1000 Terms in solid waste management. In: Skitt J (ed) International Solid Wastes Association, Copenhagen, Denmark

Kebin HE, Yao Z, Zhang Y (2010) Characteristics of vehicle emissions in China based on portable emission measurement system. 19th Annual International Emission Inventory Conference "Emissions Inventories-Informing Emerging Issues", San Antonio, TX

Lakshmikantha H (2006) Report on waste dump sites around Bangalore. Waste Manag 26:640–650

Lang DJ et al (2007) Sustainability Potential Analysis (SPA) of landfills—a systemic approach: initial application towards a legal landfill assessment. J Clean Prod 15(17):1654–1661

Warith M (2002) Bioreactor landfills: experimental and field results. Waste Manag 22(1):7–17

Chapter 16
Approaches to Selecting Sustainable Technologies for Remediation of Contaminated Sites: Case Studies

Bala Yamini Sadasivam and Krishna R. Reddy

Abstract Conventional remediation technologies are designed and implemented with the aim of achieving reduction of contaminant concentrations to meet the remedial goals in a cost-effective and timely manner. On the contrary, green and sustainable remediation (GSR) is a holistic approach to remediation which helps reduce the overall environmental impact by weighing out different technically feasible remedial options and selecting the one with minimal impact. This paper provides in-depth information pertaining to the application of GSR at three different contaminated sites in Illinois, USA. All the three sites had varied contaminant characteristics and site-specific conditions for which different remedial options were evaluated. Environmental site assessment reports were reviewed, and the final recommendations for remedial action (RA) were made based on conducting a qualitative as well as quantitative comparison between technically feasible remedial options using available tools to quantify the sustainability metrics, such as Green Remediation Evaluation Matrix (GREM), Illinois Greener Cleanups Matrix, Sustainable Remediation Technology (SRT™), and SiteWise™. Remedial Action Plans (RAPs) were developed by incorporating best management practices (BMPs), and the use of coupled-treatment techniques (remedial train) was proposed and designed for site-specific conditions along with detailed cost estimates and expected time frame to achieve the remedial goals. Long-term monitoring and maintenance plans were also included in the proposed RAPs.

Keywords Contaminated sites • Remediation • Sustainability • Soils • Groundwater

---

B.Y. Sadasivam • K.R. Reddy (✉)
Department of Civil and Materials Engineering, University of Illinois at Chicago, 842 West Taylor Street, Chicago, IL 60607, USA
e-mail: bsadas2@uic.edu; kreddy@uic.edu

G.L. Sivakumar Babu et al. (eds.), *Sustainability Issues in Civil Engineering*, Springer Transactions in Civil and Environmental Engineering,
DOI 10.1007/978-981-10-1930-2_16

16.1 Introduction

High levels of risk posed to human health and the environment from exposures to toxic contaminants in soils and groundwater require prompt attention from technical experts and regulating agencies in terms of remediating the contaminants to acceptable levels and reducing the risks. The promulgation of major environmental laws and regulations pertaining to cleanup of highly contaminated sites in the United States (e.g., CERCLA and SARA) along with prior regulatory standards, such as CWA, CAA, and RCRA (USEPA 2011), has enforced stricter norms with respect to environmentally safe practices, thereby reducing environmental impact and promoting the cleanup of polluted sites. The conventional approach to remediation warranted the cleanup goals which aimed at restoring the contaminated sites to their pristine condition. However, this resulted in no form of action taken on behalf of agencies that were required to clean up the sites especially due to the extensive costs and time involved behind rendering the site safe for use. This in turn led to the development of risk-based, tiered approach to meeting the corrective action objectives (known as TACO in Illinois) for remediating the sites. This modified approach comprises meeting more realistic cleanup goals that were devised based on comprehensive risk assessment modeling which also reduced the costs and time associated with cleanup.

The use of state-of-the-art practices in site characterization and preliminary assessments (e.g., semiquantitative mobile tools for obtaining expedited site data) has immensely contributed to more efficient cleanup activities. In addition to this, innovative remediation technologies that coupled the use of two or more technologies (treatment train) at the same time proved to be more effective in remediating a wide array of contaminants as well as reduced the time required to meet the cleanup goals. Details on different site remediation technologies can be found in Sharma and Reddy (2004). Recently, the concept of green and sustainable remediation evolved consequently with the publication of NRC's "green book" wherein the principles of sustainability were recommended for environmental cleanup decision making by the USEPA. This paved way for the use of best management practices and cleanup methods that were more environmentally friendly and caused less adverse effects to workers and surrounding community while executing the remedial action (USEPA 2008, 2010). The objective of this paper is to demonstrate the step-by-step approach in the application of sustainability concepts in remediation at three different contaminated sites in Illinois.

16.2 Green and Sustainable Remediation: Background

GSR aims at holistically achieving the goals of environmental remediation while minimizing the environmental impacts. Some critical aspects considered in this approach are as follows: (1) reduce the use of nonrenewable energy and promote the use of renewable energy for operations, (2) preserve natural resources, (3) reduce waste generation and promote waste reuse and recycle, and (4) maximize future reuse options for the site (ITRC 2011). In general, sustainability is defined broadly as meeting the needs of the present without compromising the ability of future generations to meet their needs (UN Report). Sustainable remediation encompasses a wide spectrum of environmental factors and social impacts applied throughout the entire life cycle of the remediation project so as to maximize the environmental, social, and economic benefits, otherwise known as "triple bottom line" (Ellis and Hadley 2009).

Green remediation is the practice that accounts for potential environmental hazards that are likely to arise during each phase of the cleanup process and employ suitable techniques to maximize net environmental benefit of the cleanup. Thus, in a broader perspective, green remediation involves the incorporation of sustainability into remediation (Favara et al. 2011). The prime objective of GSR is to focus on improving and sustaining the quality of life in the surrounding community without compromising on the existing standard of living for future generations (sustainability). By employing the principles of GSR in the cleanup of contaminated sites, there can be reduced impacts to the local environment, including air, water, and soil, increased energy efficiency, lower carbon emissions, acceleration of ecosystem and biodiversity restoration, reduced long-term maintenance costs, and reduced energy costs (ITRC 2011, Reddy and Adams 2001 and USEPA CLU-IN 2011).

It is critical to realize the primary goal of any remedial approach, which is achieving remediation goals that are protective of public health and the environment. In the process of achieving this goal, some core elements need to be optimized to achieve GSR which include energy requirements, water requirements and impacts to water resources, air emissions, land and ecosystem impacts, material consumption and waste generation, and long-term stewardship actions (USEPA 2008). These core elements of GSR are addressed through best management practices (BMPs) to achieve sustainable long-term net environmental benefit during all phases of site remediation (site investigation, remediation system selection and design, remediation system construction, operation and monitoring, and land end use).

16.3 Case Study 1: Villareal Building Site

16.3.1 Site Background

The site is known as the Villareal building located in Rock Island County in the City of Moline, IL, and is approximately 0.20 acres. The site is occupied by a vacant 6000 square feet single-story building, and a parking lot is also adjacent on the west side of the building.

The earliest recorded history for this site is from 1886. From 1886 to 1927, this area was a residential complex. In 1933, a restaurant was built on this site. Up to 1938, the building was a host to various business operations. From 1949 to 1970, a dry cleaner was operated on the site. Currently, the site is vacant and tax record indicated that the site belongs to the City of Moline. This site has been zoned as a B-2 industrial/commercial area by the city. B-2 industrial/commercial area is defined by the Rock Island County zoning and building safety department stated as area intended to be high-density, compact, pedestrian-oriented shopping, office, service, or entertainment area.

The site investigations identified several sources of contaminants or relevant environmental conditions (RECs) that could pose a threat to surrounding environment. Apart from contamination that could have resulted from the historical operation of the on-site dry cleaner, there were other possible sources identified, such as the adjacent garage/wrecker service on the west, leaking underground storage tank facility to the west, filling station located on the west and east side of the site, as well as a gasoline underground storage tank on the south side of the building. Within the building itself, two potential areas for chemical storage were identified. A machine pad and a closet with a floor drain suspected to be remnants of the on-site dry cleaning operation were observed.

This area has been zoned as B-2 area by the City of Moline, which restricts future site development for industrial or commercial purposes only. The city has also imposed ordinance which bars current and future owners from installing potable water wells within the municipal limits. As this site will only be developed for industrial or commercial purposes, the Occupational Safety and Health Act (OSHA) would apply. Future workers at this site must be protected against exposure to hazardous COCs. The Clean Water Act (CWA) may be also applied to this site; given the proximity to Mississippi River, it is necessary to ensure that any COC identified at this site does not migrate toward the river. In the future, if this site is repossessed by individual parties, the Small Business Liability Relief (SBLR) can be used to protect small business owners from sustaining the full cost of site remediation and can provide funding for cleanup efforts. This site can also be subjected under the Pollution Prevention Act (PPA) which can help the future owners prevent pollution by providing assistance from the United States Environmental Protection Agency (USEPA) in incorporating practices that promote waste reduction and minimization.

16.3.2 Site Characterization

Several RECs were identified in the preliminary site investigation pointing to possible VOC contamination at the site. Several detailed investigations confirmed the presence of VOCs in the groundwater at this site. The site investigations also included investigation of the site geology, site hydrogeology, and contaminant profiles. In order to characterize the site geology, four soil borings were installed on the outer perimeter of the building and three soil borings were installed inside the building near the area of suspected contaminant leakage. Results from the soil boring indicated a predominance of orthents, silty, and undulating soil with high hydraulic conductivity. Water table was found at depth of 11–14 ft below ground surface (bgs). Sandy fill layer is present just below the surface to a depth of 2 ft. Below the sandy fill, silty clay and coarse to fine sand were present to a depth of approximately 18 ft bgs. Weathered limestone bedrock was found at approximately 18 ft bgs.

Three monitoring wells were installed to collect the water samples for chemical analyses and also to estimate the hydrogeology properties of the site. The depth of water table was 10–14 ft bgs. The groundwater sloped north and the estimated groundwater gradient from the groundwater contours was very high (20%). Hydraulic conductivity test was performed at two monitoring wells and the average hydraulic conductivity was 5.71×10^{-4} cm/s. The nearest surface water is Mississippi River located approximately 1600 ft north of the site which is a concern as groundwater flow was found to slope north at high gradient (approximately 20%). The two nearest wells were found at 500 ft west and 1300 ft south, well beyond the regulated setback zone in Illinois which is 200 ft.

PNAs were detected at two soil borings that were located at the edge of the site perimeter, which could indicate that the sources of PNA contamination are potentially off-site. Benzene, ethylbenzene, and chlorinated compounds were detected more uniformly throughout the sampling locations. However, all the contaminant concentrations in the soil samples were below the most restrictive residential TACO Tier I soil remediation objective (SRO), thus requiring no further remediation. But, in the groundwater samples, concentrations of cis1,2-dichloroethene (cis-DCE), trichloroethane, and vinyl chloride (VC) were higher than the most restrictive residential Tier I groundwater remediation objective (GRO) in sampling locations within the building. Cis-DCE and VC were higher than the most restrictive industrial TACO Tier I Class II GRO. MTBEs appeared to be only detected in sampling locations on the east, indicating a potential source in the eastern side of the site. However, the levels of MTBEs met the TACO Tier I GRO for residential areas.

16.3.3 Risk Assessment

Human health risk assessment was performed guided by the soil and groundwater screening level provided in IAC Title 35 Part 742 TACO. Ecological risk assessment was not performed as the site is located in an urban area, and literature review indicated that no sensitive or threatened ecological subpopulation is located within downtown Moline, IL.

COCs are defined as contaminants that are present in excess of the TACO Tier I soil and groundwater screening levels. Among the 88 contaminants that were analyzed for in the soil and groundwater samples, only three groundwater COCs were identified, trichloroethane, VC, and cis-DCE. Since the site has been restricted for industrial or commercial use, only the TACO Tier I objectives specific for industrial screening levels are applicable. Thus, only VC and cis-DCE concentrations were above the industrial TACO Tier I Class II GRO, and these contaminants were deemed as COCs for which targeted remedial goals were set.

Groundwater ingestion is eliminated as a potential exposure pathway as the City of Moline established ordinance on the Villareal building site prohibiting installation of any drinking water well within the property boundary. The site investigations also reported that no drinking water wells are located within the regulated setback zone of 200 ft. Incidental soil ingestion pathway is also eliminated as the site has an existing building which covers over 80% of the site surface. Area not covered by the building subgrade is paved with 2 in. asphalt. Given the subgrade and asphalt cover at this site, soil contact with human via ingestion, inhalation, or dermal pathways was deemed not possible. Site soil possessed a high permeability (5.71×10^{-4} cm/s) for Villareal building site; coupled with high estimated groundwater gradient (20%) toward the north, the main concern is contamination migration toward the Mississippi River, located 1600 ft north and 15 ft lower in elevation from the site. The City of Moline obtained drinking water from a surface water intake at the Mississippi River with daily water intake reaching up to 5.3 million gallons distributed to over 43,000 people daily.

As the COCs are from the VOC class, vapor intrusion can also be a problem. Vapor intrusion can be a significant exposure pathway as the low water table (11–14 ft) was contaminated with chlorinated VOCs in excess of the conservative industrial TACO Tier I Class II groundwater screening values. Vapor intrusion can impair the indoor air quality and expose future occupants to the hazardous COCs.

Information for toxicity assessment is obtained from USEPA's Integrated Risk Information System (IRIS) and the Agency for Toxic Substances and Disease Registry (ATSDR). The USEPA classifies VC as a Class A known human carcinogen, and the critical exposure pathway is via inhalation and oral ingestion. The oral slope factor for VC is 0.72 mg/kg/day. Cis-DCE is classified by the USEPA as Class D carcinogens, indicating inadequate toxicity data or evidence to classify. Under acute exposure, cis-DCE can act as anesthetic resulting in central nervous system depression with symptoms including nausea, fatigue, vertigo, and drowsiness. Chronic exposure to cis-DCE has been linked to decrease in hematocrit and hemoglobin level. The oral reference dose is 0.002 mg/kg/day.

In determining the appropriate remediation goal, a comparison against groundwater vapor intrusion screening values was also deemed necessary as vapor intrusion can be a significant exposure pathway. IL does not have any vapor intrusion screening values; however, certain states such as NJ, MI, and CA published soil and groundwater screening values for VOCs. In most cases, the vapor intrusion screening levels are more restrictive than the TACO Tier I Class II groundwater GRO.

To obtain the site-specific vapor intrusion screening level values, a USEPA online tool for site assessment calculation was used. The tool for vapor intrusion screening level was developed by the USEPA Office of Solid Waste and Emergency Response (OSWER). In this model, the Johnson and Ettinger simplified model for subsurface vapor intrusion was incorporated. Using site-specific input parameters, the model performs sensitivity analysis to determine the less protective, best protective, and more protective screening level estimates. The remedial goals were thus set to 0.0029 mg/L and 0.426 mg/L for VC and cis-DCE, respectively.

16.3.4 Remedial Options

Several potential remedial actions (RAs) were investigated with focus given to groundwater RAs that are applicable toward extracting, immobilizing, or detoxifying the chlorinated COCs from a source zone (Reddy et al. 1999). Ex situ and in situ RAs were investigated to determine their feasibility in remediating chlorinated VOCs at the Villareal building site. Ex situ groundwater remediation technologies that were investigated include groundwater extraction and subsurface draining followed by on-site or off-site treatment and excavation. In situ groundwater remediation technologies investigated were pump and treat, in situ flushing, passive containment, active containment, permeable reactive barriers, in situ air sparging (Adams et al. 2011), in situ chemical oxidation, monitored natural attenuation, and enhanced bioremediation. While in general various groundwater remediation technologies could work to remove chlorinated VOCs, site-specific consideration is very important to ensure effective and efficient removal rate. Among the site-specific considerations are excluding any technology that requires large-scale excavation to salvage the existing structure at the site, equipments have to be small as the site is very small and only the adjacent parking area is available as open work area, contaminant migration control is necessary given the high groundwater gradient toward the Mississippi River, quick remediation time is preferred to enable building utilization, and cost should be low. Table 16.1 summarizes the

Table 16.1 Qualitative evaluation of groundwater remediation technologies based on site-specific considerations for Villareal site. Check signs indicated that the technology fits the site-specific considerations, and cross signs indicated that the technology is not suitable for the site

Strategy	In-situ	Size	Migration	Time	Cost
MNA	✓	✓	×	×	×
Ex-situ (large scale excavations)	×	✓	✓	✓	×
Passive in-situ containment	✓	✓	✓	×	✓
Active in-situ containment	✓	✓	✓	✓	✓
Ex-situ remediation	×	✓	✓	×	×
Pump and treat	✓	✓	✓	✓	×
In-situ flushing	✓	✓	×	✓	✓
Permeable reactive barrier	✓	✓	✓	×	✓
In-situ air sparging	✓	✓	×	✓	✓
Bioremediation	✓	✓	×	×	✓
In-situ chemical oxidation	✓	✓	×	✓	✓

qualitative evaluation of all available groundwater treatment technologies based on site-specific considerations.

Preliminary evaluation of technical details led to the consideration of three groundwater remedial technologies: permeable reactive barrier (PRB), enhance bioremediation (EB), and in situ chemical oxidation (ISCO). PRB was selected as it is a relatively easy technology to implement and it can immobilize the COCs and provides contaminant migration control. Enhanced bioremediation is also relatively easy to implement and should also be a very economically feasible technology. ISCO is an attractive technology to consider as it can detoxify the COCs by transforming hazardous compounds into innocuous compounds via oxidation-reduction chemical reaction (USEPA CLU-IN 2011).

The sustainability of PRB, EB, and ISCO was qualitatively assessed using the Illinois Greener Cleanups Matrix and the Minnesota Toolkit for Greener Practices. The matrix includes sustainable and best management practices (BMPs) that can be incorporated into the remedial design that can directly benefit land, air, energy, and water. BMPs identified from the IL Greener Cleanups Matrix are shown in Table 16.2. Overall, the three technologies considered are highly beneficial to the air as significant reduction in air emission can be accomplished by avoiding demolition activity on the existing structure and by incorporating in situ technologies that remove the need for off-site groundwater treatment process. PRB has the

Table 16.2 Summary of best method practices (BMPs) for IEPA Greener Cleanups Matrix Tool that can be incorporated into permeable reactive barrier (PRB), enhanced bioremediation (EB), and in situ chemical oxidation (ISCO) designs for Villareal site

BMP's	PRB	EB	ISCO	Notes
Air:				
Reduce emission from construction equipment and from waste truck	Y	Y	Y	Use retrofitted equipment with PM-reducing technology
Reduce air emission from demolition activity	Y	Y	Y	Building structure remains intact
Reduce air emission from treatment process	Y	Y	Y	Treatment process is in situ
Water:				
Reduces future migration concern	Y	N	N	Containment provided with PRB
Land:				
Reduces waste material requiring off-site disposal	N	Y	Y	Only PRB has soil waste material
Reduces future migration concern	Y	N	N	
Energy:				
Reduces fuel use in on-site construction equipment	Y	Y	Y	Use local construction company
Reduces energy use in remediation system	Y	Y	Y	Technology requires passive operation
Reduces purchase energy use	Y	Y	Y	Energy used only in construction

Table 16.3 Summary of sustainability objectives from the MN Toolkit for Greener Practices that can be achieved by choosing permeable reactive barrier (PRB), enhanced bioremediation (EB), and in situ chemical oxidation (ISCO) remedial technologies for Villareal site

Aims	PRB	EB	ISCO	Notes
Address contaminants without complex construction	Y	Y	Y	Construction is minimum for all technologies
Avoid immobilization of contaminants to different media/phases	Y	Y	Y	

added advantage of providing containment which benefits water and land; however, the trade-off is the need to dispose of excavated soil in hazardous waste landfill. In all three remedial technologies considered, BMPs will be applied to reduce and energy usage.

The MN Toolkit for Greener Practices is developed by the Minnesota Pollution Control Agency. The sustainability goals identified from the Toolkit for Greener Practices are summarized in Table 16.3. As the construction required for PRB, EB, and ISCO is minimal and nonintrusive, the sustainable goal of addressing the contaminants without complex construction is easily accomplished. EB and ISCO have the added advantage of actually transforming the contaminants into innocuous by-products instead of immobilizing the contaminants into different media or phases.

California Department of Toxic Substances Control (DTSC 2007) created a qualitative method to assess the sustainability of remedial technologies via the Green Remediation Evaluation Matrix (GREM). GREM is a simple Excel-based tool which assists comparison of different technologies based on different green parameters. GREM was used in this study to qualitatively evaluate the emissions of toxic substances (nitrogen oxides (NOxs), sulfur oxides (SOxs), and greenhouse gases (GHGs)), liquid and solid waste production, thermal release (water and vapor form), physical disturbances and disruptions (noise and traffic), and potential for resource depletion. GREM provides an opportunity for investors to evaluate the environmental impacts of each selected remediation technology. The GREM qualitative assessment results for PRB, EB, and ISCO are shown in Table 16.4. The matrix helps to identify key areas for improvement and opportunities to reduce impacts on the three chosen remedial technologies. For example, PRB has no liquid waste production, while, on the other hand, the generated solid waste needs to be properly disposed of at a hazardous landfill; ISCO does not generate any solid or liquid waste; however, emissions from construction and consumables can be very significant. Overall, it was concluded that the use of qualitative assessment such as the IL Greener Cleanups Matrix, MN Toolkit for Greener Practices, and GREM can help narrow down potential remedial action candidates, but cannot be used to definitively select the final remedial technology selection.

Quantitative sustainability evaluation tools such as Sustainable Remediation Technology (SRT™) and SiteWise™ were used to select the final remedial option. SRT™ was developed by Air Force Center for Engineering and Environment

Table 16.4 GREM for PRB, EB, and ISCO. N indicates no impact on a particular stressor or parameter, whereas Y indicates potential impact on the stressor or parameter

Stressors	Affected media	Mechanism/Effect	PRB	EB	ISCO
Substance release/production:					
Airborne NOx and Sox	Air	Acid rain/photochemical smog	N	N	Y
Chlorofluorocarbon vapors	Air	Ozone depletion	N	N	N
Greenhouse gas emissions	Air	Atmospheric warming	N	N	Y
Airborne particulates/ toxic vapors/gases/ water vapor	Air	General air pollution/toxic air/humidity increase	Y	N	Y
Liquid waste production	Water	Water toxicity/sediment toxicity	N	Y	Y
Solid waste production	Land	Land use/toxicity	Y	Y	N
Thermal releases:					
Warm water	Water	Habitat warming	N	Y	N
Warm vapor	Air	Atmospheric humidity	N	N	N
Physical disturbances/disruptions:					
Soil structure disruption	Land	Habitat destruction/soil infertility	Y	Y	N
Noise/odor/vibration/ esthetics	Env	Nuisance and safety	N	N	Y
Traffic	Land/Env	Nuisance and safety	N	N	N
Land stagnation	Land, environment	Remediation time, cleanup efficiency, redevelopment	Y	Y	N
Resource depletion/gain (recycling):					
Petroleum (energy)	Subsurface	Consumption	N	Y	Y
Mineral	Subsurface	Consumption	Y	Y	Y
Construction materials	Land	Consumption/reuse	Y	Y	Y
Land and space	Land	Impoundment/reuse	Y	Y	N
Surface water and groundwater	Water, land	Impoundment/sequester/reuse	Y	Y	Y
Biology resources (plants/trees/animals/ microorganisms)	Air, water, land/forest, subsurface	Species disappearance/diversity reduction regenerative ability reduction	Y	N	Y

(AFCEE 2010). Groundwater remediation technologies included in SRT for the Villareal site are EB, ISCO, PRB, and monitored natural attenuation. SiteWise™ is a GSR tool developed by US Navy, US Army Corps of Engineers (USACE), and Battelle (Bhargava and Sirabian 2011). Unlike SRT™, SiteWise™ focuses on activities rather than technology. Given the focus on activities, SiteWise™ can be used for environmental impact modeling of a variety of remediation technologies.

The results for SRT analysis for all three technologies are presented in Figs. 16.1 and 16.2. Carbon dioxide emission is highest for PRB, followed by ISCO and

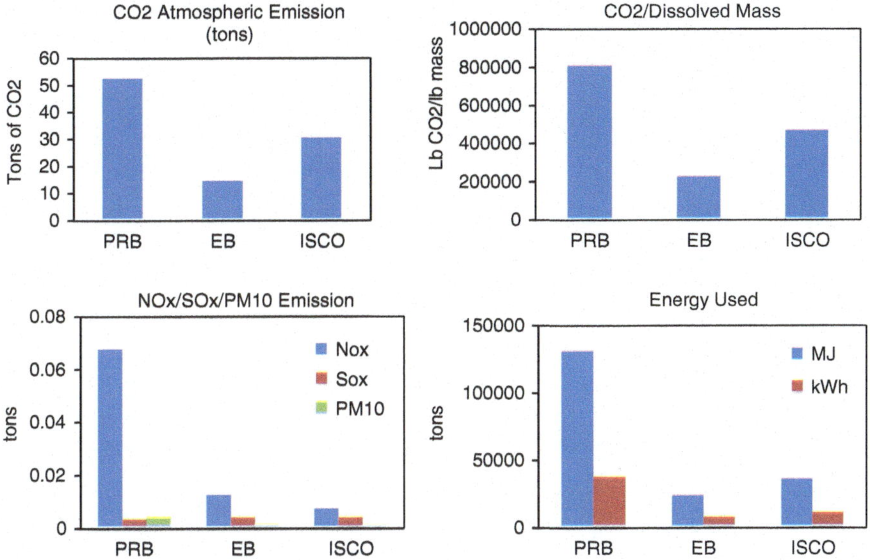

Fig. 16.1 Villareal site SRT™ results for PRB, EB, and ISCO for carbon dioxide emission, nitrogen oxides, sulfur oxides, particulate matter emissions, and energy consumption

EB. The same trend is observed when carbon dioxide emission is normalized for mass. PRB also has the highest emission in nitrogen oxides, sulfur oxides, and particulate matter emissions, followed by ISCO and EB. High emission rate for PRB could stem from the fact that PRB involves disposal of large amount of excavated soil in hazardous landfill. Emissions from ISCO are potentially from the transport and use of oxidants, whereas EB emissions are very low as only personnel and equipment transport contributed toward emission. Energy used is most intensive for PRB as it requires the most amount of construction work; EB and ISCO on the other hand only require the construction of injection wells.

In terms of lost hour and injury risk, ISCO scores the highest followed by PRB and EB (Fig. 16.2). This stems from the fact that some hazards are associated with the handling of highly reactive oxidant, and risk of personnel exposure to oxidant is significant in ISCO. In the 1-year duration used for SRT modeling, plume reduction of PRB and ISCO is comparable; however, almost no plume reduction is observed with EB. This demonstrates that PRB and ISCO have similar effectiveness in terms of COC removal; however, a much longer time is required to reduce the COC plume using EB technology. Cost is highest for PRB, followed by ISCO and EB. The higher cost in PRB is probably associated with the cost of soil disposal at hazardous landfill. In ISCO, the higher cost stems from consumables, specifically from oxidant. EB has the lowest cost as this technology requires very little construction, no associated operation cost, and low substrate or amendment cost.

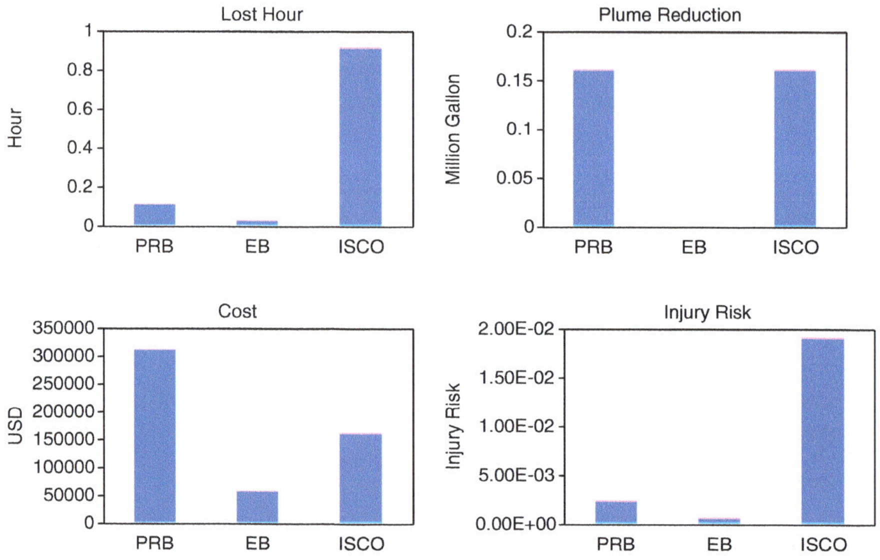

Fig. 16.2 Villareal site SRT^{TM} results for PRB, EB, and ISCO for lost hour, plume reduction, cost, and injury risk

Environmental impacts associated with the PRB, EB, and ISCO construction and monitoring were also modeled using SiteWise^{TM}. The overall SiteWise^{TM} results for construction and monitoring activities are summarized in Fig. 16.3. PRB has the highest combined GHG emission and is the most energy-intensive remedial options. SOx and PM10 emission is highest for ISCO and is associated with the residual handling of the consumable oxidants. NOx emission for PRB, EB, and ISCO is comparable. PRB is associated with the highest fatal risk accidents potentially from the earthwork construction activities which involve the use of heavy-duty equipment.

As far as sustainability, EB has the lowest environmental impacts. The trade-off with EB is a very long remediation time. Apart from remediation time concern, there is also bioavailability concern as literature review of case studies showed that EB is implemented at sites which contain 1–3 times higher concentrations of VC and cis-DCE compared to the concentrations detected at Villareal site. Given that there is not enough case studies providing evidence of EB feasibility at low contaminant concentrations, EB was not chosen as the final remedy selection. PRB and ISCO have comparable predicted effectiveness in removing the COCs and the overall environmental impacts are also similar. PRB is estimated to cost twice as much as ISCO mostly due to the cost associated with disposal of excavated soil in hazardous landfills. As cost is one of the most important drivers in decision making, ISCO was chosen as the final remedy selection.

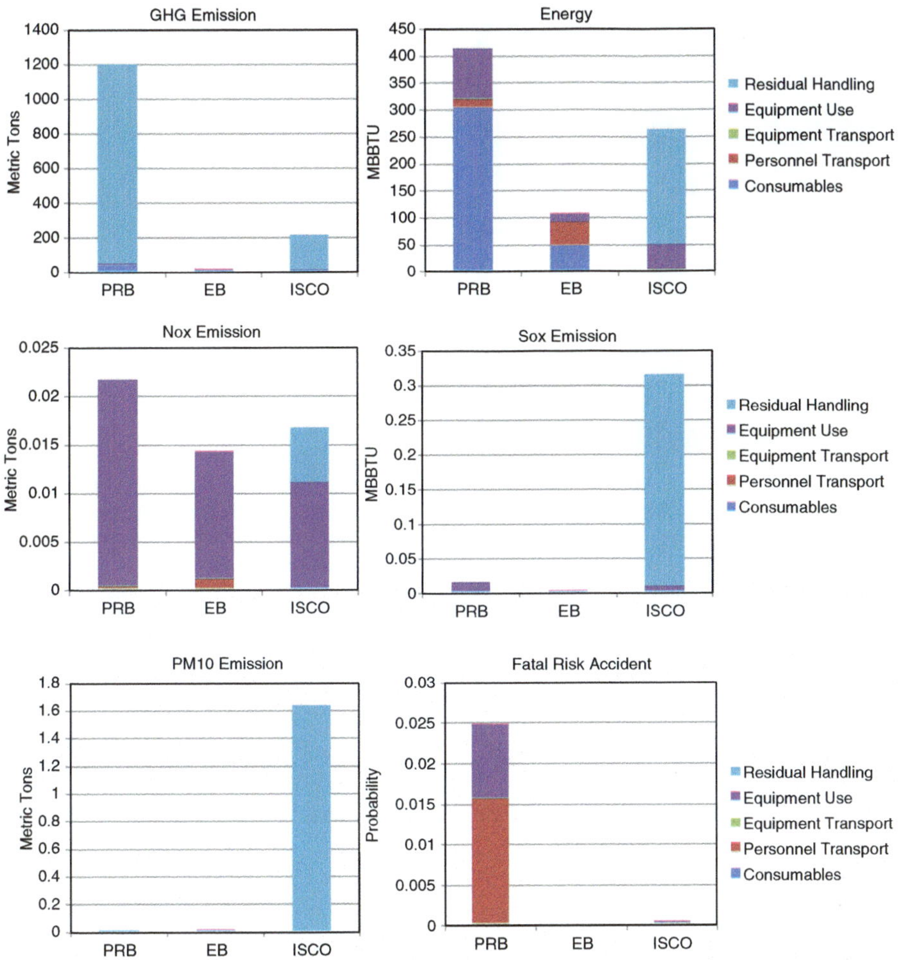

Fig. 16.3 SiteWise™ results for Villareal site for PRB, EB, and ISCO for overall construction and long-term monitoring activities

16.3.5 Remedial Design and Monitoring Requirements

To ensure adequate level of environmental and human protection as this site, design of a remedial train incorporating ISCO is necessary. In addition to this, the high groundwater gradient at the site calls for a containment system. Though ISCO can transform the chlorinated COCs into innocuous compound, this technology alone lacks the ability to contain the COCs and prevents contaminant spread and migration. The containment system however does not have to be a PRB. A soil-bentonite wall was chosen as a containment method. Soil-bentonite wall is a very common vertical barrier that is easy to implement and is economically attractive. The

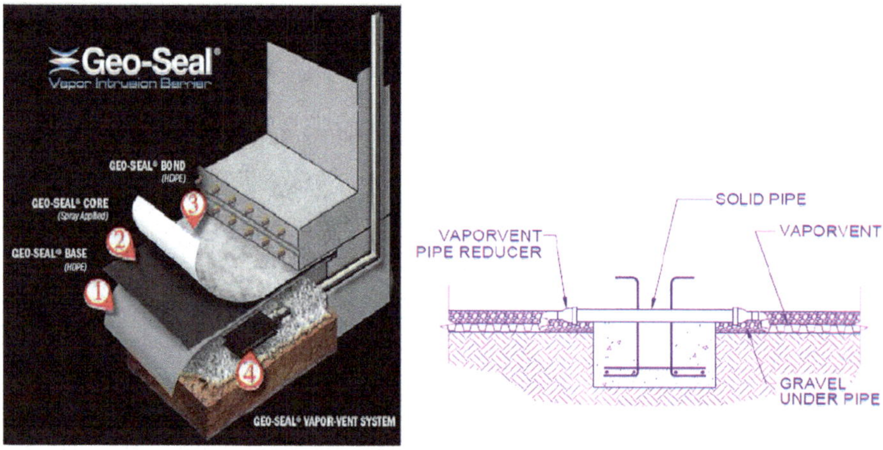

Fig. 16.4 Schematic of the GeoSeal™ vapor intrusion barrier layers is shown on the *left*. Layer 1 is the base which is a thin, low-permeability HDPE layer; layer 2 is the core, a spray on polymer adhesive; and layer 3 is the bond, a proprietary layer developed by Life Science Technologies Inc. Schematic of the GeoSeal™ low-profile vapor vent system is shown on the *right*. The vapor vent system consists of a vapor collection end and connecting pipes to transfer the collected vapor to a vapor trap

presence of highly toxic contaminants at dangerous level necessitates the implementation of an interim remedial design. One of the concerns at the Villareal building site is vapor intrusion via the building subgrade. An interim RA was taken to reduce risk of exposure by cutting off exposure pathway. To eliminate this exposure pathway, a low-permeability GeoSeal™ with vapor vent can be installed as shown in Fig. 16.4.

Two long-term remedial actions are proposed, the installation of a soil-bentonite slurry trench wall which acts as a passive containment system and ISCO which has the ability to detoxify the COCs by transforming them into innocuous compound. The soil-bentonite wall will ensure no contaminant spread or migration, whereas ISCO can lower the COC concentration to a safe level. As the groundwater flows north at a high gradient and given the limited space available around the building, it is proposed to construct the soil-bentonite slurry wall toward the northern side of the building. The trench required for the slurry should have a thickness of 2 ft, which is equal to the width of a standard backhoe. The proposed trench dimension is 125 ft in length and 18 ft in depth, which would allow the trench to reach the limestone formation.

Components considered in ISCO design include choosing the appropriate oxidant, designing an effective oxidant injection and delivery system, and ensuring safety and regulatory compliances during construction and long-term monitoring. Common oxidants used in ISCO are potassium permanganate, ozone, and hydrogen peroxide (ITRC 2001, 2005; Siegrist et al. 2011). The appropriate oxidant choice depends on the COCs. Ozone and potassium permanganate are both suitable for VC and cis-DCE remediation. Ozone is the strongest oxidant used in ISCO with 2.1 V of oxidation-reduction potential and 1.5 relative strength when compared to chlorine (ITRC 2005). For every one unit weight of contaminants, 1–10 unit weights of ozone can be required. Ozone can be generated on-site, which removes the need for chemical storage; however, cost of ozone generation can be high and specialty equipments are required. In terms of oxidizing strength, potassium permanganate is less oxidizing than ozone. However, potassium permanganate is more economically feasible than ozone. VC reacts with potassium permanganate to form carbon dioxide, manganese dioxide, potassium ions, chloride ions, hydroxide ions, and water. Carbon dioxide exists naturally in the subsurface and manganese dioxide is a natural mineral in soil; as these compounds are naturally occurring compounds, there is a need to remove the reagent posttreatment. Best management practice can be employed such that potassium permanganate is mixed and injected into the groundwater on the same day of delivery, removing the need for on-site chemical storage. Potassium permanganate is also stable in the groundwater media. Given economical consideration and the small open area at the site which may not accommodate ozone-generating equipment, potassium permanganate was chosen as the ISCO oxidant.

Oxidant injection and delivery system were designed based on ITRC guidance documents, SRT estimates, and case studies (ITRC 2001, 2005; Reddy 2011). Injection parameters are summarized in Table 16.5. Number of injection points is estimated from SRT, which assumed well spacing of 20 ft. Schedule 40 PVC pipes will be used to construct the injection well. SRT estimates 13,600 lbs of potassium permanganate is needed. ITRC guidance documents and case studies recommended potassium permanganate to be injected at 4% v/v solution; injecting at higher concentration will increase COC removal rate but will decrease removal efficiency. Twenty-one thousand gal of water is required to dilute the potassium permanganate to the desired concentration. A batch injection method will be used; this method is

Table 16.5 Summary of injection parameters for in situ chemical oxidation at Villareal building site

Item	Value
Number of injection points	7
PVC specification:	
(Use schedule 40 PVC)	
Length (ft)	112
Mass (lbs)	228
Oxidant specification:	
Mass (lbs)	13,600
Injection concentration (volume %)	4
Water required (gallon)	18,000
Injection scheme specification	
(Batch injection)	
Number of injection	3
Amount per injection (gal/well)	860
Time between injection (months)	1
Injection flow (gpm)	2

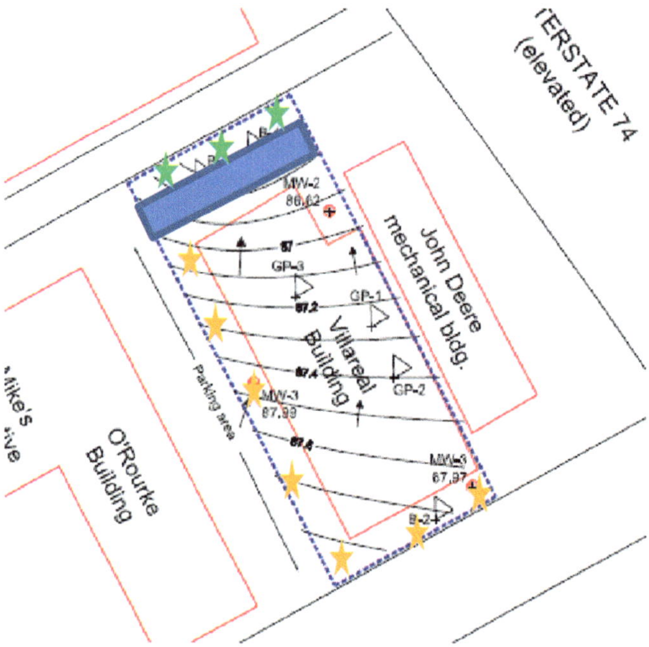

Fig. 16.5 Approximate location for proposed monitoring wells (*green* stars) and injection wells (*orange* stars) and the *blue* patch represents the location of soil-bentonite slurry wall

commonly used for ISCO potassium permanganate injection. A recirculation option for potassium permanganate injection was deemed not feasible given the constraint placed by the limited site area and existing building. It is proposed to have three injections, with each injection being 2 months apart. The ITRC guidance document suggested an injection rate of 2 gal per minute for groundwater delivery at 16 ft. Manifold injection wells can be employed for the potassium permanganate injection. Rough estimate of the locations of the injection points is shown in Fig. 16.5. The injection wells are installed to obtain a maximum radius of oxidant, and given the high hydraulic conductivity at this site, effective spreading of oxidants could be easily achieved.

To ensure successful implementation of the proposed remedial actions, a comprehensive groundwater monitoring design is included in the cost estimate. Three monitoring wells are proposed to be installed downstream of the soil-bentonite slurry wall; approximate location is shown in Fig. 16.6. Groundwater samples are to be collected every month after each ISCO injection event, every 3 months following the first year following the last ISCO injection event, and every 6 months for the next 4 years. Groundwater samples shall be analyzed for VC and cis-DCE and also for other chlorinated VOCs to detect if any other chlorinated metabolites are generated from the remedial actions.

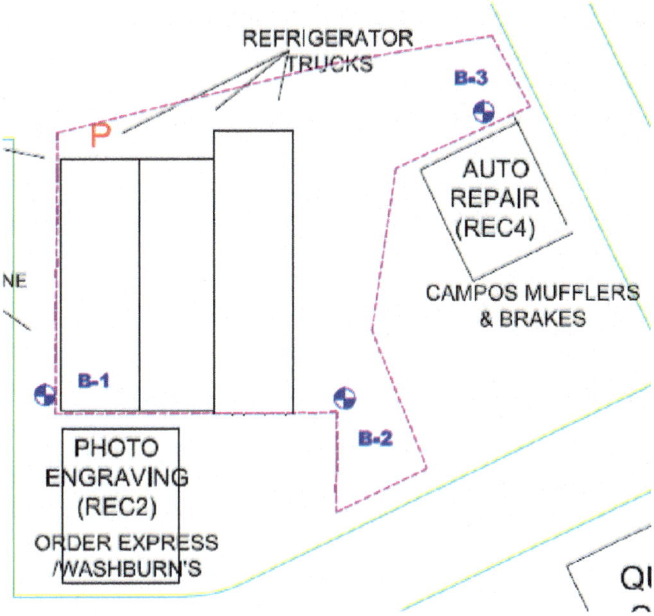

Fig. 16.6 River Bend Foodbank site location with surrounding features and approximate soil borings/well locations

16.3.6 Recommendations

Although there are only two COCs at the Villareal building, one of these contaminants is a known Class A carcinogen. ISCO injection combined with construction of a vertical barrier from soil-bentonite mixture on the northern end of the property was proposed. This ensures that contaminant from the Villareal building site will not be discharged at the Mississippi River which is the surface water intake for drinking water supply for the City of Moline. This also reduces liability for current and future owner of this site. Installation of low-permeability GeoSeal^{TM} layers helps to eliminate exposure pathway via subgrade vapor intrusion. Installation of the vapor intrusion barrier will also help current and future site owner comply with OSHA requirement for safe indoor air quality to workers or employees. To ensure effective remediation, a pilot scale study is recommended. The pilot scale study should investigate the oxidant delivery efficiency and the optimum oxidant concentration.

16.4 Case Study 2: River Bend Foodbank Site

16.4.1 Site Background

The site consists of one building occupied by River Bend Foodbank, a food storage and warehouse facility. The size of the building and surrounding parking lot was around 0.2 acres. The site is currently used as a warehouse and storage facility for a food bank. From 1886 to 1950, this site has been developed for residential purposes. From 1950 to 1958, it served as a private garage. From 1958 to 1964, the current building was under construction, and from 1964 to present, it has been occupied by River Bend Foodbank and used as a food storage and warehouse facility. Future site use has not been specified, but it is assumed that the site and surrounding area are to continue serving as commercial/industrial development area.

RECs for this site include a petroleum facility, the adjacent photo facility, an SRP listing located nearby, and an adjoining auto repair shop (Fig. 16.6). The Phase I Environmental Site Assessment (ESA) for this property indicated the possibility of heavy metal contamination in the soil and groundwater.

The applicable regulations for the site include CWA, as the gradient of the groundwater flow is toward the Mississippi River. The City of Moline has established ordinance which prohibits the installation of potable wells within the municipal limits. The River Bend site is eligible for enrollment in the IEPA SRP (Site Remediation Program) and subject to the regulations at 35 IAC Chapter 740: *Site Remediation Program*. Since the site is designated for future commercial and industrial use, regulations such as the OSHA and the PPA are also applicable.

16.4.2 Site Characterization

Three soil borings (B-1 to B-3 as shown in Fig. 16.6) were drilled to depths ranging from 10 to 12.5 ft below ground surface (bgs). Soil samples were collected using a 24 inch. split spoon sampler. The soil samples in each soil boring were analyzed with a Photo Ionization Detector (PID). Soil samples were collected from boring B-2 at the 0–3 ft bgs interval and at the 6–8 ft bgs interval and analyzed for VOCs, SVOCs, RCRA metals, and pH. Temporary groundwater monitoring wells were installed in each of the borings. Samples from GW-1 were analyzed for VOCs, PAHs, and RCRA metals; samples from GW-2 were analyzed for VOCs and RCRA metals; and samples from GW-3 were analyzed for VOCs, SVOCs, and RCRA metals.

The majority of the site soils were silty clays (CL-ML), with the depth ranging between 10 ft and 12 ft. Auger refusal occurred at depth between 10 ft and 12.5 ft, which was due to the presence of underlying limestone formation. Water table was found to be located between 9 ft and 10 ft below ground level. The site hydrogeology information indicated that the site property and surrounding land areas were

Table 16.6 Tier 1 GRO Class II GW ingestion route standards and soil contaminants detected at River Bend site

Contaminant	Class II (mg/kg)	Max. Concentration (mg/kg)	Location
Arsenic	0.2	0.45	GW-2 & GW-3
Barium	2.0	8.4	GW-2
Cadmium	0.05	0.058	GW-2
Chromium	1.0	2	GW-2
Lead	0.1	1.3	GW-2
Silver	–	0.01	GW-3D

sloping toward the north with an estimated groundwater gradient along that direction toward the Mississippi River.

Based on extensive soil and groundwater sampling and analyses, it was found that only groundwater at the site was contaminated with toxic levels of inorganic metals such as As, Ba, Cd, Cr, Pb, and Hg. The reported concentrations for these metals exceeded the GRO for Class II groundwater ingestion pathway. The extent of contamination at the site is presented in Table 16.6. The soil at the site was not found to contain toxic levels of contaminants. The type of contamination at the site was solely from heavy metals in the groundwater media, and no toxic levels of VOCs and SVOCs were detected.

16.4.3 Risk Assessment

Human health risk assessment was performed guided by the soil and groundwater screening level provided in IAC Title 35 Part 742 TACO. Ecological risk assessment was not performed as the site is located in an urban area and literature review indicated that no sensitive or threatened ecological subpopulation is located within downtown Moline, IL.

Arsenic, barium, cadmium, chromium, and lead were found to be COCs at the site. Arsenic is classified as Class 1 carcinogen, linked to cancer of the bladder, lungs, skin, kidney, nasal passages, liver, and prostate. Barium is a noncarcinogen with lungs as target organs. Cadmium is a Class B1 carcinogen, linked to pancreatic cancer. Chromium is a noncarcinogen in most cases. Finally, lead is said to be a probable carcinogen in high doses, linked to kidney cancer.

No RA was deemed necessary for soil as contaminants were below the most restrictive Tier 1 SRO. Conservative RAs were set for 5 GW contaminants (arsenic, barium, cadmium, chromium, and lead) because they had concentrations above Class II industrial GRO. Restrictions need to be placed on the use of GW at this site to avoid risk to human health. The remedial goals for the site are also listed in Table 16.6.

16.4.4 Remedial Options

Both ex situ and in situ groundwater remediation technologies were evaluated with the following site-specific considerations:

- Large-scale excavations are to be avoided in order to conserve the existing building.
- Remedial strategy should involve small and simple equipment given small area.
- Remedial time should be fast to allow building to be used.
- Technology that provides containment is preferred to control contaminant migration toward the Mississippi River, which is the drinking water source for the City of Moline.

After performing a comparative assessment of all the available remedial technologies, it was decided that PRB, Pump and Treat (P&T), and MNA are the technologies that have the best potential for the site. They are all in situ technologies, which allow for continued use of the site as industrial/commercial establishment without having to dismantle the building and pavement that is already present.

SiteWise™ analysis was performed, and the results (Fig. 16.7) indicate that GHG emissions for PRB were lower than the Pump and Treat technology. NOx emitted from construction is of concern mostly for MNA well construction, and for PRB, these emissions were the lowest among all options evaluated. PM emissions

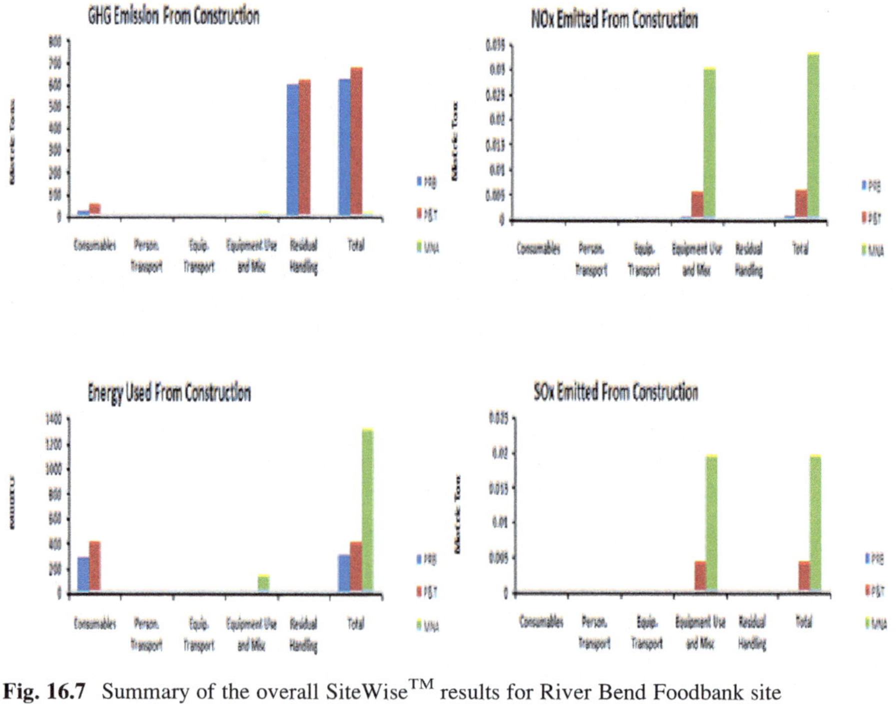

Fig. 16.7 Summary of the overall SiteWise™ results for River Bend Foodbank site

Table 16.7 Summary of investigated groundwater remediation technologies at River Bend site (SiteWise ™ results)

Technology	Total GHGs	NOx emissions	Energy used	SOx emissions
PRB	High	Moderate	Moderate	Moderate
P&T	High	High	Moderate	Moderate
MNA	Low	Moderate	High	Moderate

are highest for MNA, with PRB and P&T having comparable values. Energy used for construction is higher for MNA and P&T technologies, and SOxs are emitted in higher amounts from MNA well construction as well as from P&T method. Accident risk from construction shows higher percentages for P&T, followed by PRB. From the long-term monitoring SiteWise™ results, it was found that PRB and P&T methods have the highest GHG emissions at very comparable values. NOxs emitted during long-term monitoring are highest for P&T. PM emissions for long-term monitoring are a result of all three technologies, with PRB having the highest amounts and P&T coming in at a close second. Both P&T and PRB technologies have comparable values for the energy being used during long-term monitoring phase. SOxs emitted and accident risk also show similar values for both the aforementioned technologies. The overall SiteWise™ results for construction and monitoring activities are summarized in Table 16.7. Based solely on these SiteWise™ results, MNA and PRB prove to be desirable technologies which require further analyses.

Upon examination of case studies that have used P&T before, it was concluded that this is probably not an effective method of remediation for this site since the cost of this technology is too high ($1,000,000+). Moreover, this technology takes a long time, which is not suitable as the site was intended to be reopened by the City of Moline as soon as possible. MNA might not be the best option for remediation of this site because energy used and SOx emission are higher than with other technologies. These higher values might be due to a conservative approach in construction time and cost estimate, as well as due to installation of additional monitoring wells. Based on case studies, cost for PRB remediation technology would be reasonable ($380, 000), and taking into account that energy used and NOx and SOx emission are the lowest among these three technologies, PRB is a good remediation technology for this specific site, despite the fact that GHG emissions were relatively high.

16.4.5 Remedial Design and Monitoring Requirements

A trench approximately 15 ft deep and 150 ft wide is to be to the north of this site, mainly around borehole B-3, which was found to contain the highest levels of contaminants. A funnel-and-gate system was designed to be placed into the trench with the gate at a depth of approximately 10–12 ft. The funnel is made of

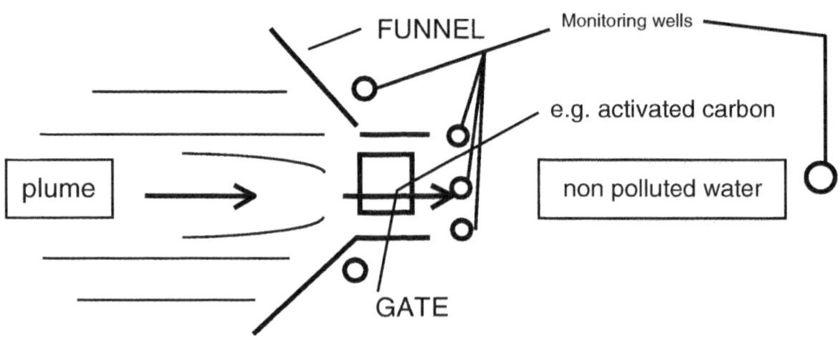

Fig. 16.8 Schematic of PRB design for River Bend Foodbank site with approximate locations of monitoring wells

non-permeable material that directs the captured contaminated groundwater flow toward the permeable gate. The permeable gate is approximately 4 ft thick and is filled with zeolites. Zeolites are microporous aluminosilicates commonly used as commercial adsorbents for metal contaminants. Zeolites have cavities, which maintain hydraulic conductivity, allowing their use in PRB.

After the PRB system is installed, a monitoring system consisting of a network of monitoring wells needs to be installed to evaluate adequate capture and treatment of the contaminated plume. A total of six monitoring wells are to be installed, and the system will be monitored for 2 years to insure that it is working properly. After that, samples need to be taken quarterly, or less depending on the stability of the system, to check if the gate has become saturated and is in need of replacement. A graphical representation of the design is shown in Fig. 16.8.

16.4.6 Recommendations

Based on the overall GSR analyses conducted for site-specific conditions, it is recommend that a PRB be installed, with zeolites in the gate, to remediate the groundwater contamination and prohibit any migration of contaminants to the Mississippi River.

16.5 Case Study 3: High School Site

16.5.1 Site Background

The proposed remediation site, high school site, is located between the Naperville Township and the City of Aurora, in unincorporated DuPage County approximately 35 miles west of the City of Chicago, Illinois. The site is 87.52 acres (3,812,371 ft^2) and is of irregular shape. Information from the preliminary site assessment report was used to perform the analysis for this site. Figure 16.9 displays the site location, shape, and surrounding features of the project site. The north half of the site is currently in a state of decommissioning the 16 peaker generators that operated on the site during the last half of the twentieth century. The south half of the site has been developed into Metea Valley High School, which opened in August 2009. A natural gas pipeline runs down the middle of the site from north to south. The pipeline is at an elevation deep enough to have minimal impact on existing infrastructure in place and any potential remediation solution. Additionally, the position of the pipeline on the site is west of the contaminated soils.

Review of historical information dating back to 1874 indicated that the site has been primarily used for agricultural purposes. A residential building was constructed on the southwest corner of the site in 1961 and remains standing to date. Commonwealth Edison (ComED) transformed the site from agricultural to electrical power generation in 1969 through a property acquisition. The property was broken into two sections with the northeast containing a peaker station and the eastern section a lineman training area. The remaining land remained agricultural. In 1999, Midwest Generation purchased the property, and St. John African Methodist Episcopal Church acquired the southern portion in 2004. The peaker station was constructed in 1969 to generate electricity on an as-needed basis during peak periods of high demand. Electricity was generated through the burning of fuel oil and/or natural gas in the peaker turbine units. A total of 16 peaker units were installed on the northeastern section of the property. The site discontinued electrical generation on September 24, 2004, and remains inactive to date. Six peaker turbine units have been completely dismantled with the other ten currently in various stages of dismantlement. The Indian Prairie School District, operators of Metea Valley High School, and other environmental groups indicate the desire to turn the contaminated portion of the site into a nature preserve and an area containing trails for walking, jogging, and biking. Current features of the site are indicated in Fig. 16.9.

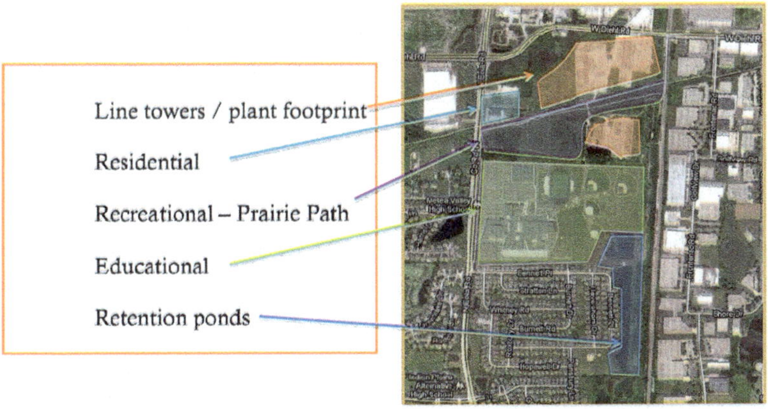

Fig. 16.9 High school site location and surrounding features

No environmental issues have been identified through the agricultural use at the site. Leaves that were placed at the site from the Lisle and Winfield Townships pose no significant environmental impacts. Interviews with the former farmers who cultivated the land indicate herbicides and pesticides were used, but at manufacturer's recommended rates. Additionally, there were no reports of any herbicide or pesticide spills. The peaker plant operated for 35 years (1969–2004) under control of ComED and Midwest Generation, during which time numerous spills occurred. Generally, it is uncertain based on the provided existing spill and cleanup information as to the amount of contaminants, the location of the contaminants, and magnitude of area that was affected at the site. Furthermore, other undocumented contaminations may have occurred. In total, five spills were documented, and one occurred in 1993, where 100 gal of fuel oil leaked influencing a volume of 20 feet by 60 feet by 6 in or 600 cubic yards of soil. As a result, 20 cubic yards of gravel was excavated to remove the contamination. The exact location of the spill was not documented. The second documented spill happened in February of 1994, when one of the peaker units leaked lubricating oil. The spill released 600 gal of lubricating oil and affected an area of 2500 square feet. The majority of the spill was cleaned by pumping the contaminants through an oil and water separator. The exact quantity of the spill that was recovered was not disclosed in the available data. The third documented spill transpired in June 1994 and 1000 gal of diesel fuel overflowed from a sump. The spill affected an area of 1000 square feet consisting primarily of gravel and concrete. However, the exact location of the spill is unavailable. The affected gravel was removed from the area and taken to a landfill for proper treatment. Through this treatment, 900 gal of diesel fuel was recovered. In 1996, 100 gal of diesel fuel and mineral oil were leaked from an oil and water separator. A hired remediation contractor used absorbents to soak the oil from the water surface. Through this process, four 55 gallons drums of oil and absorbent materials were disposed off-site. The estimated area of concern is approximately 20,000 square feet to 80,000 square feet.

The final documented spill happened in 2002 as one of the peaker units leaked an estimated 10,500 gal of fuel oil. This spill was the most severe case as the magnitude of the spill size was significantly larger than all aforementioned spill events. The severity of the spill caused an overflow of the drain tank that was able to hold 4500 gal of spilled fuel. As a result, Midwest Generation hired a remediation contractor to remediate the spill. The firm estimated that 8000 gal was recovered through a multitude of remediation technologies including flushing the drain tank, implementing an oil and water separator, using yard drains, pond discharge ditch, and utilizing underground drain lines. In addition, a biological reagent was added to the retention basin and ditch, and absorbent was added to the storm water sewer to control hydrocarbons that were still leaching in the sewer system.

IEPA indicated periodic generation of hazardous materials commonly associated with petroleum products to be at the site between 1985 and 1996. These hazardous materials include antifreeze, oil additives, and waste oils, which include F001, F002, F003, F004, and F005 (spent solvent) and D001 (ignitable), D002 (acid), D004 (arsenic), D007 (barium), D008 (lead), and D018 (benzene). Several federal and state regulations must be consulted to ensure compliance of remediation at the high school site.

16.5.2 Site Characterization

The geology of the site is mostly clay deposits with some concrete and fill from previous site activity. The saturated hydraulic conductivity of the project site soils ranges between 1.26×10^{-5} m/s and 3.17×10^{-6} m/s. It is uncertain based on provided data how deep bedrock is below surface elevation as the given soil borings only go to a depth of 15 ft. The existing site grade is negligible and nearly flat. Over the entire area, it contains an elevation change of 25 ft with an easterly gradient. The site lies on the Manhattan-Minooka Ground Moraine, which is a glacially formed accumulation of unconsolidated glacier debris consisting of soil and rock. Specifically, a ground moraine is an area of irregular topography with no ridges, generally forming rolling hills or plains. The borings illustrate that the groundwater table is between 3 feet and 12 feet bgs. The site is not in a FEMA floodplain.

During the initial site investigation, 96 borings were taken to evaluate potential subsurface environmental impacts. The site was tested for the following contaminants: BTEX, polychlorinated biphenyl (PCBs), ethylene glycol, volatile organic compounds (VOCs), semi-volatile organic compounds (SVOCs), pesticides, and priority pollutant metals. Altogether 41 chemicals were tested. Six monitoring wells were constructed to test groundwater conditions.

The test results showed that among 13 locations where samples were taken, six locations (the aboveground storage tank (AST), the oil and water separators, the maintenance shop, the oil storage shed, the septic fields, and the eastern property line) had not been impacted by previous releases. Yet, seven other locations (the detention pond, the antifreeze storage tanks, the fuel uploading pump house and hose connectors, the fuel forwarding pump house, 16 existing or former generating units, the power transformers, and the waist oil sump) had been detected to contain hazards. All seven locations were contaminated with PAHs. Areas adjacent to the 16 generating units had the most types of contaminants: BTEX, PAHs, PCBs, and metals. It is clear that previous releases of diesel and mineral oil had contaminated the site. However, neither pesticides nor ethylene glycol was detected in any samples analyzed.

The test results were compared to the TACO Tier 1 ROs with the residential standard. The ROs for soils are broken down into three separate exposure pathways: ingestion, inhalation, and the soil component of the groundwater ingestion. Table 16.8 summarizes the detected concentration of a specific chemical in a particular boring that was above the TACO Tier 1 ROs. Borings located adjacent to the 16 existing or former generating units contained benzene, naphthalene, PCB,

Table 16.8 Summary of COCs in soil exceeding TACO Tier I RO at high school site

COC (mg/l)	B-9	B-10	B-23	B-79	B-18	B-42	B-76	TACO Tier 1 RO	
Naphthalene	2.7	15.5	5.24			20.5	1.95	1.8	Construction worker inhalation
Benzo(a) anthracene					12	3.43	1.92	1.8	Residential ingestion
Benzo(a) pyrene					7	2.28		2.1	
Benzo(b) fluoranthene					5.96	2.15		2.1	
Dibenz(a,h) anthracene					0.773			0.42	
Indeno (1,2,3-c,d) pyrene					3.26			1.6	
Aroclor 1254		2.31						1	
Arsenic				19.4				13	
Benzene				0.0446		0.074		0.03	Soil component of groundwater ingestion
Total chromium				22.4				21	

and arsenic at levels above the TACO Tier 1 ROs. Boring adjacent to the fuel forwarding pump house east of the aboveground storage tank had five PAHs with unacceptable levels. Borings adjacent to the waste oil sump showed that levels of PAHs were too high, especially with naphthalene whose concentration far exceeded the TACO Tier 1 RO. Thus, these three locations were identified to require remediation. It can also be concluded that BTEX and PAH concentrations in the soil samples from the borings made to evaluate the detention pond did not exceed TACO Tier 1 ROs, although organic vapors and diesel odor were observed. Similarly, the AST hose connectors, the antifreeze storage tank, and the current and former power transformers had not been significantly impacted by the previous releases. With regard to the groundwater of the site, test results presented that PAHs were detected in test samples. However, the concentrations did not exceed the Class I remediation objectives of TACO Tier I on groundwater. Thus, it is concluded that the groundwater of the site had not been impacted.

16.5.3 Risk Assessment

Ten chemicals whose detected concentrations exceed the TACO Tier 1 ROs were recognized as hazards of concern, and they were found in the soil media. The contaminants were naphthalene, benzo(a)anthracene, benzo(a)pyrene, benzo(b) fluoranthene, dibenz(a,h)anthracene, indeno(1,2,3-c,d)pyrene, Aroclor 1254, arsenic, benzene, and total chromium, and their maximum detected concentrations are listed in Table 16.8.

A conceptual site model (CSM) was then developed to identify all potential or suspected sources of contamination, potentially contaminated media, potential routes of exposure, exposure pathways, and receptors. The exposure pathways include incidental ingestion, inhalation of particulates, and dermal contact. The potential receptors are residents, high school students, and construction workers. Based on the CSM, exposure assessment was carried out separately for both carcinogenic and noncarcinogenic effects via ingestion, inhalation, and dermal contact pathways. The toxicology data of hazards of concern was obtained primarily from USEPA's Integrated Risk Information System (IRIS). For noncarcinogens, the reference dose (RfD) or reference concentration (RfC) is used to express the daily exposure levels via oral and inhalation pathways, respectively. For carcinogens, the slope factor (SF) or the inhalation unit risk (IUR) is used to express the potential risk level associated with the exposure.

Baseline risk assessment on human health was performed based on the exposure assessment and toxicity data. For individual noncarcinogens, the non-cancer hazard quotient (HQ) was calculated. Hazard index (HI) was then obtained by adding up each individual HQ values for each exposure pathway. For individual carcinogenic risks, the probability of a person developing a cancer was calculated. The total carcinogenic risk was the sum of the carcinogenic risk for each hazard of concern. For HQ and HI, chemicals whose values are greater than one are considered to possibly cause adverse health effects. For pathway risk and total risk, values greater than 10^{-6} are considered unacceptable by USEPA, while values greater than 10^{-4} indicate severe health risks.

The baseline risk assessment shows that the total HI for the potential receptors is 3.19, while the total risk is 0.00074. Among individual chemicals, Aroclor 1254 causes most non-cancer health concerns, followed by inorganic arsenic. For carcinogenic health concerns, benzo[a]pyrene is the most problematic chemical, accounting for more than half of the total risk, while naphthalene is the only chemical that causes unacceptable inhalation risk. Based on this risk assessment, the site needs remediation to reduce the risks they expose the potential receptors.

In addition to human risk assessment, the ecology risk assessment was also carried out on the same hazards of concern. Chemicals whose concentrations exceed the screening ecological benchmarks were identified as affecting ecology. By comparison, inorganic arsenic, benz[a]anthracene, benzo[a]pyrene, total chromium, and naphthalene detected on the site would affect ecology. Among those five chemicals, the concentration of naphthalene detected was more than 200 times higher than the benchmark value. In addition, total chromium also exceeded the benchmark more than 50 times.

Finally, based on the risk assessment performed, remediation goals were set for the site to reduce the negative impacts of chemicals on human and environment. Although TACO Tier 1 ROs were used to identify hazards of concern, they were not selected to be the remedial goals for the site because the values in ROs are given as if it were the only chemical on the site, while on the Metea Valley High Site, there were multiple chemicals of concern. In order to be rendered acceptable, the total HI of all chemicals shall be less than one and total risk shall be less than 10^{-6}. Based on these two criteria and the same toxicity data, preliminary remediation goals (PRG) were set (Table 16.9).

16.5.4 Remedial Options

Many remediation alternatives where identified for the project site, as shown in Table 16.10. Ultimately, three technologies were selected for further analysis: electrokinetic (EK) remediation (Chinthamreddy and Reddy 1999) with ISCO injections, phytoremediation (Hinchman et al. 1997), and excavation. The alternatives were assessed against a variety of criteria which include cost, remediation time, environmental analysis, and final design considering site-specific conditions.

Results of the GREM analysis (not shown) indicate that EK is the most environmentally friendly option, followed by phytoremediation and then excavation, respectively. Site-specific assumptions for SiteWise™ analysis are listed in Table 16.11. The remediation time for the three alternatives varies largely. EK has a preliminary schedule taking 3 months of remediation followed by 2 years of monitoring. Phytoremediation is estimated to take between 6 and 10 years, with unknown expected results due to the clay soil and limited groundwater movement. The phytoremediation alternative provides the greatest unknown schedule as there cannot be a definite determination of remediation time. Excavation is estimated to take approximately 2 months to complete. For the shortest overall time duration, excavation will release many contaminants into the air, which is not acceptable around the high school and the residential neighborhoods, which needs to be accounted for if this alternative is selected. The results for SiteWise™ analysis are shown in Figs. 16.10 and 16.11.

After thorough consideration of all of available options for remediating the site using the analysis tool previously discussed, the method selected as the best option was EK with ISCO injection. Other in situ options were eliminated due to the clay soils present at the project site, and ex situ options were eliminated primarily due to cost.

16.5.5 Remedial Design and Monitoring

Based on site characterization, contaminant locations, and contaminant concentrations, seven hot spots were identified as shown in Fig. 16.12. A geometric square of 10,000 ft^2 is reasonably assumed for each hot spot. On each area of 10,000 ft^2, the design for EK remediation requires six rows of cathodes and five rows of anodes to

Table 16.9 Remediation goals for high school site

Chemical	Ingestion PRG TR = 1.0E-6 (mg/kg)	Dermal PRG TR = 1.0E-6 (mg/kg)	Inhalation PRG TR = 1.0E-6 (mg/kg)	Carcinogenic PRG TR = 1.0E-6 (mg/kg)	Ingestion PRG HQ = 1 (mg/kg)	Dermal PRG HQ = 1 (mg/kg)	Inhalation PRG HQ = 1 (mg/kg)	Non-carcinogenic PRG HI = 1 (mg/kg)
Aroclor 1254	0.319	0.723	5790	0.221	1.56	3.99	–	1.12
Arsenic, inorganic	0.426	4.5	769	0.389	23.5	279	21,300	21.6
Benz[a] anthracene	0.204	0.532	11,900	0.148	–	–	–	–
Benzene	11.6	–	1.19	1.08	313	-	119	86.2
Benzo[a] pyrene	0.0204	0.0532	1190	0.0148	–	–	–	–
Benzo[b] fluoranthene	0.204	0.532	11,900	0.148	–	–	–	–
Chromium, total	–	–	–	–	–	–	–	–
Dibenz[a,h] anthracene	0.0204	0.0532	1090	0.0148	–	–	–	–
Indeno [1,2,3-cd] pyrene	0.204	0.532	11,900	0.148	–	–	–	–
Naphthalene	–	–	3.57	3.57	1560	4300	156	137

Table 16.10 Evaluation of remedial options for high school site

	Technology	Disqualifier	Cost
In situ	Electrokinetics	None (or cost, depending on analysis)	\$90–\$130/ton
	Phytoremediation	None (or cost, depending on analysis)	\$100/ton
	Vitrification	None (or cost, depending on analysis)	\$350–\$700/ton
	Vapor extraction	Contaminants must be heterogeneously ineffective in low K soils	<\$100/ton
	Soil flushing	Ineffective in low K soils	\$80–\$165 yd^3
	Bioremediation	Ineffective in low K soils	\$27–\$310/ton
		Ineffective for metal contaminants	
	Soil heating	Ineffective in low K soils	\$50–100/ton
	Solidification	Ineffective in low K soils (EPRI 2003)	\$100–150/yd^3
	Thermal desorption	Ineffective in low K soils	\$74–\$184/ton
Ex situ	Soil washing	Not ideal for soil with greater than 20% fines	\$100–\$300/ton
	Incineration	Very expensive	\$500–\$1500/ton
	Air sparging	Contaminants must be heterogeneous ineffective in low K soils	\$3/gallon of groundwater

Table 16.11 Input for SiteWise™ analysis for high school site

General assumptions:		
7 hot spots (10,000 ft^2 each)		
50 mile commute for site workers in SUVs		
Diesel fuel for equipment		
Electrokinetics with ISCO injections	*Phytoremediation*	*Excavation*
2″ wells	Mulch, 4″	6 f. of removal
Soil water content (16% Moisture)	Fertilizer, 60 lb/acre	Haul
Borings:	Shrubs for shallow contaminants	15 yd^3 per dump truck
5 min per bore	20 site personnel	120 miles to authorized dump site
5 teams boring	1 safety	60 working days
8 days of boring	2 engineers	10 trucks
Energy:	9 laborers	2 daily trips per truck
44 kw/h for 120,680 ft^2		
3417.6 kw/h for 2 months (60 days)		

be installed, and for all seven hot spots, 462 cathodes and 385 anodes are required for remediation. Both cathodes and anodes are to be installed at a depth of 4 feet. The cathodes in each row are to be spaced 10 feet apart, and the rows are to be spaced 18 feet apart. In between the rows of cathodes, 55 anodes are to be installed

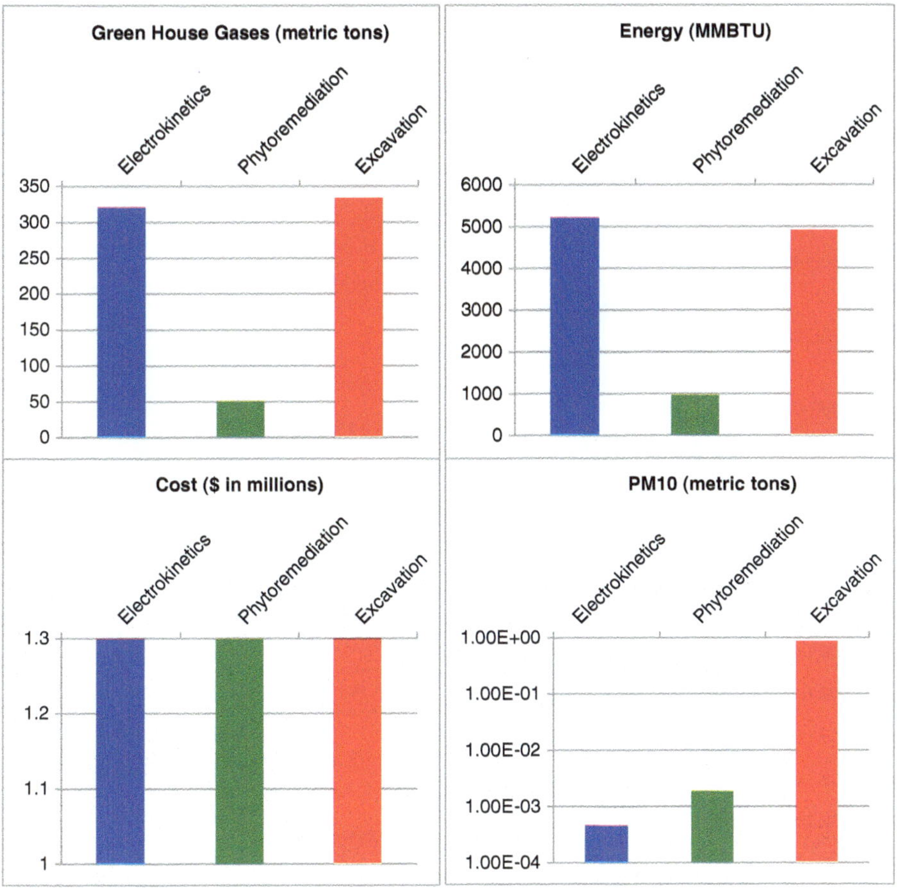

Fig. 16.10 SiteWise^{TM} analysis results for Metea Valley High School site

in five rows evenly spaced. In each row, the anodes are to be placed 10 ft apart. Figure 16.13 shows a plan view of the EK design for one hot spot, and Fig. 16.14 is a cross-sectional view of the design. Graphite was selected as the material for both cathodes and anodes as it is inert and thus will not add more contamination to the site. Compared to other materials, graphite has shown to be a serviceable replacement with limited to no efficiency loss.

The operating voltage will operate at 300 V and a current of 30 amps will be applied across the electrodes. Typically, field and pilot tests have indicated an operating voltage from 100 to 600 V and 15 to 50 amps to be implemented. While generally higher currents lead to better electroosmotic flow conditions, a middle number of 300 V and 30 amps were employed for the reason that it is deemed the lowest voltage possible to achieve cleanup objectives given the spacing of the electrodes. A balance between the operating voltage and the spacing of electrodes is necessary to maximize contamination removal and site remediation

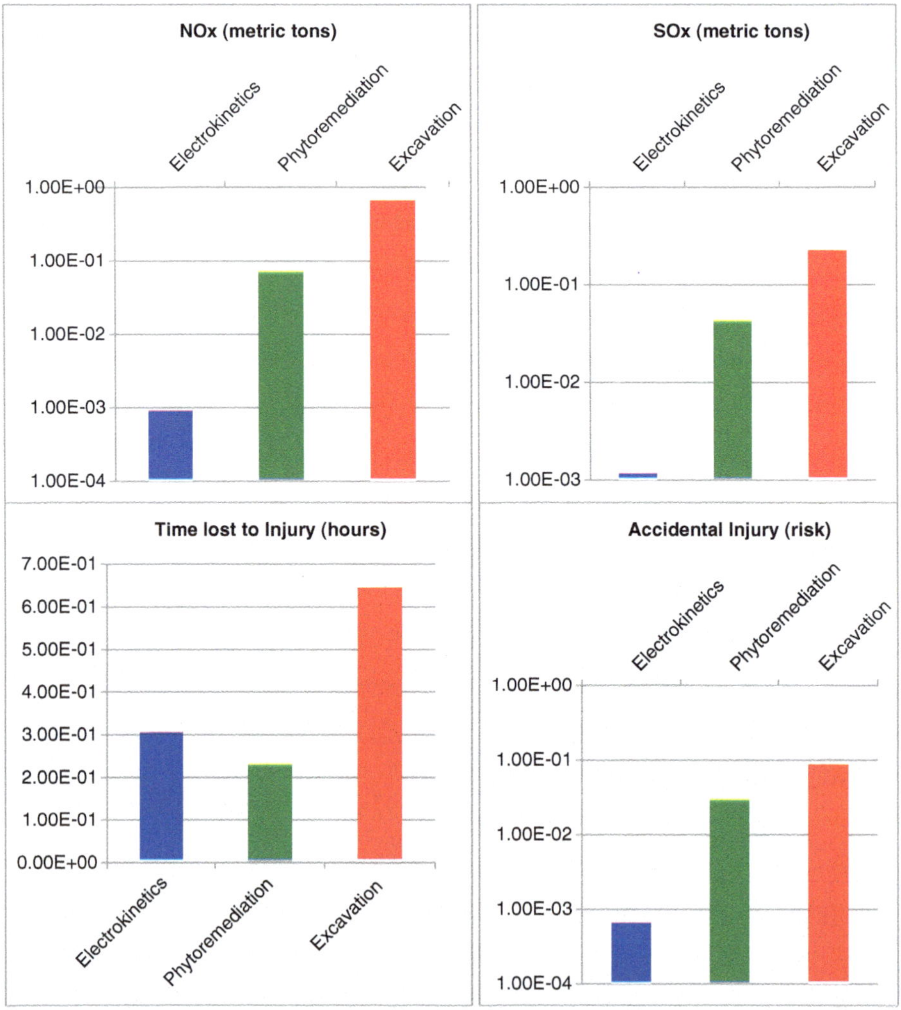

Fig. 16.11 SiteWise™ analysis results for Metea Valley High School site

economics. The closer the electrodes are spaced, the more wells will need to be constructed and more cost will be incurred. Although it may be easier to deliver currents of one amp per square foot of cross-sectional area between electrodes, it may potentially reduce the operating voltage required in a negative fashion. The recommended design has also taken into consideration the predictable current drop during the operation period; hence, it was determined that 300 V and 30 amps is sufficient and most economical.

Given that the site is contaminated with not only heavy metals but also polycyclic aromatic hydrocarbons (PAHs), in situ chemical oxidation is recommended along with electrokinetics to simultaneously oxidize organic

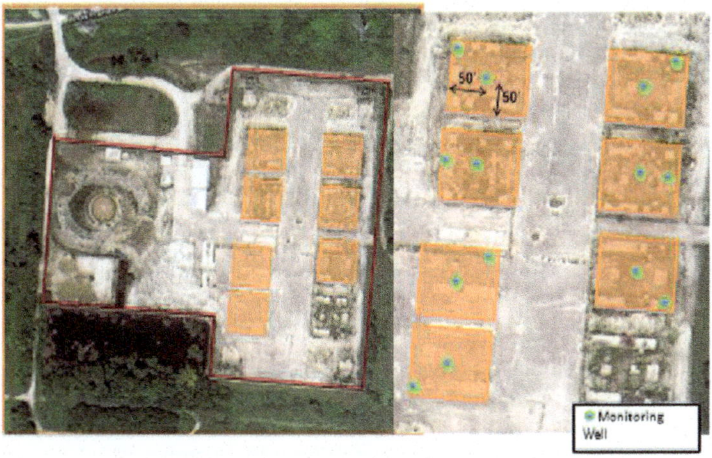

Fig. 16.12 Contaminated areas indicating hot spots and the locations of monitoring well at M

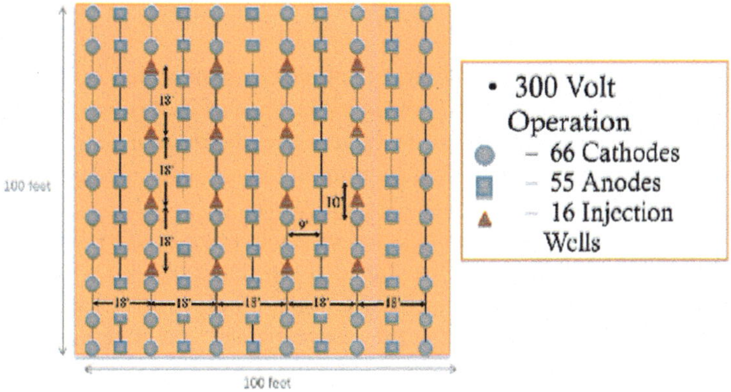

Fig. 16.13 Specifications for EK remediation design for on sample hot spot

hazardous materials and remove heavy metals. Based on research studies, Fenton's agent, composed of hydrogen peroxide and native iron catalyst, is recommended for chemical oxidation (Reddy and Karri 2008; Reddy and Chandhuri 2009). The desirable concentration of peroxide is 30%, as it was demonstrated to oxidize PAHs effectively at this level.

Sixteen ISCO injection wells are to be placed in a 4 foot × 4 foot square on each hot spot. Each injection well is spaced 18 feet from another horizontal or vertical injection well. EDTA, a chelate agent, is recommended to be the enhancement solution in the EK system to avoid premature precipitation of metals at the cathodes when pH changes in soil. Two injections are recommended, one at the beginning of the project and one 30 days after the initial injection. After the EK system is installed on the site, operation can begin during month 2, by applying the

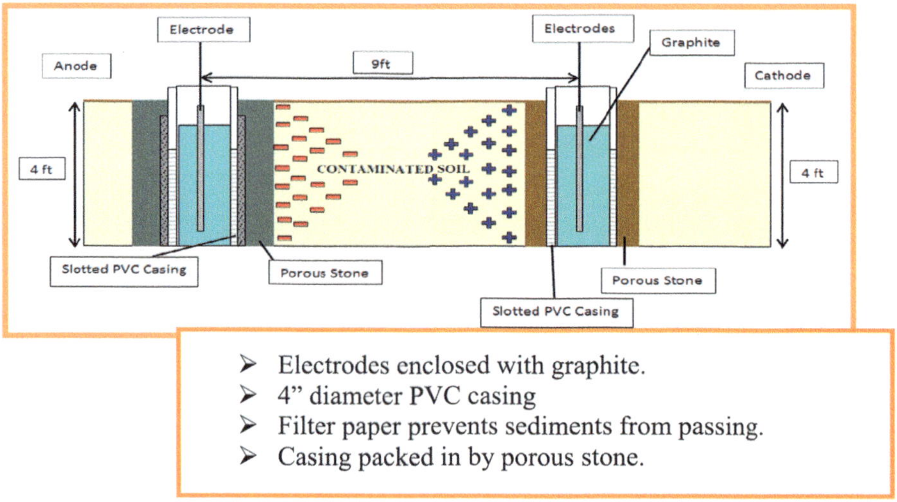

Fig. 16.14 Cross-sectional view of EK remediation design proposed for Metea Valley High School site

aforementioned designed voltage and Fenton's reagent added in simultaneously. The reservoirs shall be monitored during this process to observe the remediation effectiveness. If the removal efficiency is less than the desired goal (the operational time is set to be 90 days), considerations should be given to changing electrolytes, changing electric potential, or even electrode spacing.

Two reservoirs are also recommended on each hot spot to monitor the remediation process. One reservoir is to be placed at the center of each hot spot, while the other reservoir is placed at the peripheral as shown in Figs. 16.12 and 16.13. In this fashion, samples may be taken from the reservoirs and will be representative of the entire hot spot to determine removal efficiency. The EK system can be turned off when ROs are met. Continued monitoring will be applied for 2 years to determine if any rebound has occurred in the soils. The probability of rebound is negligible as groundwater movement is minimal and contaminants will not be moving. The preferred alternative is designed to remove all contaminants in the given schedule of 3 months and further remediation should not be required. Once the site is clean, electrodes will be removed and wells abandoned. Post-remediation monitoring shall continue for 2 years after the completion of the EK and ISCO treatments to the soil. If any rebound is observed, though highly unlikely, additional remedial technologies should be applied. However, the recommended system is designed to effectively remove all the contaminants; therefore, the potential for additional remediation is minimal. Based on the site characteristics, primarily a clay soil structure, unaffected groundwater, shallow contaminant depths, and a quick remediation time required, EK with ISCO injections is recommended for the high school site.

16.6 Conclusions

Overall, this paper clearly demonstrates the application of GSR techniques at three different contaminated sites in Illinois. It is critical to understand the technicalities of GSR and employ the right methodologies for selecting, designing, and implementing the most suitable technology for the site. In some cases, the combined use of more than one remedial option (remedial train) can prove to be more sustainable as opposed to a traditional approach. BMPs have to be carefully developed to minimize the environmental, economic, and social impacts that are likely to arise during the various stages of the remedial project. In terms of future direction, more effective remediation technologies that can be employed to clean up a broad spectrum of mixed contaminants and can address the current challenges pertaining to their application in clayey soils and in complex geological settings need to be developed. More focus needs to be placed on the advancement of cleaner technologies that currently prove to be sustainable (e.g., bio-augmentation, bio-stimulation, bio-cementation, phytoremediation, electrokinetic remediation). New approaches of combining different technologies based on site-specific conditions and reducing the overall time invested in the cleanup have to be further researched. Additionally, the absence of guidance on standardized sustainability metrics and assessment tools for GSR implementation can limit its effectiveness and application. The development of a sustainable remediation framework will allow the selection and optimization of all phases of remedial action, leading to a net benefit to environment, society, and economy. Industry and government stakeholders should continue to promote sustainability in site remediation.

Acknowledgment The assistance of graduate students in Environmental Remediation Engineering course is gratefully acknowledged.

References

Adams JA, Reddy KR, Tekola L (2011) Remediation of chlorinated solvent plumes using in-situ air sparging. Int J Environ Res Public Health 8(6):2226–2239

AFCEE (Air Force Center for Environmental Excellence) (2010) Sustainable remediation tool user guide. United States Air Force

Bhargava M, Sirabian R (2011) SiteWise^{TM} version 2 user guide. NAVFAC Engineering Service Center, Port Hueneme

Chinthamreddy S, Reddy KR (1999) Oxidation and mobility of trivalent chromium in manganese enriched clays during electrokinetic remediation. J Soil Contam 8(2):197–216

DTSC (California Department of Toxic Substances Control) (2007). Green remediation. www.dtsc.ca.gov/OMF/Grn_Remediation.cfm

Electric Power Research Institute (2003) Evaluation of the effectiveness of in situ solidification/stabilization at the Columbus, Georgia manufacturing gas plant site, EPRI, Palo Alto, CA: 2003. 1009095: pp 8–1 and 8–2

Ellis DE, Hadley PW (2009) Integrating sustainable principles, practices, and metrics into remediation projects. Remediat J 19(3):5–114

EPA (2010) Green remediation. www.clu-in.org/greenremediation

Favara PJ, Krieger TM, Boughton B, Fisher AS, Bhargava M (2011) Guidance for performing footprint analyses and life-cycle assessments for the remediation industry. Remediat J 21 (3):39–79

Hinchman RR, Negri MC, Gatliff EG (1997) Phytoremediation: using green plants to clean up contaminated soil, groundwater, and wastewater. Submitted to the U.S. department of energy, assistant secretary for energy efficient and renewable energy under contract W-31-109-Eng-38

Interstate Technology Regulatory Cooperation (ITRC) (2001) In-situ chemical oxidation work team, technical and regulatory guidance for In Situ chemical oxidation of contaminated soil and groundwater, 1st Edn. Washington DC

Interstate Technology Regulatory Cooperation (ITRC) (2005) In-situ chemical oxidation work team, technical and regulatory guidance for In Situ chemical oxidation of contaminated soil and groundwater, 2nd Edn. Washington DC

Interstate Technology & Regulatory Council (ITRC) (2011) Green and sustainable remediation: state of the science and practice. GSR-1. Washington DC: Interstate Technology & Regulatory Council, Green and Sustainable Remediation Team. www.itrcweb.org

ITRC (Interstate Technology & Regulatory Council) (2011) Green and sustainable remediation: a practical framework, *GSR-2*. Washington DC

Reddy KR (2011) Green and sustainable remediation. Proceedings 26th international conference on solid waste technology and management, Philadelphia, PA, 316–334

Reddy KR and Adams JA (2001) Cleanup of chemical spills using air sparging. In: Fingas M (ed) Chapter 14 in handbook of chemical spill technologies, McGraw-Hill, New York

Reddy KR, Chandhuri KS (2009) Fenton-like oxidation of PAHs in clayey soils using electrokinetics. J Geotech Geoenviron 135(10):1429–1439

Reddy KR, Karri MR (2008) Effect of oxidant dosage on integrated electrochemical remediation of contaminant mixtures in soils. J Environ Sci Health 43(8):881–893

Reddy KR, Adams JA and Richardson C (1999) Potential technologies for remediation of brownfields. *Practice Periodical of Hazardous, Toxic, and Radioactive Waste Management*. ASCE 3(2):61–68

Sharma HD, Reddy KR (2004) Geoenvironmental engineering: site remediation, waste containment, and emerging waste management technologies. Wiley, Hoboken

Siegrist RL, Crimi M, Simpkin TJ (2011) In situ chemical oxidation for groundwater remediation. Springer, New York

USEPA (2008) Green remediation: incorporating sustainable environmental practices into remediation of contaminated sites. EPA 542-R-08-002, Office of Solid Waste and Emergency Response, Washington DC

USEPA (2011) RCRA laws and regulations, USEPA Website, www.epa.gov/waste/laws-regs/index/htm

USEPA CLU-IN (2011) In-situ chemical oxidation overview, USEPA CLU-IN Website, www.clu-in.org

Chapter 17
Disinfection of Water Using Pulsed Power Technique: Effect of System Parameters and Kinetic Study

Raj Kamal Singh, Vigneshwar Babu, Ligy Philip, and Sarathi Ramanujam

Abstract The present study investigated the use of pulsed power technique for disinfection of water under different operating and environmental conditions. Final concentrations of reactive oxygen species (ROS) like hydroxyl radical, hydrogen peroxide, ozone and superoxide radicals generated in the system were found to be 56 mg/L, 17 mg/L, 1 mg/L and 18 mg/L, respectively, for an applied voltage of 23 kV, frequency of 25 Hz and a streamer discharge time of 12 min. It was observed that disinfection efficiency was high with sequential stress compared to continuous stress. The disinfection efficiency increased with increasing applied voltage and frequency. Disinfection efficiency was high when pH was less than 7. The presence of alkalinity, natural organic matter and turbidity reduced the disinfection efficiency significantly. For 7 log reduction of *E. coli*, the treatment time was increased from 6 min to 10 min, when pH was increased from 4 to 9. Complete disinfection of *E. coli* was achieved in a short treatment time of 4 min to 10 min, with an energy consumption of 0.0056 kWh to 0.014 kWh for 50 mL of contaminated water. An empirical model for optimum disinfection efficiency was developed using Box-Behnken design (BBD). As per the model, applied voltage, time of treatment and alkalinity were found to be the most significant factors affecting the disinfection efficiency. The model predicted values were in good agreement with the experimental values. Rate constant for disinfection and ROS formation was also evaluated. Rate of disinfection was between 0.59 and 1.68 log(cfu/mL)/min.

Keywords Disinfection • Pulsed power technique • Streamer • Reactive oxygen species • Alkalinity • Natural organic matter

---

R.K. Singh • L. Philip (✉)
Department of Civil Engineering, Indian Institute of Technology Madras, Chennai 600036, India
e-mail: ligy@iitm.ac.in

V. Babu • S. Ramanujam
Department of Electrical Engineering, Indian Institute of Technology Madras, Chennai 600036, India

G.L. Sivakumar Babu et al. (eds.), *Sustainability Issues in Civil Engineering*, Springer Transactions in Civil and Environmental Engineering,
DOI 10.1007/978-981-10-1930-2_17

17.1 Introduction

About 80 % of the diseases in developing countries are caused due to the consumption of poor-quality water (Thewaterproject.org et al. 2015). Around 11 % of the world population lacks access to safe drinking water, and more than 500,000 children die every year due to diarrhoea (World Health Organization 2012). Drinking contaminated water leads to several diseases such as cholera, jaundice, typhoid, malaria, dengue, yellow fever and river blindness. Some researchers have projected that almost 135 million people would die by the year 2020 because of pathogen-contaminated drinking water (Gleick 2002).

Chlorination is the most commonly used technique for disinfection. However, the application of this technique is reducing because of the formation of carcinogenic disinfection by-products like trihalomethanes, haloacetic acids and organochlorine compounds in water (Tian et al. 2013; Harvey 2011; Zhang et al. 2005). Other disinfection processes like ultraviolet (UV) irradiation, ozonation and gamma irradiation are expensive. Also, it is reported that these processes are inefficient in inactivating pathogens like *Cryptosporidium* (Szewzyk et al. 2000). Among several other methods, solar water disinfection (SODIS) is a simple and low-cost disinfection technique, mainly used for the removal of bacteria and viruses (McGuigan et al. 2012). However, this technique is not effective for highly turbid water with high inorganic and organic matter (Sichel et al. 2007). In addition, it cannot inactivate protozoan cysts and spores of some bacteria (McGuigan et al. 2012). Advanced oxidation processes (AOPs) are emerging as alternative technologies for the treatment of water contaminated with various pollutants. Several AOPs like Fenton, photo-Fenton (Ortega-Gómez et al. 2012), H_2O_2/UV (Penru et al. 2012), O_3/UV (Summerfelt et al. 2009), O_3/UV/Ag-TiO_2 (Wu et al. 2011), electrochemical process (Jeong et al. 2009; Schaefer et al. 2015) and photocatalysis (Marugán et al. 2008; Gao et al. 2013) have been applied for disinfection. Most of the AOPs have certain advantages and disadvantages in terms of efficiency and energy consumption (Ribeiro et al. 2014). Also, the presence of natural organic matter (NOM) like humic acid, bicarbonates, carbonates and other ions in water affects the performance of AOPs (Crittenden et al. 2012). Therefore, assessing the effects of these water constituents on the performance of AOPs is indeed important.

In recent times, plasma technology has been gaining importance because it exhibits a rapid disinfection potential (Rossi et al. 2009; Roth et al. 2010). Pulsed power technique (PPT) is a process by which energy is accumulated over a relatively long time and then it is released rapidly, thereby increasing the instantaneous power. Pulsed high-voltage streamer discharge in air and water generates plasma, that is, a mixture of highly reactive atoms, free radicals, electrons and certain photons (Locke et al. 2006). Plasma generated by streamer discharge diffuses into the water and forms reactive oxygen species (ROS), which efficiently degrades the organic pollutants and inactivates the pathogens. Shock waves and UV light generated along with the ROS, in turn, increase the disinfection efficiency (Locke et al. 2006). Study of bactericidal effects of the corona discharge was initiated by Kuzmichev et al. (2000). Several studies on the application of plasma for bacterial inactivation have been recently reported (Odic et al. 2002; Wang et al. 2008; Julák et al. 2012). Inactivation of vegetative forms and spores of *B. subtilis* using corona discharge was reported by Joubert et al (2013). Inactivation of viruses such as bacteriophage MS2 by streamer corona (Lee et al. 2011) and inactivation of *Cryptosporidium parvum* by pulsed UV with pulsed-plasma gas discharge have also been reported (Hayes et al. 2013). The information of the effect of applied pulse, wave characteristics and energy estimates of these plasma processes for water disinfection is limited.

Information on disinfection of water in the presence of high turbidity, alkalinity, NOM and different pH conditions by plasma treatment processes is scanty. It is essential to understand the effect of these parameters on the formation of ROS during the treatment. Optimisation of operating conditions of PPT such as voltage, frequency and time, along with environmental parameters such as pH, organic content and alkalinity, for effective disinfection is seldom carried out. The primary objective of the present work is to study the disinfection efficiency of PPT in various environmental conditions and quantification of ROS during the disinfection process. In order to determine optimal conditions for disinfection, an attempt was made to develop an empirical model for disinfection using Box-Behnken Design (BBD) by incorporating various operating parameters (voltage, frequency, time) and environmental parameters (pH, alkalinity and organic matters) simultaneously. The effect of intermittent and continuous discharge on disinfection efficiency was also investigated.

17.2 Materials and Methods

17.2.1 Growth Media and Bacterial Cultures

Escherichia coli was used as the model organism in all bacterial disinfection studies. A single colony of *E. coli* was isolated from tryptone bile glucuronic agar (Himedia, India) plate and inoculated in 100 mL of Luria-Bertani (Himedia, India) liquid media in a 250-mL conical flask. The culture was incubated overnight at 35 °C in an orbital shaker at 150 rpm. 10 mL of culture was then taken and centrifuged at 5000 × g for 10 min at 4 °C. The pellet was washed three times using physiological saline and resuspended in physiological saline (0.85 % NaCl) and diluted to 50 times for performing the experiment. Initial cell concentration of 10^7 cfu/mL was used for all experiments.

17.2.2 Chemicals

All the chemicals used in the study unless specifically mentioned were procured from Rankem, India. Bentonite clay was used for the preparation of turbid water. N, N Diethyl - 1, 4 Phenylenediamine Sulfate (DPD) kit (Prerana Laboratories, Pune, India) was used for ozone quantification. Titanium oxysulphate and potassium superoxide were procured from Sigma-Aldrich, India. Nitro blue tetrazolium salt (NBT) (Loba, India) and ethanol (Jiangsu Huaxi International, India) were used for quantification of superoxide. Uridine (Loba, India) was used for UV detection.

17.2.3 Experimental Setup and Reactor Configuration

The experimental setup was divided into two sections: (a) high-voltage source (Fig. 17.1a) and (b) reactor cell (Fig. 17.1b). The discharge-free test transformer (100 kV, 5 kVA) was used for the generation of high AC voltage. The generated AC voltage was converted to DC voltage by the use of high-voltage diode (140 kV, 20 mA, 100 kΩ). DC voltage was subsequently measured using resistor divider. The constant DC voltage was converted to square pulse using rotating spark gap (RSG). Also, the duration of pulse was controlled by varying the speed of the RSG. Shape and magnitude of generated square wave pulse were recorded using high-voltage probe connected to the oscilloscope (HP 54645A, 100 MHz). The measured rise time of the pulse was 0.4 μs and duration of the pulse was 14 ms, 17 ms and 20 ms for frequency of 30 Hz, 25 Hz and 20 Hz, respectively. A typical shape of generated voltage wave is shown in Fig. 17.1c.

In the present study, streamer discharge was achieved by using needle-plate configuration. Multiple needles were attached to the circular plate, which is connected to the high-voltage terminal. The circular-plane bottom electrode was connected to the ground terminal. The high voltage and ground electrode were fixed in a cylindrical container filled with water. 50 mL of water was used in all studies. The needle was made up of tungsten electrode. The radius of curvature of the tip was about 50 μm. This ensured continuous streamer propagation in the liquid. The gap between needle tip and water surface in the container was adjustable between 3 mm to 8 mm. The applied voltage was fixed in such a way that the visual streamer propagation was observed during the test. A water jacket was provided to avoid temperature rise, and it was maintained between 30 °C and 40 °C for all the experiments. A thermal vision camera was fixed to observe any temperature increase during the test period. Two ports were provided in the top electrode: one for injecting sample and the other for extracting the sample for characterisation.

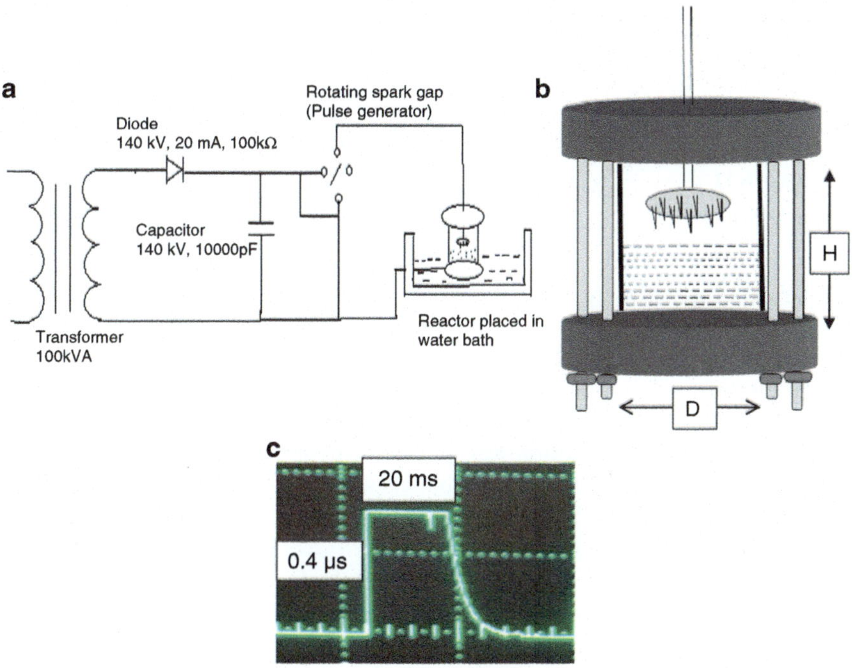

Fig. 17.1 Diagrammatic sketches for pulsed streamer discharge setup configuration. (**a**) High-voltage source, (**b**) reactor cell ($D = 60$ mm – effective diameter of reactor; $H = 70$ mm – effective height of reactor; needle spacing = 15 mm), (**c**) a typical pulse characteristics

17.2.4 Batch Studies

In order to study the effect of different operating parameters (voltage, frequency and time) and environmental parameters (pH, alkalinity, turbidity, humic acid and easily degradable organic matter), batch disinfection studies were performed. The details of experiments are presented in Table 17.1. Experiments were carried out at an initial cell concentration of 10^7 cells/mL in physiological saline (0.85 % NaCl) of conductivity 45 mS/cm to 47 mS/cm. In this study, initial *E. coli* concentration was 10^7 cells/mL, which was much higher than the concentration found in the source of drinking water (10^1–10^2 cells/mL). From the preliminary experiments, we found that the complete disinfection was achieved within 2 min of streamer discharge for the initial cell concentration of 10^1–10^2 cells/mL. Therefore, it became difficult to collect enough samples and analyse to plot proper trend line. Moreover, effects of relevant system variables might not be justified clearly with very less treatment time. This was the reason for employing high alkalinity values also. After 2, 4, 6, 8 and 10 min discharge times, treated samples were analysed by plating on tryptone bile glucuronic agar. Plates were incubated at 37 °C for 24 h, and colonies appeared on the plates were manually counted. Appropriate control experiments were carried out to assess the adverse effects of pH and alkalinity.

Table 17.1 Three levels for each of the parameters

Parameters	Lower level	Medium level	Higher level
pH	4	7	9
Alkalinity (mg/L as $CaCO_3$)	200	400	600
Glucose (mg/L)	10	50	100
Turbidity (NTU)	2	5	10
Humic acid (mg/L)	1	3	5

Similar experiments were also conducted to evaluate the effect of voltage on ROS formation. These experiments were performed at 23 kV and a frequency of 25 pulse/s.

The formation rate of ROS and energy efficiency is of importance in evaluating the performance of a reactor and for the industrialisation of the technology. Initially, the order of formation was determined for all ROS. The rate of reaction was then estimated by fitting a curve between the concentration of ROS and time. The rate constants of the reactions (formation) were determined by the slope of graphs at different applied voltages.

The energy consumption for the present technique was determined using Eq. 17.1 and Eq.17.2 (Kim et al. 2014). By integrating the voltage (U) and current profile (I) over one pulse cycle, deposited energy per cycle can be calculated (Eq. 17.1). Energy per second can be calculated using Eq. 17.2, and multiplying with treatment time provides total energy consumption for the treatment period.

$$\mathrm{E_c/pulse} = \int_{I=0}^{I=2\pi} U(t)I(t)dt \tag{17.1}$$

$$\mathrm{E_c/second} = \int_{I=0}^{I=2\pi} U(t)I(t)dt \times f \tag{17.2}$$

where E_c is the energy consumption (J), $U(t)$ and $I(t)$ are the instantaneous voltage and current, respectively, and f is the frequency applied (Pulse/s).

17.2.4.1 Evaluation of Disinfection Efficiency for Different Operating Cycles

Three modes of operations for streamer discharge were adopted for the disinfection study. In the first mode of intermittent operation (2 + 1), initially 2 min of continuous discharge was given, followed by 1 min without any discharge. This cycle was repeated five times for a total discharge time of 10 min. Similarly, for the second intermittent mode of operation (5 + 2), the bacterial culture was exposed to 5 min of continuous discharge and 2 min without discharge, for two cycles. In the third mode of operation (10 + 0), the bacterial culture was exposed to 10 min of continuous discharge. All the experiments were performed at 20 kV and a frequency of 25 pulse/s. Further, the effect of voltage was also studied by conducting the experiment with three different input voltages of 17 kV, 20 kV and 23 kV.

17.2.5 Analytical Procedures

17.2.5.1 Reactive Oxygen Species Quantification

Estimation of Hydroxyl Radical

Quantification of hydroxyl radical was carried out using derivatisation as per Tai et al. (2004) and Sahni and Locke (2006). Derivatisation is the technique used to transform a chemical compound into its derivative compound of similar chemical structure. The transformed product may have different chemical properties, which can be used for the separation or elution of the derivatives. Dimethyl sulfoxide (DMSO) is highly water soluble and able to trap most of the generated OH radicals and form methyl radicals and methanesulfinic acid as primary intermediate products (Eq. 17.3). The methyl radicals react further with oxygen molecules to form formaldehyde (Eqs. 17.4 and 17.5). In the present study, the reaction of OH radicals with DMSO leads to the formation of formaldehyde that was unstable at room temperature. Therefore, derivatisation process was used to impart stability to formaldehyde. From the reaction stoichiometry, Eqs. 17.3, 17.4 and 17.5 indicate that 2 mol of $OH^{\cdot}$ yields 1 mol of formaldehyde.

$$(CH_3)_2SO_2 + \cdot OH \rightarrow CH_3SOOH + CH_3 \tag{17.3}$$

$$CH_3 \cdot + O_2 \rightarrow CH_3COO\cdot \tag{17.4}$$

$$2CH_3OO \cdot \rightarrow HCHO + CH_3OH + O_2 \tag{17.5}$$

Dimethylsulfoxide (DMSO) having 28.2 mM concentration was dissolved in double-distilled water to ensure sufficient availability of DMSO for the trapping of OH radicals which were formed during streamer discharge. To avoid error in estimation, enough DMSO concentration should be available to trap OH radicals and should avoid the side reactions of the OH radicals with other species in the system (Sahni and Locke 2006). Also, the optimum pH for the reaction of formaldehyde and dinitrophenylhydrazine (DNPH) is 2–6 (Tai et al. 2004). In the present study, the solution pH was fixed in acidic range (~ 5.5), for the derivatisation reaction. Initial solution conductivity was 120 μS/cm. For the derivatisation process, 5 mL of the aqueous solution from the reactor was withdrawn at 0, 2, 4, 6, 8 and 10 min intervals and was mixed with 5 mL of pH 4 buffer ($NaH_2PO_4 + H_3PO_4$) and 1 mL of saturated solution of dinitrophenylhydrazine (DNPH). All the reaction mixtures were kept in an oven at 40 °C for 1 h to ensure that complete reaction took place to form hydrazone. Reaction mixtures were then equilibrated to room temperature and analysed by the HPLC–UV/vis (Dionex, ultimate 3000, USA) at 360 nm. Analysis was performed on a reverse phase high-performance liquid chromatography (HPLC) unit with an LC-18 column using 70:30 ratios of acetonitrile and water with 0.5 % acetic acid solution.

Estimation of Hydrogen Peroxide (H_2O_2)

Colorimetric determination of hydrogen peroxide concentration was performed by mixing 3 mL of aqueous solution from reactor and 1.5 mL of titanium(IV) - oxysulfate-sulphuric acid solution and analysing the concentration of resulting perititanic acid complex at 410 nm in spectrophotometer (Shimadzu, Japan) (Sahni and Locke 2006).

Estimation of Ozone (O_3)

Ozone is very unstable in water. Hence, ozone test kit was used for colorimetric determination of dissolved ozone in water by DPD (N, N-diethyl-p-phenylenediamine) method (Hayes et al. 2013). Distilled water of conductivity 5 μS/cm was used for the hydrogen peroxide and ozone quantification.

Estimation of Superoxide Radicals ($O_2^{\bullet\,-}$)

The analytical method described by Goto (2004) was used to detect the superoxide ion by nitro blue tetrazolium salt (NBT). NBT reacts with superoxide radicals to form purple formazan. During the study, NBT solution was prepared by dissolving 408.82 mg of NBT in 1 L of Millipore water along with 10 % ethanol. 0.1 M of 1-butanol was added to ensure the scavenging of hydroxyl radicals produced during the study. Also, hydrogen peroxide (generated) does not react with NBT (Goto 2004). NBT can degrade by UV during the discharge, but UV results show that the UV incident flux is as low to degrade NBT. Also, hydrogen radicals ($H^{\cdot}$) generated during streamer discharge can also react with NBT but it was shown that the probability of hydrogen radical diffusing into the bulk is quite low. Also, it is highly likely that in the oxygenated environment in and around the discharge, it will be converted to $HO_2^{\bullet}$ and subsequently to $O_2^{\bullet-}$ that diffuses and reacts in the bulk (Sahni 2006).

Initial conductivity for the reaction mixture was 116 μS/cm. After streamer discharge, 2 mL sample was drawn from the reactor at regular intervals and mixed with 2 mL of methanol to completely dissolve the formazan. At the end, the reaction mixtures were analysed using a spectrophotometer (Shimadzu, Japan) at 530 nm.

17.2.5.2 UV Quantification

In the study, uridine was used as a probe to determine UV incident flux, and its absorption maxima is 262 nm, which falls in UVC region (100–280 nm) (Sonntag and Schuchmann 2002). Therefore, only UVC was determined from the experiments that can cause cellular damage. Uridine molecule interacts with UV radiation and is converted to its photohydrate. Hence, the concentration of uridine decreases with an increase in UV dose. From the preliminary experiments conducted, it was found that the OH radicals also reacted with uridine and decreased the uridine concentration. In order to avoid ˙OH reaction with uridine, a quartz cuvette filled with a uridine solution was placed at a position inside the reactor such that it could receive a large fraction of the UV radiation from plasma channels. After exposure for a specified time, the absorbance of the solution in the cuvette was measured and compared with the absorbance before treatment. UV incident flux was calculated using the method suggested by Sonntag and Schuchmann (2002).

17.2.6 Model Development for Disinfection

To study the combined effect of different environmental (pH, alkalinity, NOM) and operating parameters (voltage, time and frequency) on bacterial deactivation, response surface methodology (RSM) was adopted. Among several environmental factors, alkalinity and pH are chosen as parameters for model development because the presence of CO_3^{2-} and HCO_3^- ions in water always contributes to alkalinity, whereas variation in pH significantly affects the water quality. Apart from these, glucose was also chosen as a model parameter to represent the easily biodegradable organic matter present in water. Turbidity was not included because it usually gets removed during secondary water treatment process. Moreover, the concentration of turbidity in tap water is negligible. Among several RSMs, Box-Behnken design (BBD) was used in the present study for optimisation of disinfection. As compared to other designs such as centre composite design (CCD) and three-level full factorial design, BBD is better because it is more efficient and requires less number of experiments (Ferreira et al. 2007). In the present study, BBD was developed by incorporating three environmental factors, i.e. pH, alkalinity and glucose concentration, and three operating factors, i.e. voltage rating, frequency and time. The factor labels required for the BBD optimisation is given in Table 17.2. Different levels of factors (parameters) such as "higher level", "moderate level" and "lower level" for each parameter were denoted by "+1", "0" and "−1", respectively. Different levels were fixed for each parameter, so that its effects on the disinfection efficiency could be studied. Treatment combinations and observed responses are presented in Table 17.A.1 (supplementary material). Statistical analysis was carried out to determine and predict disinfection efficiency (log reduction of CFU/mL) of the system in the presence of different operating and environmental conditions. The

Table 17.2 Actual factor level corresponding to coded factor level for BBD

		Actual factor levels at coded factor level of		
Factor	Symbol	+1	0	−1
Voltage (kV)	A	23	20	17
Frequency	B	30	25	20
Time (min)	C	9	6	3
Alkalinity (mg/L)	D	9	7	5
pH	E	600	400	200
Glucose (mg/L)	F	20	10	0

Note: Different factor levels such as "+1", "0" and "−1" were used to denote "higher level", "moderate level" and "lower level" of each parameter

data were analysed by the Design-Expert 9 software and a regression model was developed. The regression model permitted evaluation of the effect of linear, quadratic and cubic interaction terms of the independent variables (voltage, frequency, time, pH, alkalinity and glucose concentration) on the response. The level of significant α-value (0.05) was set for every term in the model.

17.2.7 SEM Imaging of Bacterial Kill and DNA Damage Study

Samples for the bacterial imaging were prepared by taking 2–5 μL of the sample at different time intervals of the treatment, at a voltage of 23 kV. The sample was made as a thin smear and air dried prior to the imaging. The glass slide was then coated with gold for good conduction. SEM images were obtained using Quanta FEG 200. The DNA damage was confirmed by performing the DNA extraction and analyses of bacteria before and after treatment. Bacterial genomic DNA isolation kit from Helini Biomolecules, Chennai, India, was used for the extraction by following the supplier's instructions. Gel electrophoresis was performed at 100 V, 45 min. The extracted DNA was then loaded in a 1 % agarose gel (Pronastar agarose plus, India) and viewed using the gel documentation system (Gelstan, India).

17.3 Results and Discussion

17.3.1 Effect of Pulsed Voltage Magnitude on Bacterial Disinfection

Magnitude of pulsed voltage has a significant effect on bacterial disinfection. Figure 17.2a shows the variation in log reduction of *E. coli* culture treated with

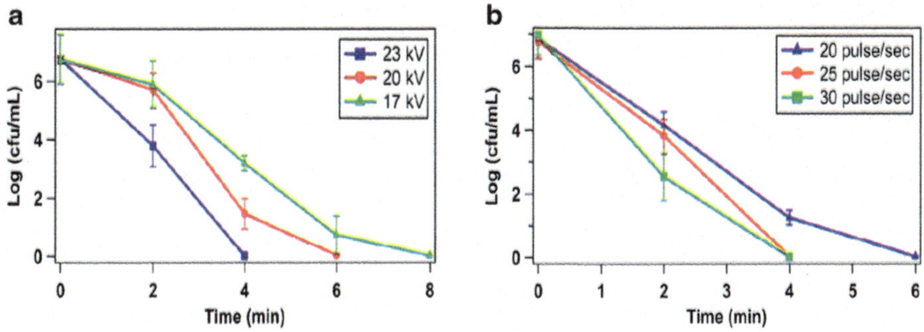

Fig. 17.2 *E. coli* degradation in physiological saline water as a function of time (**a**) at different applied voltages and (**b**) at different frequency

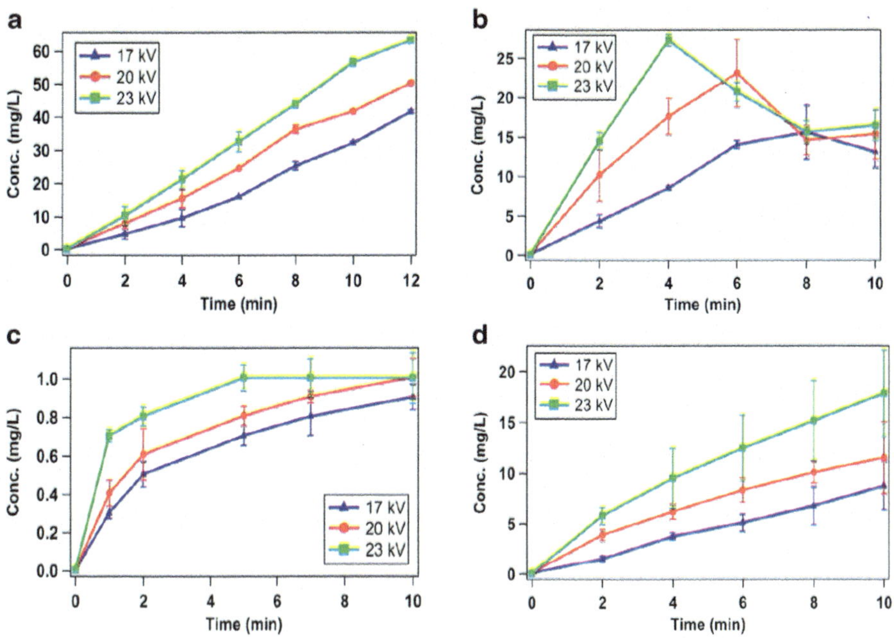

Fig. 17.3 Reactive oxygen species concentration in Millipore water as a function of pulsed streamer discharge time at different applied voltages. (**a**) Hydroxyl radicals, (**b**) hydrogen peroxides, (**c**) ozone, (**d**) superoxide radicals

the PPT using different magnitudes of applied voltage. An increase in applied voltage reduced the treatment time significantly. This may be due to the increase in ROS formation and increased UV dosage (Fig. 17.3 and supplementary material: Fig. 17.A.1). Increasing the applied voltage enhances the applied electric field which in turn produces micropores in the bacterial cell membrane (Jayaram 2000,

Neumann and Sugar 1983), which could in turn alter the cell membrane permeability. This alteration of membrane permeability might cause cell death. Moreover, ROS being a strong oxidant causes atom etching in the cell components of bacteria (Moisan et al. 2001). Apart from high voltage, UV radiation can also damage the genetic material of bacterial cell. Thus, it is evident that the cumulative effect of ROS, high electric field and UV enhanced the bacterial disinfection during streamer discharge process. This result from the present study is in good agreement with results obtained by Wang et al. (2008), who studied the deactivation of *E. coli* using different input voltages.

Figure 17.2b shows the variation of disinfection efficiency under different applied frequencies (pulse/sec) when the voltage was 23 kV. The treatment time reduced significantly when number of pulses was increased. In the present study, square pulses of a fast rise time of 0.4 μs were applied, which is considered as a better choice over the other wave forms (Chen et al. 2010), (Quin et al. 1994). Characteristics of the applied wave forms decide the energy efficiency of the pulse. Therefore, disinfection efficiency can be improved significantly by optimum selection of wave forms, by increasing pulse number and voltage magnitude.

For cases of 2, 4, 6, 8 and 10 min streamer discharge, energy inputs were determined as 0.0028, 0.0056, 0.0085, 0.011 and 0.014 kWh, respectively, at 23 kV applied voltage and 25 pulse/s frequency. When pulse frequency was increased to 30 pulse/s, the energy cost also increased to 0.017 kWh for 10 min streamer discharge. The applied voltage also has an appreciable effect on energy input. The energy input decreased from 0.014 kWh to 0.011 kWh, when the applied voltage was reduced from 23 kV to 17 kV. From the above observations, it can be concluded that the energy cost of the plasma treatment largely depends on applied voltage, frequency and time. In present form the process seems to be energy intensive like other AOPs. However, the energy cost can be further minimised by reducing the rise time, the duration of applied pulse, optimising the electrode geometry, scaling up the reactor and facilitating the transfer of reactive species which are formed during streamer discharge from air to aqueous phase. These studies need to be conducted in the future to optimise the process and minimise the cost.

17.3.1.1 Quantification of Reactive Oxygen Species Formed by Pulsed Streamer Discharge

In any advanced oxidation process (AOPs), the oxidative species play an important role in the degradation of organic substances present in water. More importantly, degradation efficiency of organic compounds greatly depends on the characteristics of the compounds. Some processes show high degradation for one compound, less or no effect on another compound. Therefore, quantitative determination of the generation rate and the concentrations of oxidative species in the advanced treatment processes are necessary for a better understanding of degradation efficiency of different systems, to what extent the oxidative species involve in the degradation of pollutants, and it may also provide a great input for the modelling of oxidative species generation in plasma processes.

ROS formation by streamer discharge process in aqueous solution is a function of input electrical power. The formation of ROS such as hydroxyl radical, hydrogen peroxide, ozone and superoxide radical in distilled water was quantified. Figure 17.3a shows the kinetics of hydroxyl radical production under different voltages. A linear increase in hydroxyl radical production was observed with increase in applied voltage. These results from the present study were compared with those of Sahni and Locke (2006) and Joshi and Locke (1995). It is evident that hydroxyl radical production is more in the present study compared to the values reported by Sahni and Locke (2006) and Joshi and Locke (1995). From Fig. 17.3b, it could be seen that the initial rate of production of hydrogen peroxide increased linearly with time. However, after certain duration, it gradually decreased and finally approached a constant state of generation with respect to time. The possible reason for such variation in hydrogen peroxide could be explained by Eqs.17.6 and 17.7 proposed by Haber and Weiss (1934). As the reactor generates the reaction mixture of ROS including H_2O_2 and OH radicals, when hydrogen peroxide is present at high concentrations, it can scavenge itself (Eqs. 17.6 and 17.7). Along with dissociation, hydrogen peroxide also forms simultaneously (Eq. 17.8). As equilibrium is attained with time, a constant concentration of hydrogen peroxide was maintained in the system, after the drop.

$$\cdot OH + H_2O_2 \rightarrow H_2O + HO_2\cdot \;;\; k = 3.3 \times 10^7 M^{-1}s^{-1} \tag{17.6}$$

$$HO_2\cdot + H_2O_2 \rightarrow H_2O + O_2 + \cdot OH \tag{17.7}$$

$$\cdot OH + \cdot OH \rightarrow H_2O_2 \;;\; k = 4 \times 10^{10}\ M^{-1}s^{-1} \tag{17.8}$$

Similarly, the rate of production of ozone increased exponentially and then reached equilibrium state (Fig. 17.3c). Similar trend was observed during the production of superoxide radical also, which increased linearly with applied voltage (Fig. 17.3d) and attained a final concentration of 18 mg/L at 23 kV.

In the present study, most of the values for $^{\bullet}OH$, H_2O_2 and O_3 were observed to be higher than those in the above-cited references (Sahni and Locke 2006, Joshi and Locke 1995), which could be due to the difference in the electrode geometry and electrode materials. Also factors such as the gap between the high-voltage electrode and water surface (ground electrode) have high influence on the generation of OH radical. The gap distance was optimised in the present study by varying the gap distance from 3 mm to 8 mm. The rate of hydroxyl radical production was maximum at a gap distance of 5 mm (Fig. 17.4). The possible reason might be that at shorter gap (3 mm), the available air thickness for the streamer discharge is less, resulting in lesser ionisation of the air. The ionisation of air is a critical factor because the reactive O and O_3 are formed only through the ionisation of air. O and

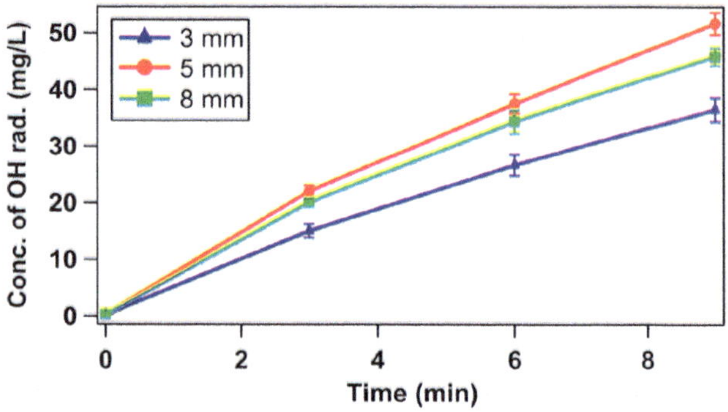

Fig. 17.4 Effect of gap distance between high-voltage electrode and water surface on hydroxyl radical formation with respect to time

O_3, when dissolved in water, result in the formation of ROS by a series of reactions as shown in Fig. 17.A.2 (supplementary material). The sequence of ROS formation is given in Eqs.(17.9), (17.10), (17.11), (17.12), (17.13), (17.14), (17.15), (17.16) and (17.17) (Joshi and Locke 1995; Fridman 2008).

$$O_3 + OH^- \rightarrow O_3^{\cdot -} + \cdot OH \tag{17.9}$$

$$O_3^{\cdot -} \rightarrow O^{\cdot -} + O_2 \tag{17.10}$$

$$O^{\cdot -} + H_2O \rightarrow \cdot OH + OH^- \tag{17.11}$$

$$e^- + H_2O \rightarrow \cdot OH + O^{\cdot} + e^- \tag{17.12}$$

$$O^{\cdot} + H_2O \rightarrow \cdot OH + \cdot OH \tag{17.13}$$

$$e_{aq}^- + O_2 \rightarrow O_2^- \tag{17.14}$$

$$O_2^{\cdot -} + H^+ \rightarrow HO_2^{\cdot} \tag{17.15}$$

$$\cdot OH + \cdot OH \rightarrow H_2O_2 \tag{17.16}$$

$$HO_2^{\cdot} + HO_2^{\cdot} \rightarrow H_2O_2 + O_2 \tag{17.17}$$

When the gap distance is increased to 8 mm, most of the ionised forms of O_3 or O get diffused in the air phase rather than getting dissolved in the aqueous phase. As a result, the concentration of ROS in the aqueous phase decreases. Hence, an optimum gap distance of 5 mm was adopted in all disinfection studies.

It is observed that the formation of hydroxyl radical, hydrogen peroxide and superoxide radical followed zero-order kinetics in the reactor, whereas ozone production was found to be following pseudo first-order kinetics. This can be explained by Eq. (17.18), where rate constant for ozone formation mainly depends on O atom, as O_2 molecule is always present in excess compared to O atom. The rate constant for ROS formation is provided in Table 17.3.

$$O + O_2 + M \rightarrow O_3 + M \tag{17.18}$$

Table 17.3 Kinetics of ROS at different applied voltages

Voltage (kV)	Rate of reaction for ·OH (mol L^{-1} s^{-1})	Rate of reaction for H_2O_2 (mol L^{-1} s^{-1})	Rate of reaction for $O_2^{\cdot-}$ (mol L^{-1} s^{-1})	Rate of reaction for O_3 (s^{-1})
17	3.1	1.0	0.4	0.195
20	4.0	2.0	0.6	0.225
23	5.3	3.4	1.0	0.28

Note: The unit for rate of reaction; $r_{\cdot OH}$, $r_{H_2O_2}$, $r_{O_2^{\cdot-}}$ and r_{O_3} are 10^{-6}mol L^{-1} s^{-1}, 10^{-6}mol L^{-1} s^{-1}, 10^{-6}mol L^{-1} s^{-1} and 10^{-6} s^{-1}, respectively

17.3.1.2 Effect of Voltage on Incident UV Energy Flux Generated by Pulsed Streamer Discharge

The energy flux of generated UV due to the streamer discharge was quantified for different voltages by dissolving uridine in distilled water. Figure 17.A.1 (supplementary material) shows the variation in total incident UV energy flux at different applied voltages. It was observed that the level of UV irradiation dosage is much lower to deactivate *E. coli* cell. UV dose required for complete *E. coli* inactivation is about 140–450 mW/cm^2 (McGuigan et al. 2012). Thus, the generated UV flux did not significantly influence the disinfection.

17.3.2 Effect of Different Modes of Streamer Discharge on Bacterial Disinfection

The variation in log reduction of *E. coli* cells under different modes of applied voltages is shown in Fig. 17.5. The disinfection efficiency was more in the cyclic stress condition as compared to the continuous stress condition. It is observed that 2 + 1 mode is the most effective process for deactivation of *E. coli* cells. This could be due to the impact of hydroxyl radical formed during the discharge process. During continuous discharge, the hydroxyl radical production would have been relatively more, which eventually got converted to hydrogen peroxide (Eq. 17.19) bringing down the levels of hydroxyl radicals needed for the bacterial destruction.

$$\cdot OH + \cdot OH \rightarrow H_2O_2\ ; k = 4 \times 10^{10} M^{-1} s^{-1} \tag{17.19}$$

In (2 + 1) mode of operation, most of the hydroxyl species were available for the bacterial disinfection, and there was less possibility for the formation of hydrogen peroxide. However, in (5 + 2) and (10 + 0) modes of operation, the availability of hydroxyl radicals for bacterial disinfection decreased and more of hydrogen peroxide was produced. It is evident that the oxidation potential of hydroxyl radical (2.3 eV) is much higher than hydrogen peroxide (1.8 eV). Hence, the bacterial disinfection decreased for these operating conditions. The sequential generation of

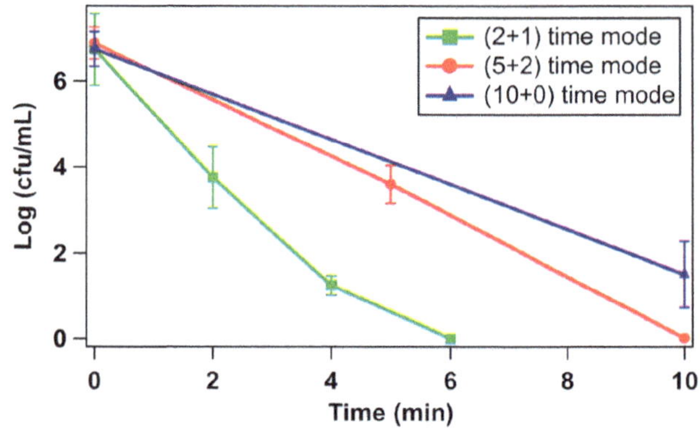

Fig. 17.5 *E. coli* degradation in physiological saline water by adapting different mode of streamer discharge

streamer resulted in an elevated production of OH radicals that enhanced the bacterial disinfection.

17.3.3 Effects of pH, Alkalinity, Biodegradable Organic Matter, Humic Acid and Turbidity on Bacterial Disinfection

17.3.3.1 pH

In the present study, the effect of pH on disinfection efficiency was studied by conducting experiments at different pH values (17.4, 17.7 and 17.9). Approximately 7 log reduction of bacterial concentration was obtained in 6 min at pH 5, whereas same efficiency was achieved in 8 min and 10 min at pH 7 and pH 9, respectively (Fig. 17.6a). Similar results were reported by Zhang et al. (2014) for the inactivation of *Bacillus atrophaeus* spores by UV/H_2O_2 treatment, where a better inactivation of bacteria was achieved in acidic pH. The possible reason for this could be explained using Eqs. 17.20, 17.21, 17.22, 17.23, 17.24, 17.25, 17.26, 17.27 and 17.28 (Chen et al. 2010, Joshi and Locke 1995). In acidic conditions, more $H^{\bullet}$ are generated (Eq. 17.20). $H^{\bullet}$ radicals formed are subsequently converted to ROS like $^{\bullet}OH$ and H_2O_2 as per Eqs. (17.21), (17.22), (17.23) and (17.24). However, as mentioned in Eq. (17.25), H_2O_2 also acts as a scavenger of $^{\bullet}OH$, but its rate constant is lower than the other equations which favour the ROS formation. Also, the intermediate products $HO_2^{\bullet}$ and $^{\bullet}OH$ also form H_2O_2 by Eqs. (17.26) and (17.27). Moreover, high pH may also strongly favour the self-decomposition of H_2O_2 through Eq. (17.28). Hence, the disinfection efficiency was more in acidic pH followed by neutral pH and alkaline pH conditions.

$$e_{aq}^{-} + H^{+} \rightarrow H^{\cdot} + H_2O\,; k = 2.3 \times 10^{10} M^{-1} s^{-1} \quad (17.20)$$

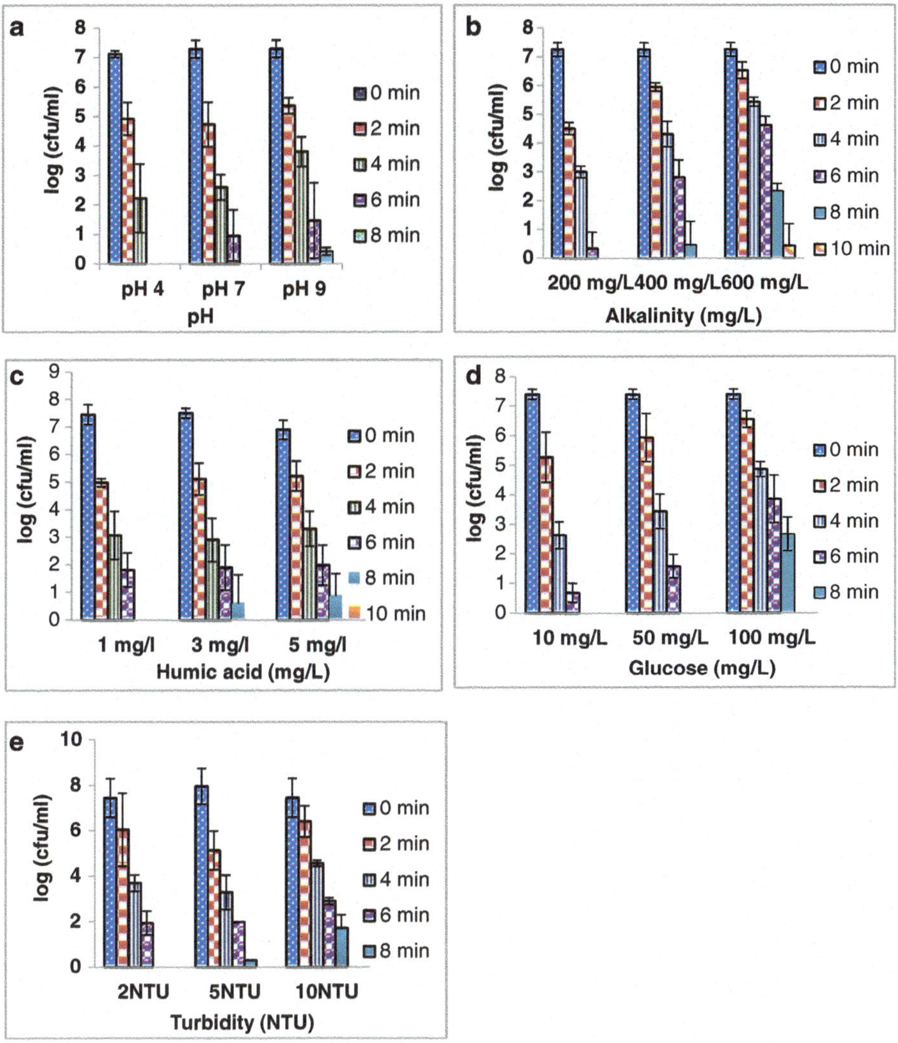

Fig. 17.6 Effect of different environmental conditions on *E. coli* disinfection. (**a**) Effect of pH, (**b**) alkalinity, (**c**) humic acid, (**d**) glucose, (**e**) turbidity

$$H^{\cdot} + H_2O_2 \rightarrow H_2O + \cdot OH\,;k = 5 \times 10^7\,M^{-1}s^{-1} \tag{17.21}$$

$$H^{\cdot} + H_2O \rightarrow H_2 + \cdot OH\,;k = 1 \times 10^{10}\,M^{-1}s^{-1} \tag{17.22}$$

$$H^{\cdot} + O_2 \rightarrow HO_2^{\cdot}\,;\,k = 1.2 \times 10^{10}\,M^{-1}s^{-1} \tag{17.23}$$

$$H^{\cdot} + HO_2^{\cdot} \rightarrow H_2O_2\,;\,k = 1 \times 10^{10}\,M^{-1}s^{-1} \tag{17.24}$$

$$\cdot OH + H_2O_2 \rightarrow H_2O + HO_2^{\cdot} ; k = 2.7 \times 10^7 M^{-1}s^{-1} \quad (17.25)$$

$$\cdot OH + \cdot OH \rightarrow H_2O_2 ; k = 4 \times 10^{10} M^{-1}s^{-1} \quad (17.26)$$

$$HO_2^{\cdot} + HO_2^{\cdot} \rightarrow H_2O_2 + O_2 ; k = 2 \times 10^6 M^{-1}s^{-1} \quad (17.27)$$

$$2H_2O_2 \rightarrow H_2O + O_2 \quad (17.28)$$

Blank experiments were performed without streamer discharge and the results are presented in Table 17.A.2 (supplementary material). The disinfection at slightly acidic pH was negligible. Only 0.21 log reduction of bacterial concentration was observed at pH 4.

17.3.3.2 Alkalinity

The effect of alkalinity on bacterial disinfection was studied by varying the alkalinity from 200 mg/L to 600 mg/L. During this study, the reactor was operated in (2 + 1) mode and pH was kept at 7. From Fig. 17.6b, it can be seen that 7 log reduction was obtained in 8 min at an alkalinity of 200 mg/L, and the time for complete disinfection increased to 10 min and 12 min, as the alkalinity increased to 400 mg/L and 600 mg/L, respectively. Similar results were reported by Marugan et al. (Jeong et al. 2009), where the increase in alkalinity (HCO_3^- ions) decreased the disinfection efficiency. As alkalinity increases, higher concentrations of HCO_3^- and CO_3^{2-} scavenge the $^{\bullet}OH$ radicals produced (Wang and Chen 2011). Thus the disinfection efficiency decreases with increase in alkalinity.

17.3.3.3 Effect of Natural Organic Matter

The effect of natural organic matter (NOM) like humic acid and an easily biodegradable organic matter like glucose on disinfection efficiency was studied by varying their concentrations. Humic acid concentration was varied from 1 mg/L to 5 mg/L, and the glucose concentration was varied from 10 mg/L to 100 mg/L to represent the actual conditions. When the humic acid concentration was 1 mg/L, a 7 log reduction was obtained in 8 min (Fig. 17.6c). With an increase in humic acid concentration to 3 and 5 mg/L, the deactivation time also increased to 10 min. A similar effect was observed in the presence of glucose. From Fig. 17.6d, it is evident that the 7 log reduction was obtained in 6 min in the presence of 10 mg/L glucose, and it increased to 8 min and 10 min as the glucose concentration increased to 50 mg/L and 100 mg/L, respectively. Though NOM (humic acid) is complex and recalcitrant, the concentration (1 mg/L to 5 mg/L) employed in the present study is relatively less than the concentration of glucose (10 mg/L to 100 mg/L). The high concentration of glucose thus had a significant effect on disinfection. Thus, the presence of organic matter in raw water lowers the disinfection efficiency by PPT (Keen et al. 2014). Also, the complexity of NOM can affect the disinfection efficiency significantly.

Table 17.4 Rate of disinfection in the presence of different operating and environmental conditions

Parameters	Levels of parameters	Rate of disinfection (log CFU/ml)/min
Effect of voltage (kV)	17	0.841
	20	1.122
	23	1.68
Effect of pH	5	1.187
	7	0.912
	9	0.858
Effect of turbidity (NTU)	2	0.93
	5	0.796
	10	0.745
Effect of alkalinity (mg/L)	200	0.906
	400	0.725
	600	0.681
Effect of glucose (mg/L)	10	1.12
	50	0.97
	100	0.59
Effect of humic acid (mg/L)	1	0.931
	3	0.75
	5	0.69

17.3.3.4 Turbidity

The effect of turbidity on the performance of PPT was also studied. The presence of turbidity affected PPT in a similar way as that of NOM. From Fig. 17.6e, it could be seen that a 7 log reduction was achieved in 8 min when turbidity was 2 NTU, and it increased to 10 min for turbidity of 10 NTU. Similar result was reported by Arya and Philip (2014), where the disinfection efficiency decreased with increase in turbidity. This can be explained as follows: as turbidity increases, the optical penetration of UV decreases. Thus, the presence of turbid materials decreases the efficiency of disinfection. The presence of components like NOM, turbidity and alkalinity in water scavenges the ROS (Crittenden et al. 2012). Therefore, the disinfection efficiency decreased with the increase in the concentration of these components commonly occurring in natural aqueous systems. Rate constants for disinfection were calculated under different operating and environmental conditions presented in Table 17.4.

17.3.4 Combined Effects of Operating and Environmental Factors on Bacterial Deactivation and Disinfection Regression Model

The individual effect of pH, alkalinity, turbidity and presence of NOM on disinfection efficiency was discussed in Sect. 17.3.3. In order to study the combined effect of important environmental factors (pH, alkalinity and organic matter) and operating conditions (voltage, time and frequency) on disinfection, BBD was employed. Using BBD, a multinomial regression model containing six linear, fifteen interactions and six interaction terms was employed in ANOVA table using Design-Expert 9 software. In order to assess the dependency (effect) of environmental factors on response (disinfection efficiency), the multinomial regression model was tested for linear, two-factor interaction (2FI), quadratic and cubical fits. The best fit of the model was evaluated using R^2, adjusted R^2 and lack of fit test. The results obtained using quadratic model were better compared to other fits, with an $R^2 = 0.9751$ and adjusted $R^2 = 0.9493$. The results are presented in Table 17.A.3 (supplementary material). The lack of fit was found to be insignificant (*P*-value > 0.05) for the quadratic model (supplementary material: Table 17.A.4). The results suggested that the quadratic model accurately represented the experimental data.

From the ANOVA (supplementary material: Table 17.A.5), it was observed that the P-values for voltage (A), time (C), alkalinity (D) and quadratic term voltage^2 (A^2) are much lower than 0.05. This suggests that the effect of these factors is very significant. P-values for frequency (B), pH (E), glucose concentration (F) and quadratic interaction term voltage × pH (AE) and alkalinity × pH (DE) were also lower than 0.05, thereby indicating that these factors are also of significant importance. However P-value for other quadratic terms was more than 0.05, and hence it indicates that these factors are not significant. Considering the factors of significance as discussed above, a functional equation interrelating different operating and environmental parameters developed from the model is given in Eq. (17.29).

$$\begin{aligned}\text{Disinfection efficiency (log reduction)} = {} & 4.982 + 0.91\ \text{voltage} + 0.078\ \text{frequency} \\ & +0.602\ \text{time} - 0.196\ \text{alkalinity} \\ & -0.112\ \text{pH} - 0.094\ \text{glucose} \\ & +(0.174\ \text{voltage} \times \text{pH}) \\ & + (0.134\ \text{pH} \times \text{alkalinity}) \\ & - 0.29\ (\text{voltage})^2\end{aligned} \tag{17.29}$$

The partial effect plots (supplementary material: Fig. 17.A.3) were used to determine the effect of each factor graphically. The partial effect plot of a factor illustrates the variation in response move with the various factors at different levels under an optimum condition (Lim et al. 2007). From the partial effect plot, it was evident that the factors such as the voltage (A), time (B) and alkalinity (D) were most prominent in the disinfection process. From Eq. (17.29) and Fig. 17.A.3 (supplementary material), it can be observed that the voltage, frequency and time positively affect the disinfection efficiency, whereas pH, alkalinity and organic matter (glucose) negatively affect the disinfection efficiency.

The quadratic interaction term in the model can be understood by additive or nonadditive effect. Additive effect of the two-factor effects means that the effect of one factor on the response is independent of the level of the other factor (Lim

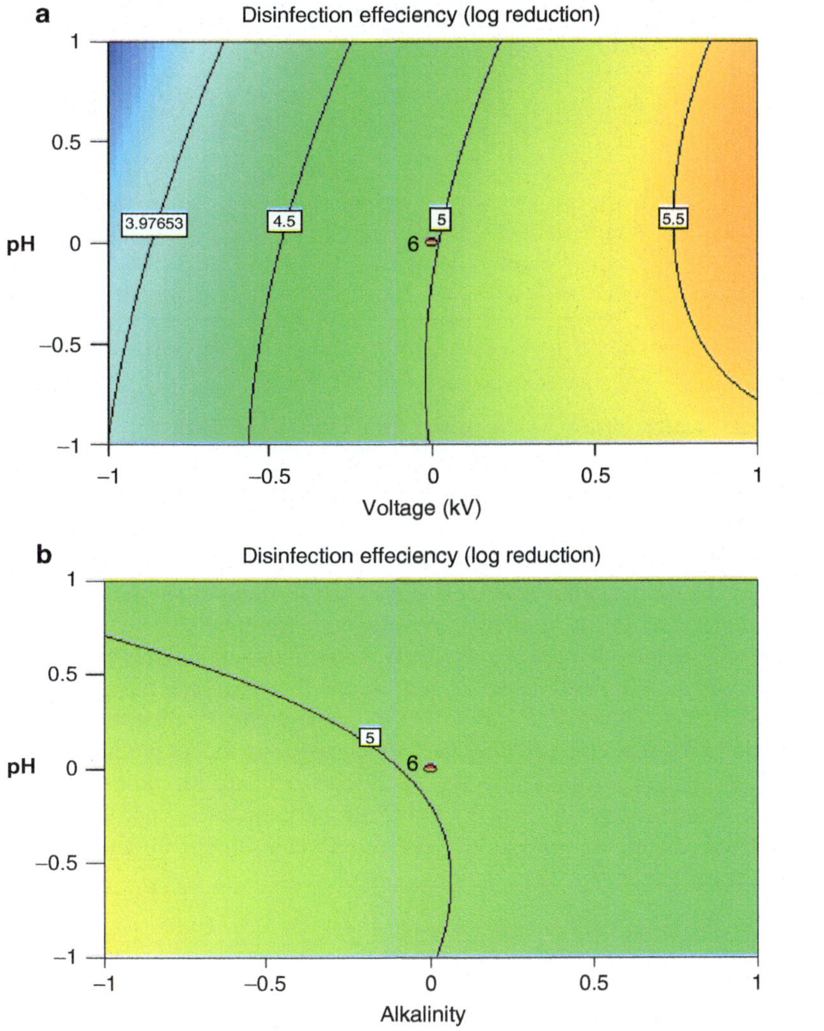

Fig. 17.7 Contour plot for response surface for the effect of different factor combinations on *E. coli* log reduction. Disinfection efficiency increases from *blue*- to *red*-colour region in contour plot area. (**a**) Voltage and pH nonadditive effect, (**b**) pH and alkalinity nonadditive effect

et al. 2007). Figure 17.7a explains the nonadditive effect between voltage and pH observed due to the significant interaction between them. This can be explained by the following possible process. As the input voltage increases, more ionisation of atmospheric nitrogen takes place by corona discharge, and nitrogen molecule gets converted to nitrogen radicals. With the presence of oxygen, reactive atoms or radicals such as $N^{\bullet}$, $O^{\bullet}$, $H^{\bullet}$ and $^{\bullet}OH$ are formed, and they recombine among themselves or with the initial molecules (N_2, O_2, H_2O, etc.), thereby inducing the

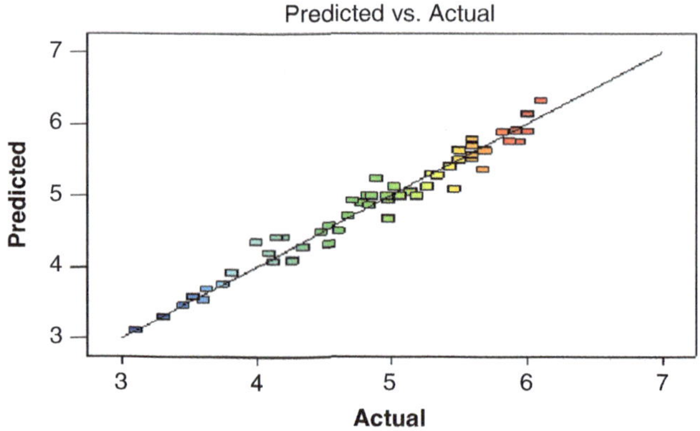

Fig. 17.8 Model prediction with respect to actual experimental results

formation of new products such as NO_2, HNO_2 and HNO_3 (Jiang et al. 2014). These HNO_2 and HNO_3 then induce a steep decrease in the pH. Thus, a nonadditive interaction was observed between voltage and pH. Similarly nonadditive effect was observed between alkalinity and pH (Fig. 17.7b). Alkalinity in the aqueous solution is pH dependent. As mentioned above, the increase in voltage changes the solution pH and this in turn changes the alkalinity. Other interaction terms are not significant (as shown in ANOVA Table 17.A.4: supplementary material), and they show additive effect. Figure 17.8 shows that the experimental values are evenly distributed along the model prediction line, which suggests model prediction is good for the experimental result. Thus, Eq. 17.29 can estimate the time required for the desired disinfection efficiency by knowing the parameters such as pH, alkalinity and organic matter concentration in the given water sample, for a given voltage and frequency.

17.3.5 Scanning Electron Microscope (SEM) Imaging of Bacterial Cells During Disinfection

Scanning electron microscope (SEM) was used to determine the morphological damage of *E. coli* before and after streamer discharge (Fig. 17.9). The deformity in the morphology of *E. coli* indicated the capability of the streamer discharge to destruct the bacteria completely. Different stages of the bacterial destruction are depicted in Fig. 17.9a–c. ROS generated during streamer discharge can rupture the cell membrane and genomic DNA. Gel electrophoresis images showed that the genomic DNA was completely degraded in 6 min of streamer discharge (result not shown).

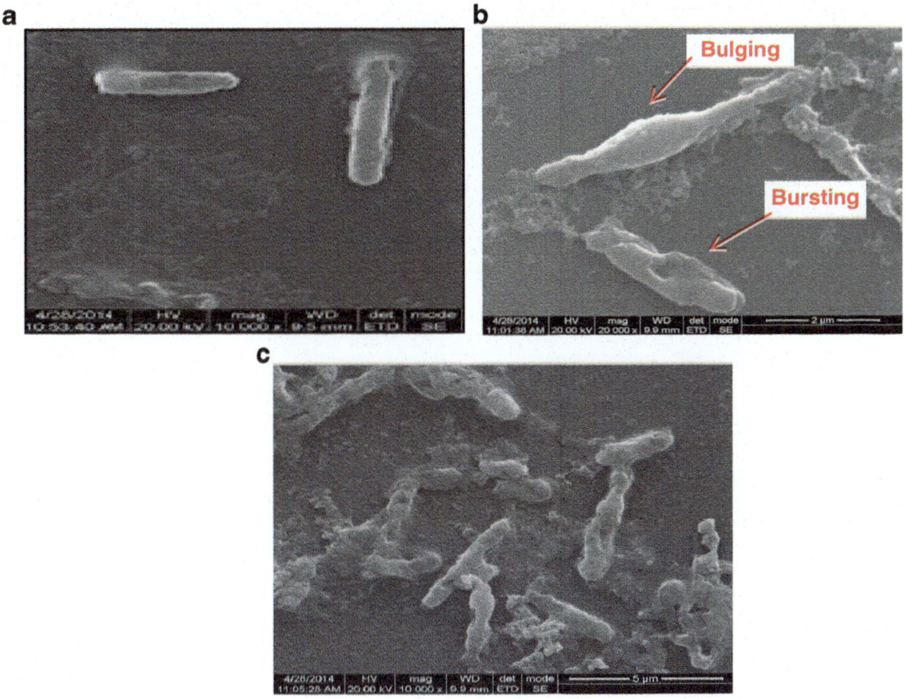

Fig. 17.9 SEM image of *E. coli* (**a**) before streamer discharge, (**b**) after streamer discharge (bulging and bursting of *E. coli* cell) and (**c**) after discharge (deformed and broken cells)

17.4 Conclusion

The efficiency of pulsed power technique for the disinfection of water has been demonstrated. It was observed that ROS concentration increased linearly with the increase in applied voltage magnitude. With the increase in pH, alkalinity, turbidity and organic matter, the rate of disinfection decreased because of the decrease in the net availability of ROS. The complete disinfection (7 log reduction) was achieved in 4 min to 12 min of plasma treatment, depending on the energy input. The combined effect of operating and environmental conditions was studied, and an empirical model was developed to predict the disinfection efficiency under different operating and environmental conditions. Based on the experimental results, it could be concluded that generated UV during the streamer discharge process has less impact on inactivation of *E. coli* cells in water. It could be also concluded that the voltage magnitude, number of pulses and alkalinity affect the disinfection efficiency significantly. However, further investigation is needed to understand the disinfection mechanism completely during the process.

Acknowledgement The authors place on record their gratitude to the Department of Science and Technology (DST), Government of India, for the financial support for this study.

Appendix A: Supplementary Data

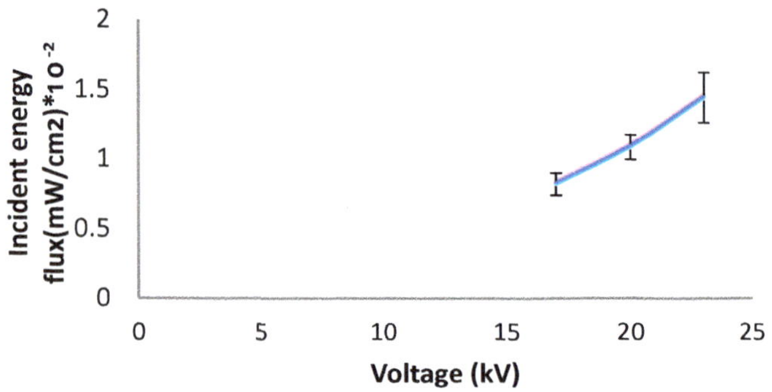

Fig. 17.A.1 UV incident energy flux quantification by uridine probe at different applied voltages

Fig. 17.A.2 Ionisation of air by pulsed streamer discharge and initiation of formation of ROS in water and reactive oxygen species formation

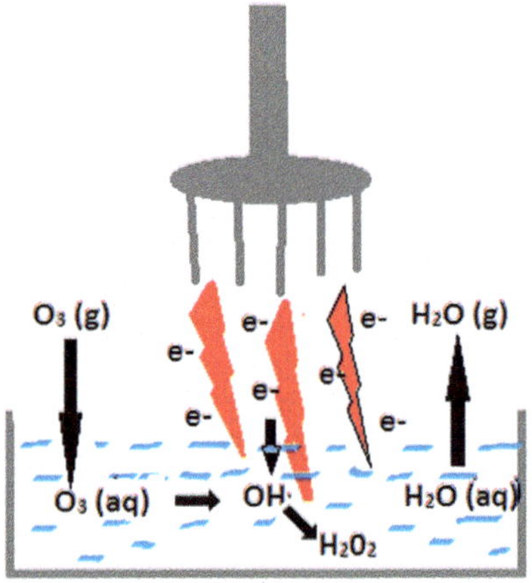

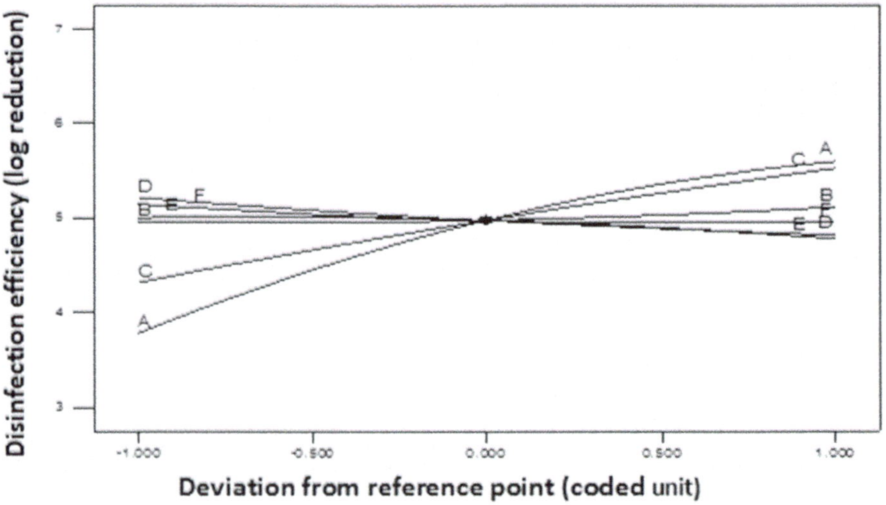

Fig. 17.A.3 Partial effect plot showing the individual effect of *A* voltage, *B* frequency, *C* time, *D* alkalinity, *E* pH and *F* glucose on the variation of disinfection efficiency

Table 17.A.1 Treatment combinations and responses

Run	Coded variable level						Response *Y*
	*X*1	*X*2	*X*3	*X*4	*X*5	*X*6	
1	0	0	−1	1	0	−1	4.35
2	1	0	−1	0	0	−1	5.02
3	−1	0	0	1	−1	0	3.75
4	0	−1	0	0	−1	1	4.71
5	1	0	−1	0	0	−1	4.98
6	−1	0	0	1	1	0	3.46
7	0	1	0	0	1	1	4.84
8	1	−1	0	−1	0	0	5.94
9	0	−1	−1	0	−1	0	4.21
10	0	−1	1	0	−1	0	5.69
11	0	0	−1	−1	0	−1	4.98
12	1	1	0	−1	0	0	5.92
13	0	1	0	0	−1	−1	5.34
14	0	0	0	0	0	0	4.97
15	0	−1	0	0	−1	−1	5.27
16	0	−1	−1	0	1	0	4.1
17	0	1	0	0	−1	1	5.46
18	0	0	1	1	0	1	5.3
19	−1	−1	0	−1	0	0	3.82
20	0	0	0	0	0	0	4.86
21	0	0	1	−1	0	−1	5.82
22	0	0	−1	1	0	1	4.27
23	0	0	1	−1	0	1	5.6
24	0	0	0	0	0	0	4.82
25	1	0	0	1	1	0	5.5
26	1	0	1	0	0	1	6
27	−1	0	1	0	0	−1	4.61
28	−1	0	1	0	0	1	4.54
29	1	1	0	1	0	0	5.6
30	0	0	−1	−1	0	1	4.48
31	0	1	−1	0	−1	0	4.54
32	0	0	0	0	0	0	5.19
33	0	1	1	0	−1	0	5.6
34	0	1	0	0	1	−1	5.14
35	−1	−1	0	1	0	0	3.61
36	1	0	0	−1	1	0	5.87
37	−1	0	−1	0	0	1	3.1
38	−1	0	0	−1	1	0	3.53
39	−1	0	−1	0	0	−1	3.31
40	0	0	0	0	0	0	5.07
41	0	0	1	1	0	−1	5.5

(continued)

Table 17.A.1 (continued)

Run	Coded variable level						Response Y
	$X1$	$X2$	$X3$	$X4$	$X5$	$X6$	
42	0	−1	0	0	1	1	4.68
43	1	0	0	1	−1	0	4.89
44	1	0	1	0	0	−1	6.1
45	1	0	0	−1	−1	0	6
46	−1	0	0	−1	−1	0	4.15
47	0	0	0	0	0	0	4.98
48	−1	1	0	−1	0	0	4.13
49	1	−1	0	1	0	0	5.67
50	0	1	−1	0	1	0	4
51	0	−1	1	0	1	0	5.43
52	−1	1	0	1	0	0	3.63
53	0	1	1	0	1	0	5.6
54	0	−1	0	0	1	−1	4.78

Table 17.A.2 Effect of pH on bacterial disinfection without streamer discharge

Parameters	Initial bacterial conc. (cfu/mL)	Final bacterial conc. (cfu/mL)
Physiological saline	1.51×10^7	1.46×10^7
pH 4	1.23×10^7	7.6×10^6
pH 7	1.45×10^7	1.31×10^7
pH 9	1.25×10^7	1.00×10^7

Note: Plating was done after 10 min incubation time

Table 17.A.3 Model summary statistics

Source	Std. Dev.	R-squared	Adjusted R-squared	Predicted R-squared	PRESS	
Linear	0.24	0.9203	0.9102	0.8931	3.50	
2FI	0.25	0.9393	0.8995	0.8074	6.31	
Quadratic	0.18	0.9751	0.9493	0.8807	3.91	Suggested
Cubic	0.15	0.9947	0.9649	0.3615	20.94	Aliased

Table 17.A.4 Lack of fit tests

Source	Sum of squares	df	Mean square	F value	p-value Prob > F	
Linear	2.52	42	0.060	3.25	0.0944	
2FI	1.90	27	0.070	3.81	0.0705	
Quadratic	0.72	21	0.034	1.87	0.2533	Suggested
Cubic	0.081	3	0.027	1.47	0.3295	Aliased
Pure Error	0.092	5	0.018			

Table 17.A.5 ANOVA for response surface quadratic model for bacterial disinfection

Source	Sum of squares	df	Mean square	F value	p-value Prob > F	
Model	31.97	27	1.18	37.76	<0.0001	Significant
A-Voltage	19.89	1	19.89	634.32	<0.0001	Significant
B-Frequency	0.15	1	0.15	4.75	0.0386	Significant
C-Time	8.70	1	8.70	277.42	<0.0001	Significant
D-Alkalinity	0.92	1	0.92	29.47	<0.0001	Significant
E-pH	0.30	1	0.30	9.54	0.0047	Significant
F-Glucose conc.	0.21	1	0.21	6.79	0.0150	Significant
AB	0.022	1	0.022	0.70	0.4094	
AC	0.051	1	0.051	1.63	0.2126	
AD	0.050	1	0.050	1.58	0.2201	
AE	0.24	1	0.24	7.70	0.0101	Significant
AF	2.450E-003	1	2.450E-003	0.078	0.7821	
BC	2.813E-003	1	2.813E-003	0.090	0.7670	
BD	0.014	1	0.014	0.46	0.5033	
BE	0.014	1	0.014	0.44	0.5128	
BF	0.029	1	0.029	0.92	0.3467	
CD	6.050E-003	1	6.050E-003	0.19	0.6641	
CE	0.019	1	0.019	0.61	0.4432	
CF	3.600E-003	1	3.600E-003	0.11	0.7375	
DE	0.14	1	0.14	4.56	0.0422	Significant
DF	0.024	1	0.024	0.77	0.3877	
EF	2.000E-004	1	2.000E-004	6.377E-003	0.9370	
A^2	0.87	1	0.87	27.66	<0.0001	Significant
B^2	0.036	1	0.036	1.15	0.2938	
C^2	0.034	1	0.034	1.08	0.3073	
D^2	0.016	1	0.016	0.51	0.4798	
E^2	0.078	1	0.078	2.49	0.1269	
F^2	0.056	1	0.056	1.78	0.1932	
Residual	0.82	26	0.031			
Lack of fit	0.72	21	0.034	1.87	0.2533	Not significant
Pure error	0.092	5	0.018			
Cor total	32.79	53				

References

Arya V, Philip L (2014) Visible and solar light photocatalytic disinfection of bacteria by N-doped TiO_2. Water Sci Technol Water Supply 14:924

Chen J, Zhang R, Xiao J, Li J, Wang L, Guan Z, MacAlpine M (2010) Influence of pulse rise time on the inactivation of Staphylococcus aureus by pulse electric fields. IEEE Trans Plasma Sci 38:1935–1941

Crittenden JC, Trussell RR, Hand DW, Howe KJ, Tchobanoglous G (2012) Water treatment principles and design, 3rd edn. Willy, Hoboken, pp 1415–1484
Ferreira SLC, Bruns RE, Ferreira HS, Matos GD, David JM, Brandão GC et al (2007) Box-Behnken design: an alternative for the optimization of analytical methods. Anal Chim Acta 597:179–186
Fridman A (2008) Plasma chemistry. Cambridge university press, New York
Gao P, Liu J, Sun DD, Ng W (2013) Graphene oxide-CdS composite with high photocatalytic degradation and disinfection activities under visible light irradiation. J Hazard Mater 250–251:412–420
Gleick PH (2002) Dirty water: Estimated deaths from water-related disease 2000–2020. Pacific Institute Research report
Goto H (2004) Quantitative analysis of superoxide ion and hydrogen peroxide produced from molecular oxygen on photoirradiated TiO2 particles. J Catal 225:223–229
Haber F, Weiss J (1934) The catalytic decomposition of hydrogen peroxide by iron salts. Proc Royal Soc Lond 147:332–351
Harvey RG (2011) Historical overview of chemical carcinogenesis, chemical carcinogenesis, Penning TM (ed) Humana, New York, pp 1–26
Hayes J, Kirf D, Garvey M, Rowan N (2013) Disinfection and toxicological assessments of pulsed UV and pulsed-plasma gas-discharge treated-water containing the waterborne protozoan enteroparasite Cryptosporidium parvum. J Microbiol Methods 94:325–337
Jayaram SH (2000) Sterilization of liquid foods by pulsed electric fields. IEEE Electr Insul Mag 16:17–25
Jeong J, Kim C, Yoon J (2009) The effect of electrode material on the generation of oxidants and microbial inactivation in the electrochemical disinfection processes. Water Res 43:895–901
Jiang B, Zheng J, Qiu S, Wu M, Zhang Q, Yan Z et al (2014) Review on electrical discharge plasma technology for wastewater remediation. Chem Eng J 236:348–368
Joshi AA, Locke BR (1995) Formation of hydroxyl radicals, hydrogen peroxide and aqueous electrons by pulsed streamer corona discharge in aqueous solution. J Hazard Mater 41:3–30
Joubert V, Cheype C, Bonnet J, Packan D, Garnier, J-P, Teissié J et al (2013) Inactivation of Bacillus subtilis var. niger of both spore and vegetative forms by means of corona discharges applied in water. Water Res 47:1381–1389
Julák J, Scholtz V, Kotúčová S, Janoušková O (2012) The persistent microbicidal effect in water exposed to the corona discharge. Phys Med 28:230–239
Keen OS, McKay G, Mezyk SP, Linden KG, Rosario-Ortiz FL (2014) Identifying the factors that influence the reactivity of effluent organic matter with hydroxyl radicals. Water Res 50:408–419
Kim HS, Lee DH, Fridman A, Cho YI (2014) Residuals effects and energy cost of gliding arc discharge treatment on the inactivation of Escherichia coli in water. Int J Heat Mass Trans 77:1075–1083
Kuzmichev AI, Soloshenko IA, Tsiolko VV, Kryzhanovsky VI, Bazhenov VY, Mikhno IL, Khomomich VA (2000) Features of sterilization by different type of atmospheric pressure discharges. In: Proceedings of the international symposium on high pressure, low temperature plasma chemistry, Hakone VII, Greifswald, Germany, pp 402–406
Lee C, Kim J, Yoon J (2011) Inactivation of MS2 bacteriophage by streamer corona discharge in water. Chemosphere 82:1135–1140
Lim BA, Rahim CH, Ho RA, and Arbakariya YW (2007) Optimization of growth medium for efficient cultivation of Lactobacillus salivarius i 24 using response surface method responses. Malays J Microbiol 3:41–47
Locke BR, Sato M, Sunka P, Hoffmann MR, Chang J-S (2006) Electrohydraulic discharge and nonthermal plasma for water treatment. Ind Eng Chem Res 45:882–905
Marugán J, van Grieken R, Sordo C, Cruz C (2008) Kinetics of the photocatalytic disinfection of Escherichia coli suspensions. Appl Catal B Environ 82:27–36

McGuigan KG, Conroy RM, Mosler H-J, du Preez M, Ubomba-Jaswa E, Fernandez-Ibañez P (2012) Solar water disinfection (SODIS): a review from bench-top to roof-top. J Hazard Mater 235–236:29–46

Moisan M, Barbeau J, Moreau S, Pelletier J, Tabrizian M, Yahia LH (2001) Low-temperature sterilization using gas plasmas: a review of the experiments and an analysis of the inactivation mechanisms. Int J Pharm 226:1–21

Neumann E, Sugar I (1983) Electroporation – stochastic model for electric field-induced membrane pores. J Electroanal Chem Interfacial Electrochem 156:479

Odic E, Goldman A, Goldman M, Delaveau S, Hegarat FL (2002) Plasma sterilization technologies and processes. High Temp Mater Processes 6:385–396

Ortega-Gómez E, Fernández-Ibáñez P, Ballesteros Martín MM, Polo-López MI, Esteban García B, Sánchez Pérez JA (2012) Water disinfection using photo-Fenton: effect of temperature on Enterococcus faecalis survival. Water Res 46:6154–6162

Penru Y, Guastalli AR, Esplugas S, Baig S (2012) Application of UV and UV/H2O2 to seawater: disinfection and natural organic matter removal. J Photochem Photobiol A Chem 233:40–45

Quin BL, Zhang Q, Barbosa Canovas GV (1994) Inactivation of microorganisms by pulsed electric fields of different waveforms. IEEE Trans Dielectr Electr Insul 1:1047–1057

Ribeiro AR, Nunes OC, Pereira MFR, Silva AMT (2014) An overview on the advanced oxidation processes applied for the treatment of water pollutants defined in the recently launched Directive 2013/39/EU. Environ Int 75C:33–51

Rossi F, Kylian O, Rauscher H, Hasiwa M (2009) Low pressure plasma discharges for the sterilization and decontamination of surfaces. New J Phys 11:115017

Roth S, Feichtinger J, Hertel C (2010) Characterization of *Bacillus subtilis* spore inactivation in low-pressure, low-temperature gas plasma sterilization processes. J Appl Microbiol 108:521–531

Sahni M (2006) Analysis of chemical reactions in pulsed streamer discharges: an experimental study. PhD thesis, Florida state university

Sahni M, Locke BR (2006) Quantification of Hydroxyl Radicals Produced in Aqueous Phase Pulsed Electrical Discharge Reactors. Industrial & Engineering Chemistry Research 45:5819–5825

Schaefer CE, Andaya C, Urtiaga A (2015) Assessment and by-products forming during electrochemical treatment of surface water using a Ti/IrO2 anode. Chem Eng J 264:411–416

Sichel C, Blanco J, Malato S, Fernández-Ibáñez P (2007) Effects of experimental conditions on E. coli survival during solar photocatalytic water disinfection. J Photochem Photobiol A Chem 189:239–246

Sonntag CV, Schuchmann HP (2002) UV disinfection of drinking water and by-product formation-some basic considerations. J Water SRT-Aqua 41:67–74

Summerfelt ST, Sharrer MJ, Tsukuda SM, Gearheart M (2009) Process requirements for achieving full-flow disinfection of recirculating water using ozonation and UV irradiation. Aquac Eng 40:17–27

Szewzyk U, Szewzyk R, Manz W, Schleifer KH (2000) Microbiological safety of drinking water. Ann Rev Microbiol 54:81–127

Tai C, Peng JF, Liu JF, Jiang GB, Zou H (2004) Determination of hydroxyl radicals in advanced oxidation processes with dimethyl sulfoxide trapping and liquid chromatography. Analytica Chemica Acta 527:73–80

Thewaterproject.org, The water crisis, statistics and figures. Retrieved on January 4, 2015 from http://thewaterproject.org/water_stats_2#stats

Tian C, Liu R, Liu H, Qu J (2013) Disinfection by-products formation and precursors transformation during chlorination and chloramination of highly-polluted source water: significance of ammonia. Water Res 47:5901–5910

Wang H, Chen X (2011) Kinetic analysis and energy efficiency of phenol degradation in a plasma-photocatalysis system. J Hazard Mater 186:1888–1892

Wang C, Wu Y, Li G (2008) Inactivation of E. coli with plasma generated by bipolar pulsed discharge in a three-phase discharge plasma reactor. J Electrostat 66:71–78

World Health Organization (2012) Water fact sheet. Retrieved on November 4, 2014, from http://www.who.int/mediacentre/factsheets/fs391/en/

Wu D, You H, Zhang R, Chen C, Lee D-J (2011) Ballast waters treatment using UV/Ag–TiO2 + O3 advanced oxidation process with Escherichia coli and Vibrio alginolyticus as indicator microorganisms. Chem Eng J 174:714–718

Zhang X, Echigo S, Lei H, Smith ME, Minear RA, Talley JW (2005) Effects of temperature and chemical addition on the formation of bromoorganic DBPs during ozonation. Water Res 39:423–435

Zhang Y, Zhang Y, Zhou L, Tan C (2014) Factors affecting UV/H2O2 inactivation of Bacillus atrophaeus spores in drinking water. J Photochem Photobiol B 134:9–15

Part IV
Sustainable Structural Systems

Chapter 18
Pre-engineered Bamboo Structures: A Step Towards Sustainable Construction

Suresh Bhalla, Roger P. West, Diwaker Bhagat, Mukul Gupta, and Aarti Nagpal

Abstract This paper presents the new developments undertaken by the bamboo research group (BRG) at IIT Delhi in the area of sustainable construction. Bamboo is a natural grass, which grows fully in much less time and with much less resources as compared to timber, which is fast depleting across the world. A scientific design approach has been formulated recently by BRG along with detailed material characterization (which includes both strength and elastic characterization) of the common Indian species *Dendrocalamus strictus*. This is followed by the development of modular structural components, which, when connected, can pave way for pre-engineered construction. In order to achieve easy connectability, special steel shoes have been developed which easily fix at the ends of the bamboo culms, hold up there through friction and impart weldability as well as bolting. The paper also briefly describes the development of high-capacity flexural/compression members, which can pave way for the construction of full portal frame of multistory buildings.

Keywords Engineered bamboo structures • Sustainability • Pre-engineered • Design approach • Bamboo-steel connections

18.1 Introduction

Bamboo is a versatile, strong, renewable and environment-friendly material. It is a member of the grass family and the fastest-growing woody plant on earth. Most bamboo species produce mature fibre in 3–4 years after sowing, much sooner than

S. Bhalla (✉) • D. Bhagat • M. Gupta
Department of Civil Engineering, IIT Delhi, New Delhi 110016, India
e-mail: sbhalla@civil.iitd.ac.in

R.P. West
Department of Civil, Structural and Environmental Engineering, Trinity College, University of Dublin, Dublin, Ireland

A. Nagpal
School of Planning and Architecture, IP Estate, New Delhi 110002, India

G.L. Sivakumar Babu et al. (eds.), *Sustainability Issues in Civil Engineering*,
Springer Transactions in Civil and Environmental Engineering,
DOI 10.1007/978-981-10-1930-2_18

any tree species. Some bamboos grow up to 1 m a day, with many reaching culm lengths of 25 m or more. Bamboo can be grown quickly and easily and sustainably harvested in 3–5-year cycles. It can grow on marginal and degraded lands, elevated ground and along field bunds and river banks. It adapts to most climatic conditions and soil types, acting as a soil stabilizer and an effective carbon sink and helping to counter the greenhouse effect (Bhalla et al. 2008). Although throughout the past, (especially in India) it has been used in rural buildings, it needs to be characterized for strength and other properties. Also, scientific design approach needs to be developed on lines of concrete and steel before it can be used in an economical fashion for widespread modern urban as well as rural construction. Recently, there has been rapid progress in this regard as a result of the groundbreaking work done by the bamboo research group (BRG) at IIT Delhi. This paper briefly presents an overview of the recent developments in this front, starting from material characterization, development of scientific design methodology and development of pre-engineered modular components, both moderate and high capacity.

18.2 Mechanical Characterization

A comprehensive test plan was undertaken by BRC to carry out the complete mechanical characterization of two varieties of *Dendrocalamus strictus* grown in Assam, red type and black type. The detailed test methodologies and results can be found in Bhalla and Shaw (2013). Figure 18.1 shows the failure pattern of four

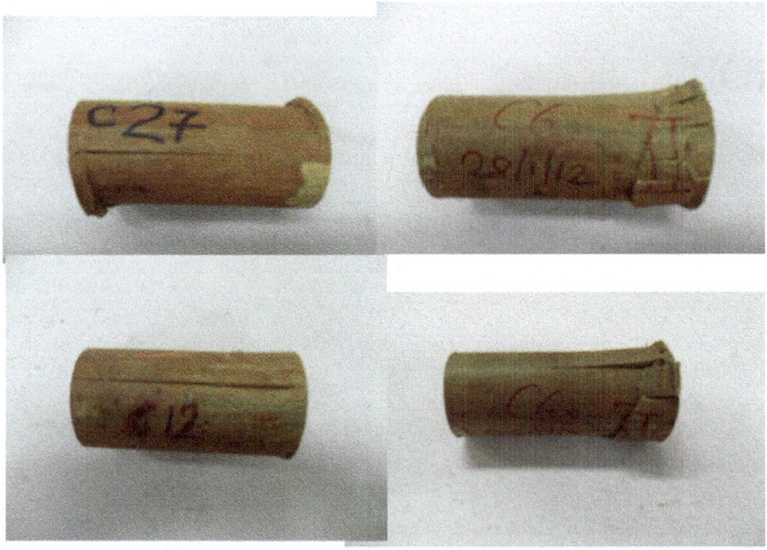

Fig. 18.1 Representative bamboo specimens tested for compressive strength

typical specimens tested as part of the project. Based on a large number of tests, characteristic strengths were worked out in line with the standard practice in concrete industry. In summary, the typical characteristic compressive strengths of black and red bamboo were found to be 63 MPa and 44 MPa, respectively. The respective characteristic tensile strengths were measured to be 76 MPa and 62 MPa, respectively.

18.3 Development of Moderate-Capacity Built-Up Bamboo Beams and Columns

The next step after material characterization is the fabrication of built-up beams and columns. In this endeavour, the BRC came out with a new kind of built-up beams/columns consisting of bamboo culms tied together by half-split bamboo battens using steel fasteners. A complete view of one such column is shown in Fig. 18.2. These columns were tested at the Structures Lab, Department of Civil, Structural and Environmental Engineering, Trinity College, University of Dublin. The details of the test can be found in Kajjam et al. (2012).

Suitable steel fixtures were developed and characterized, as shown in Fig. 18.3a, to impart weldability to the built-up columns. The striking feature of the fixture is that it does not involve any drilling in the culm. The fixture is attached to the culm by tightening two half-split pipe pieces using bolts tightened to a specific torque level. The fixture was experimentally characterized and details can be found in Gupta et al. (2013). The fixtures enable fixing of base plate at the bottom of the column as shown in Fig. 18.3b. This makes the columns easily connectable to the foundations, resulting in the erection of the total bent up portal frame in

Fig. 18.2 A full-scale built-up bamboo column

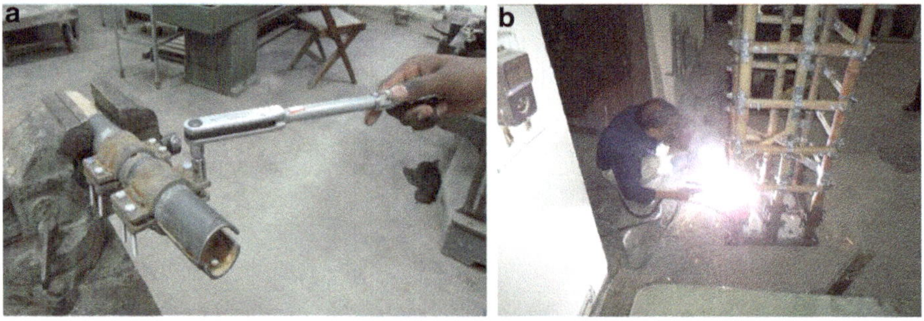

Fig. 18.3 (**a**) Typical fixture for bamboo. (**b**) Welding of column to base plate

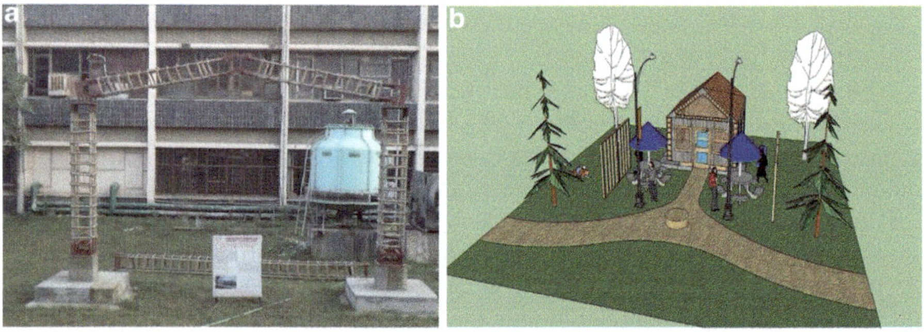

Fig. 18.4 (**a**) Prototype pre-engineered bamboo portal frame. (**b**) Artistic impression of a bamboo-based student activity centre

pre-engineered fashion, as shown in Fig. 18.4a. The complete erection process can be visualized through a video link available at http://web.iitd.ac.in/~sbhalla/video.html.

Possible applications of this system include construction of cattle sheds, warehouses and rural industrial sheds. They can also possibly be employed to construct stylish activity centres in colleges and commercial hubs; a typical artistic impression of one such centre is shown in Fig. 18.4b. Wind load analysis of a typical frame can be accessed from the related project reports of Gupta (2014) and Chauhan (2012). The above developments took place as part of the R&D project bamboo as a green engineering material in rural housing and agricultural structures for sustainable economic growth (Gupta et al. 2012).

18.4 Development of High-Capacity Flexural Members

The above developments are suitable for moderate capacity members, such as battened either beams or columns. However, these cannot be employed for multistory constructions where the frame members are likely to encounter large

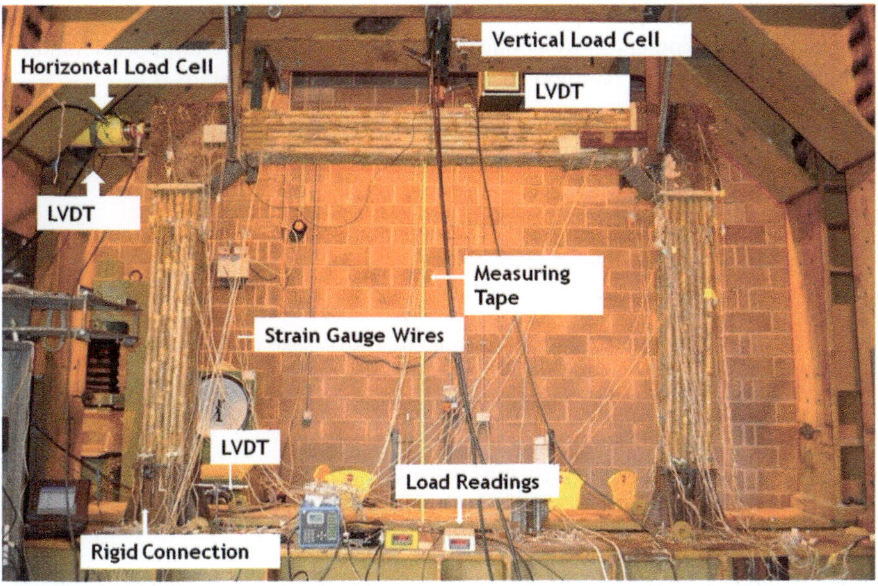

Fig. 18.5 High-capacity bamboo portal frame under test at Trinity College, University of Dublin

loads, including lateral loads. For such structures, the BRC has come out with novel fibre reinforced composite bamboo (FRCB) members (Bhagat et al. 2013; Bhagat 2014). These members can be easily joined together to result in a rigid portal frame with the help of steel rebars cast with epoxy concrete. Figure 18.5 shows one such frame, about 3 × 3 m in dimensions, ready for testing at the Structures Lab at the Department of Civil, Structural and Environmental Engineering, Trinity College, University of Dublin. Three such prototype frames were tested for combined vertical and lateral loads. The test data is still being analysed. In overall, the tests clearly established the feasibility of such high-capacity rigid frame being employed in modern multistory buildings.

18.5 Conclusions

This paper has presented the recent research work carried out by BRC at IIT Delhi towards realization of pre-engineered bamboo structures for modern sustainable construction. A scientific design methodology has been developed based on structural engineering principles and the mechanical properties, including strength and stiffness available from rigorous testing of the common Indian species *Dendrocalamus strictus*. Suitable steel fixtures have been developed to impart connectability to the members, including welding. The entire fabrication can be carried out in a workshop in rural settings and the final fabrication be done with fast speed at the site using the principles of pre-engineered construction. The technology has immense potential for countries like India which have large wastelands and suitable climate where bamboo species can be easily cultivated.

References

Bhagat D (2014) Engineered bamboo structures: fabrication and testing of high capacity bamboo fibre reinforced composite members, Ph. D. synopsis, Department of Civil Engineering, IIT Delhi

Bhagat D, Gupta M, Bhalla S (2013) Composite bamboo junction elements for structural applications. In: Proceedings of the Innovative Techniques in Civil and Environmental Engineering (SITCEE-2013), 05–06 June, Jawaharlal Nehru University, New Delhi (Published in special edition in the Int J Civil Engg Appls.) 3(7):36–41

Bhalla S, Shaw A (2013) Engineering characterization of bamboo as a sustainable alternative to steel. In: Proceedings of the UKIERI congress on innovations in concrete construction, 5–8 March, B. R. Ambdekar National Institute of Technology, Jalandhar, paper no UCC-490

Bhalla S, Gupta S, Puttaguna S, Suresh R (2008) Bamboo as green alternative to concrete and steel for modern structures. J Env Res 3(2):362–370

Chauhan (2012) Analysis and design of bamboo based cowshed. B. Tech. Project, Department of Civil Engineering, IIT Delhi.

Gupta M (2014) Engineered bamboo structures for sustainable construction. B. Tech. Project, Department of Civil Engineering, IIT Delhi

Gupta S, Mohanty S, Bhalla S, Satya S, Shahani DT, Babu R, Subromony T (2012) Bamboo as a green engineering material in rural housing and agricultural structures for sustainable economic growth. Final Project Report, Indian Council of Agricultural Research, Under National Agricultural Innovation Project

Gupta M, Bhagat D, Bhalla S (2013) Design and strength analysis of steel-bamboo fixtures for pre-engineered bamboo portal frame. In: Proceedings of the Innovative Techniques in Civil and Environmental Engineering (SITCEE-2013), 5–6 June, Jawaharlal Nehru University, New Delhi (Published in special edition in the Int J Civil Engg Appls) 3(7):42–46

Kajjam S, Chaudhry DK, Bhalla S, West R (2012) Fabrication and testing of built-up bamboo columns for structural applications. In: Proceedings of the third international Conference on Mechanics of Functional Materials and Structures (ACMFMS 2012), 5–8 December, IIT Delhi, pp 433–436

Chapter 19
Sustainability: Way Forward in Architectural Engineering

S.P. Anchuri and N.V. Ramana Rao

Abstract "What's the more sustainable alternative?" "How does these substitutes work?" "Where has that been done earlier?" "Who can source it?" Some of the major sustainable development challenges that threaten the construction industry include issues that have global as well as local importance, such as resource exhaustion, protecting biodiversity, and climate change. This paper gives practical information about the components that have key role in meeting the challenges of sustainable construction and development. It is based on the simple concept of ensuring a better quality life for everyone, now and for generations to come.

Keywords Paperless project • Indoor air quality • Net-zero energy • Day lighting • Building form

19.1 Introduction

Sustainability requires eco-conscious thinking at every stage in the building life cycle starting with the decision to build. Let us understand critical decisions and the steps it takes to meet targets for energy and water efficiency, achieve cost savings, and minimize overall impact on the environment exploring each project phase for opportunities to create and deliver more sustainable projects. We need to cover architectural engineering profession which includes architectural or design services, civil engineering, MEP engineering, structural engineering, construction, energy services, and above all owner.

S.P. Anchuri (✉)
Anchuri & Anchuri, Architects, Industrial Designers and Structural Engineers, 1-11-248/3, Shamlal Building, S.P. Road, Begumpet, Hyderabad, TS 500016, India
e-mail: spanchuri@gmail.com

N.V. Ramana Rao
Department of Civil Engineering, JNTUH, Hyderabad, India

G.L. Sivakumar Babu et al. (eds.), *Sustainability Issues in Civil Engineering*, Springer Transactions in Civil and Environmental Engineering,
DOI 10.1007/978-981-10-1930-2_19

This chapter prepares us to meet the basic needs of all those concerned with planning and building activities within the building and construction industry. This document will be of specific interest to environmental coordinators, developers, builders, manufacturers, housing associations, clients, planners, and development officers.

19.1.1 Requirements Developments

19.1.1.1 Energy and Environmental Goals

Set goals for net-zero energy or carbon-neutral operations. Consider available renewable energy sources and determine potential for on-site wind or solar electric energy generation and solar thermal opportunities. Consider building form and orientation for optimal energy efficiency. Look for possibilities for natural ventilation given site constraints, climate, security, air quality, and occupancy requirements.

19.1.1.2 Building Use

Establish building space requirements. Explore floor area allocation and assign usages.

19.1.1.3 Regulatory and Certification Targets

Conduct preliminary analysis to set achievable goals for meeting or exceeding regulatory or certification requirements for energy, water, and land impacts.

19.1.1.4 Paperless Project

Require electronic document management and sharing to reduce paper and shipping impact. ATG Technical Knowledge Center specially comes up with requirements giving emphasis on paperless work and will revolutionize the process across the global use of paper system and transform into paperless working strategy.

The typical example of requirements development is shown in Fig. 19.1, and the same example will be used to show all the points in the process.

Fig. 19.1 Requirements developments

19.1.2 Conceptualization

19.1.2.1 Envelope Type and Performance

Consider regionally available materials and regulatory requirements. Determine thermal and mass requirements of materials that maximize performance.

19.1.2.2 Daylighting

Consider building form, openings, glass amounts, and glazing types. Consider required or desired views, ventilation, and comfort. Consider automatic lighting controls and energy cost benefits.

19.1.2.3 Site Planning

Plan location of all elements of project design, such as building, parking, and landscaping. Consider man-made and natural features of site and adjacent areas (e.g., structures, geology, and hydrology).

19.1.2.4 Energy and Environmental

Conduct whole building energy, water, and carbon analysis of early conceptual models to estimate potential performance and evaluate alternatives.

19.1.2.5 Building Form and Orientation

Balance design to achieve energy requirements, views, lighting, ventilation, and comfort. Plan to maximize natural energy benefits of site, such as passive solar or natural ventilation.

Fig. 19.2 Conceptualization

Figure 19.2 shows us the conceptualization stage of the typical example project.

19.1.3 Detailed Design

19.1.3.1 Building Design

Visualize, design, and analyze potential performance, and evaluate costs of the projects before construction begins.

19.1.3.2 Lighting Design

Design lighting while adhering to light level and performance requirements, and minimize embodied measures. Design lighting after daylighting has been optimized.

19.1.3.3 Envelope Assemblies

Meet defined performance criteria with design of envelope-specific assemblies, while minimizing embodied measures.

19.1.3.4 Indoor Air Quality

Integrate natural ventilation. Consider night ventilation options and the indoor/outdoor pollutant issues it may raise.

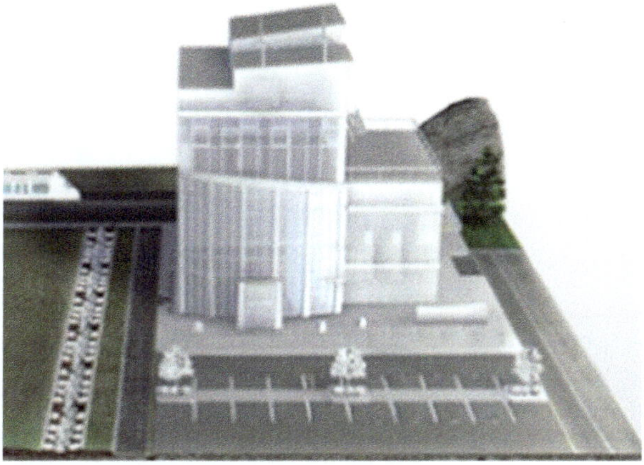

Fig. 19.3 Detailed design

19.1.3.5 Minimizing Lighting Pollution

Ensure indoor lighting does not appear externally, or plan for auto-control lighting for after hours. Optimize outdoor and façade/landscape lighting.

Figure 19.3 shows us the adoption of detailed design for the typical example project.

19.1.4 Implementation Documentation

19.1.4.1 Building Component Fabrication

Consider manufacturing in controlled environments to reduce waste and minimize site environmental impact.

19.1.4.2 Final Code Compliance

Review plans for code compliance with local agencies—avoid rework and waste.

19.1.4.3 Paperless Project

Continue digital sharing of documentation among all stakeholders, including external agencies. Enable electronic document management and sharing to reduce paper and shipping impact.

Fig. 19.4 Implementation documentation

19.1.4.4 Project Delivery

Many of the decisions on sustainability have already been made in the earlier phases of design, and now is the time to ensure successful implementation.

19.1.4.5 Material Quantities

Plan carefully—estimate materials needed to avoid waste. Anticipate problems and resolve conflicts before committing.

Figure 19.4 explains us about the implementation of document for the typical example.

19.1.5 Construction

19.1.5.1 Fabrication

Communicate with fabricators to ensure quality and determine arrival of various building components.

19.1.5.2 Regulatory and Certification Targets

Monitor progress toward goals for meeting or exceeding regulatory or certification requirements for energy, water, and land impacts.

Fig. 19.5 Construction

19.1.5.3 Paperless Project

Continue digital sharing of documentation among all stakeholders, including external agencies. Enable electronic document management and sharing to reduce paper and shipping impact.

Figure 19.5 gives us clear picture of construction process shown for a typical example.

19.2 Operation and Maintain

19.2.1 Environmental Impact

Buildings and infrastructure should continue to serve the community usefully over a long life span while leaving minimal environmental footprint.

19.2.1.1 Paperless as Built

Employ the digital record of the building or infrastructure, enabled by the latest techniques including BIM-like process used throughout the project.

19.2.1.2 Certification

Document necessary energy efficient building information to obtain desired certifications.

Fig. 19.6 Operation and maintain

19.2.1.3 Occupant Health and Comfort

Interview building occupants to determine how the building is performing and if the building is operating as expected.

Figure 19.6 shows us the typical example of operation and maintenance of the project.

As per Hounslow's of London, the above-described steps should be considered as early as possible in the development process. When the concept of a development is first discussed, we should see which of the different elements of architectural engineering referred to can be accurately employed in the new project to enable a reduced overall environmental impact and making buildings more cost-effective in the long run.

19.3 Measuring Sustainability in Construction

Environmental assessment method can be adopted to measure sustainability to assess the performance of buildings in a number of areas and issues, which are briefly outline below.

Management: overall management policy, contracting site management and procedural issues
Pollution: air and water pollution issues
Materials: environmental implication of building materials, including life cycle impacts
Energy use: operational energy and carbon dioxide issues
Transport: transport-related CO_2 and location-related factors
Ecology: ecological value conservation and enhancement of the site
Health and well-being: indoor and external issues affecting health and well-being
Land use: greenfield and brownfield sites

It is widely accepted and recognized that assessment tools/techniques should be used by all agencies involved in building construction, including client/users, planners, and developers, so that sustainability can be assessed quickly and easily.

Based on the techniques used in decontamination method, use of aggregates, water conservation, prefabrication, sustainable building materials, energy efficient designs, renewable energy systems, landscaping, etc., the assessment can be done as poor, average, good, or excellent.

Our commitment shall be to a Green Tomorrow leading the way to a sustainable future and more comfortable homes for everyone, with the goals of Zero Energy, Zero Emissions, and Green IT, perfecting techniques and advanced technologies that provide the highest possible value to owners. Use no fossil fuels in buildings, cut greenhouse gas emissions by 50 % in terms of costs, and try to replace up to 10 % of the energy used in residential complexes with renewable energy sources.

19.4 Conclusions

An important fragment of sustainable construction is the careful planning of the development. More front-loading is required for sustainability to work well. The end result is more wanted and more cost-efficient.

The sustainable construction of buildings is achievable and worthwhile. It is achievable if we take the initiative and be ambitious.

The world is moving, the economy is varying, and our practice is changing. Many firms are looking for the best way to integrate immerging technologies with sustainable design and analysis tools for new construction projects and also renovation projects.

Pre-sustainability and the way forward in architectural engineering and technologies for every phase of the project life cycle with the study of the following points

- What decisions affect the sustainability of the project?
- When to make those decisions?
- What tools and data do we need to make better up-to-date decisions?

Making decisions early in the design process can help deliver important results in the efficient use of energy, water, materials, and land in a final project realization.

References

Adams WM (2006) *The future of sustainability: re-thinking environment and development in the twenty first century*. Report of the IUCN Renowned Thinkers Meeting, 29–31 January 2006

ATG Technical Knowledge Center (2010) *Autodesk tech Source Document*, April 29, 2010

Chen TY, Burnett J, Chau CK (2001) *Analysis of embodied energy use in the residential building of Hong Kong*. Energy 26:323–340

Guthrie A (2008) *Tall buildings: sustainability from the bottom up*. Proceedings of the CTBUH 8th World Congress, "Tall & Green: Typology for a Sustainable Urban Future", Dubai, March 193 3–5, 2008, pp 95–101

Hammond GP, Jones CI (2011) *Inventory of Carbon and Energy (ICE)*, Version 2.0, Annex B: methodologies for recycling. University of Bath, Bath Industry solutions

Industry solutions. http://www.autodesk.com/solutions/building-information-modeling/overview

Kestner D (2010). *Sustainability: thinking beyond the checklist*. Structural design Magazine, June 2009, p 5

London Borough of Hounslow. www.hounslow.gov.uk/index/housing.htm, information

Chapter 20
Harnessing Sustainable Solutions Through Challenges: A Case Study of World Record Long-Distance Pumping of Concrete

Chetan Hazaree, Viswanath Mahadevan, Sunil Bauchkar, and Shankar Kottur

Abstract This paper depicts some of the experiences gained while solving challenges of pumping concrete through a world record distance of 2432 m in a sustainable manner. The choice of method and materials made the overall scheme have a better sustainable footprint. Pumping of concrete is more of an art requiring synergistic association of assured quality materials, equipments, plant and machinery, weather, and most importantly manpower. Holistic planning, minute detailing of methodology, and above all disciplined execution made such an effort successful. Full-scale field trials, safety planning, monitoring of weather, and maintaining adequate control points play an integral role.

Keywords Sustainability • Pumping • Fly ash • Tunnel lining • Concreting

20.1 Introduction

Pumping of concrete – vertical, horizontal, or a combination – is a well-established and widely accepted method of concrete placement for many applications. Discernible advantages of pumping concrete, experience gained over the years, and amalgamation of the state of the research with ever-increasing extreme construction needs have been driving concrete pump and pipeline technology to newer scales. Flawless, clean, safe, and economic pumping of concrete is a specialized job requiring both relevant skills and experience – from materials' selection through the final ball passing in the pipeline. Albeit the demand for pumping for longer distances is increasing, the know-how for designing and controlling such concrete

C. Hazaree (✉)
Hindustan Construction Co. Ltd, Mumbai, Maharashtra 400083, India
e-mail: Chetan.Hazaree@hccindia.com

V. Mahadevan • S. Bauchkar • S. Kottur
BASF Construction Chemicals, Mumbai, Maharashtra 400705, India

G.L. Sivakumar Babu et al. (eds.), *Sustainability Issues in Civil Engineering*,
Springer Transactions in Civil and Environmental Engineering,
DOI 10.1007/978-981-10-1930-2_20

mixtures, establishing a steady pumpable regime, and troubleshooting of related issues do not seem to exist in general. This eventually culminates into incompatible concrete specifications for a selected construction methodology.

20.2 Pumpability

Pumpability of a concrete can be defined as the "aptitude" of a concrete to be placed by pump and relates itself to the formation of blockages (Kaplan et al. 2005a, b). Published research and state of the practice suggest that a pumpable concrete must remain stable and homogenous (nonsegregating) throughout the pumping operations (external stresses) and should be deformable enough to homogenously maneuver over elbows, joints, and tapers (Kaplan et al. 2005a, b; Gray 1962; Browne and Bamforth 1977) under operating pressures. Although test methods are devised to test specific aspects of pumpability of concrete, concrete pumping remains to date an empirical trial and error process, involving frequent troubleshooting on construction sites (Kaplan et al. 2005a, b). Ideally for concrete to be pumpable, a plug flow is desired wherein there is a solid core of concrete with a slippery interface between concrete and steel. See Fig. 20.1.

20.3 Objectives

The objective of this article is to narrate experiences gained from arriving at sustainable method and materials required for long-distance pumping (LDP) (500–2450 m) of concretes, where:

(i) Concrete was to be placed in relatively smaller tunnel (finished diameter of 3.76 m) for a distance of close to 2500 m.
(ii) Difficulties were experienced due to incompatible and general concrete specifications.

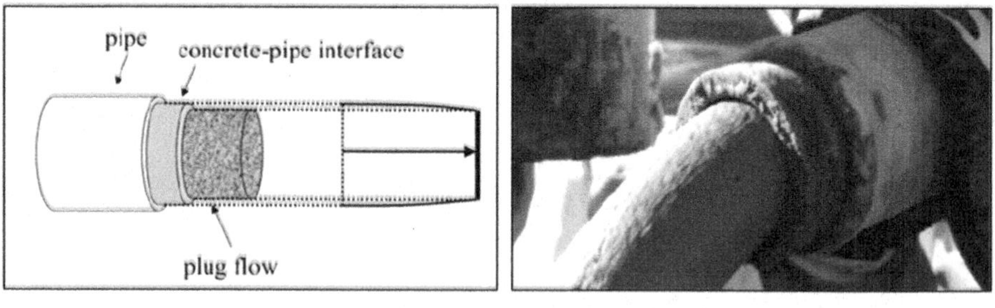

Fig. 20.1 Ideal plug flow of concrete (L) (Kaplan et al. 2005a, b). Picture showing solid-like core and slippery interface layer (R) (Rio et al. 2011)

(iii) General methodology for commencing and continuously pumping couldn't work.
(iv) Sustainable materials' selection proved to be rather deterministic.
(v) Establishing a steady and stable concrete pumping operations was a challenge.

20.4 Selecting Long-Distance Pumping (LDP)

The project under discussion is a typical hydropower project comprising of a diversion barrage, intake arrangement for drawing water, two underground de-silting chambers, a 6300 m long 3.76 m diameter head race tunnel (HRT), an underground surge shaft, and a pressure shaft, turbine assembly located in an underground power house. The finished diameter of HRT is 3.76 m with an average lining thickness of 0.25 m. The distances for lining included 1.360 km on upstream side and 2.432 km on the downstream side. In such smaller diameter tunnel, the vehicle movement is unidirectional and simultaneous, and safe carrying out of multifarious activities poses serious challenge due to congested working space. Completing the concrete lining activity within a stipulated (and challenging) time frame is critical for timely project completion.

Carrying concrete in a one-way traffic, with longer wait periods and unloading times, the implied environmental pollution inside the tunnel, and above all risking the safety of the working teams were some of the factors that lead to thinking about an alternative solution. After substantial deliberations, an alternative method for concrete conveying and placing was zeroed down, viz., long-distance pumping. Comparing in terms of fuel consumptions and time requirements for completing the tunnel lining, it was inferred that the pumping of concrete is a more sustainable option, thus justifying the method further from a deeper perspective. Subsequent to detailed discussion with equipment manufacturer, a pump suitable for matching the concreting requirements was selected while considering the overall layout of pipeline in the tunnel (bends, curvatures, tapers, and straight lengths), pipe diameter (150 mm), and pumping distances (up to 2500 m). This paper presents experiences regarding the materials' development involved in this project.

20.5 Concrete and Concreting Considerations

The desired output is a crucial factor in deciding the retention period of concrete. See Fig. 20.2. The workability retention should take into account contingency time of around 2 h in addition to the batching, transporting, and testing time requirements. This ensures safe cleaning of pipeline in case of disruption of concreting. Admixture selection must be guided by requirements for retention, setting time, weather, pump characteristics, and binder combinations. Based on considerations for these factors, it was inferred that a flowable concrete with slump flow ranging

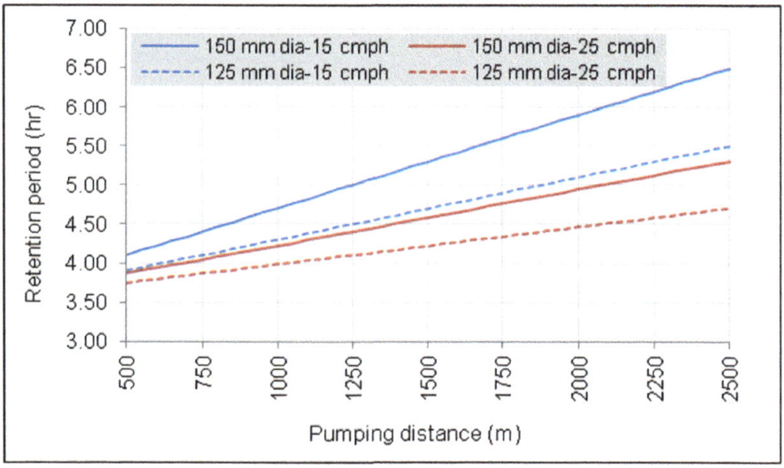

Fig. 20.2 Influence of delivery rate and pipe diameter on retention time of concrete

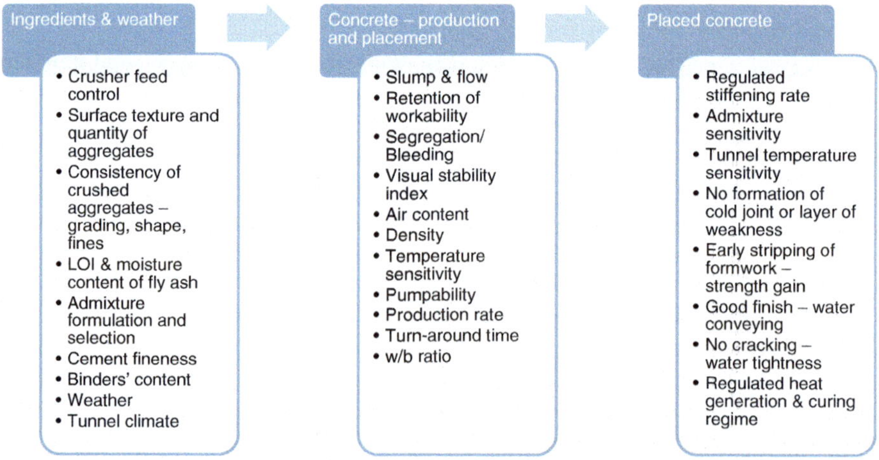

Fig. 20.3 Various parameters needing accounting during such long-distance pumping

between 600 and 700 mm flow, able to retain pumpable workability (flow value > 450 mm) through 12 h, set within 20 h, and able to gain stripping strength (4–7 MPa) by 30 h, is required. To handle the operations with better control, certain control points were established and are summarized in Fig. 20.3.

20.5.1 Pipeline Layout

As the resistance to concrete flow varies linearly to the pumping distance, any hindrance, change of direction, and impediment created in the path of concrete would increase the drop in the pressure head, thus increasing the pump back pressure. Hence it is absolutely essential to plan the layout of the pipeline with minimum possible bends and tapers. The amount of pressure head drop depends on the degree of bend, the wear and tear of the pipeline, bend location, and nature of concrete being pumped. It also needs to be noted that the wearing in pipelines is more nearer to the pump than at the farther end, in the bends and in the tapers.

20.6 Initial Considerations

The tunnel lining operation requires a well-coordinated effort in balancing the cycle time and completing the scheduled target per month. The following points summarize how these requirements were captured while deciding the concrete performance requirements.

(i) The use of OPC alone was leading to non-pumpable concrete, often disrupting concreting. Incorporating fly ash dramatically improved the mix rheology. Additionally, fly ash helped improve the sustainability, finishibility, and durability performance of concrete.
(ii) The use of relatively higher proportions of fine aggregates (44–48 %) improved dense packing and cohesion in the mix. The consistency of grading of fine aggregates was controlled at the aggregate crushing and screening plants.
(iii) The placement of concrete required vibrating of formwork at the pouring point. In order to ensure full compaction, it was decided to use self-compacting concrete.
(iv) Setting time measurements were done during the lab trials and also during execution in order to get an estimate of admixture behavior and strength development. It is very critical to have an idea regarding the setting behaviors of concrete at various temperatures.
(v) The adopted construction methodology required formwork stripping of first panel after 30 h of commencing concreting. Refer to Fig. 20.4 for a typical view of tunnel formwork. Hence, the concrete was proportioned to have strength in the range of 4–7 MPa during the stripping time. This requirement was judicially balanced with setting time requirements juxtaposed with the range of retention times. Although the target strength requirement was 30 MPa, it was actually the strength that governed the mixture proportioning.
(vi) Segregating and bleeding tendencies and stability of the mix were visually estimated and were critically considered during concrete mix proportioning.
(vii) Strength development was monitored from 18 h through 28 days.

Fig. 20.4 Tunnel lining formwork

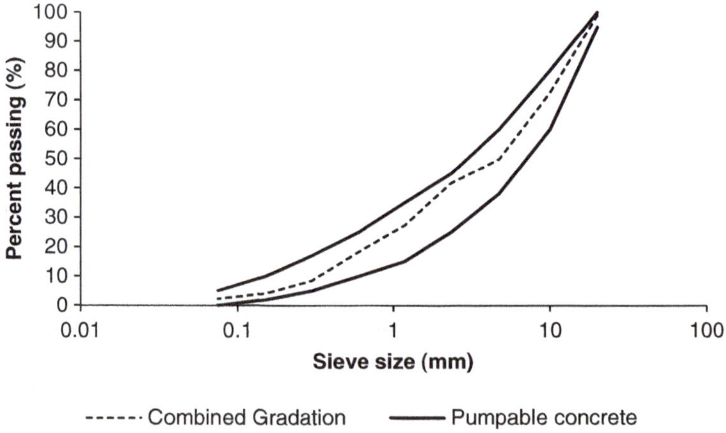

Fig. 20.5 Sample of combined aggregate grading

20.7 Materials' Development

OPC conforming to IS 8112 (BIS 2005), fly ash conforming to IS 312 (BIS 2003), and aggregates conforming to IS 383 (BIS 2002) were selected. 20 mm MSA was used in the mix and the paste volume was kept considering the pumping distance. Fly ash used was class F fly ash with a specific gravity of 2.1. The specific gravity and water absorption of coarse and fine aggregates were 2.745 and 1.26 % and 2.710 and 2.9 %, respectively. Figure 20.5 shows a sample of combined aggregate grading.

20.7.1 Formulating the Admixture

The custom-made admixture needed to be designed takes the following performance requirements into consideration:

(i) Ability to pump continuously without blockages, while retaining the workability for required retention periods
(ii) Able to set within a stimulated time window in order to have no joints in the poured concrete
(iii) Achieve required stripping strength meeting the time cycle requirements

The admixture components required choosing a low-efficiency dispersant and a workability retaining component that did not retard the setting times of concrete. The low-efficiency dispersant is allowed for the robustness during the batching plant operations reducing sensitivity to moisture variations or to dosage variations from the dosing pump. It also helped maintain better cohesiveness of the concrete mixture. These components also helped balance rheology with other critical parameters such as workability retention. Control of retention was done with minimal or no retarder to enable balance retention with early strength development for reusing the formwork.

20.7.2 Typical Mixture

Table 20.1 summarizes the evolved concrete mixture and the properties. This mix was evolved considering the required robustness and flexibility with changes in the distance of concrete pumping. Since the use of only OPC mixture was leading to

Table 20.1 Typical concrete mixture proportions and properties

Parameter	Value
OPC	340 kg/m^3
Fly ash	135 kg/m^3
Coarse aggregate	950 kg/m^3
Fine aggregate	777 kg/m^3
Water	200 kg/m^3
Admixture	0.50–1.10 %, by weight of binder (distance dependent)
Slump flow – initial, 15 min	550–650 mm
Slump flow – terminal	450–500 mm
Setting time – initial	15.0–17.5 h
Setting time – final	19.0–21.0 h
30-h strength	4–8 MPa

frequent problems, although fly ash procurement was logistically difficult, in order to ensure speedy completion of the project, it was decided to use fly ash in the mixture. Moreover, fly ash also helped improve the rheology, setting behavior, and pumpability of the concrete mix, besides improving the durability of the structure. The fly ash used was a class F fly ash.

20.8 Results and Discussions

There are two parameters that one needs to balance, viz., stickiness and cohesiveness. As a result, the admixture was designed with polymeric dispersants that imparted the best balance. Using a dispersant that is robust to variations in admixture dosages – expulsion of bleed water over time or the propensity to segregate due to moisture variations during batching can be hugely detrimental for pumping of concrete. As such, the dispersants used in the admixture were less sensitive to small moisture variations. Typically, retention properties can be achieved by the use of a retaining type polymeric dispersant and/or a retarding admixture. Therefore to maximize strength gain, a flow-retaining polymer was used over a retarder component. This helped maintain 10–12 h of flow retention while meeting required stripping strength requirements. Refer to Fig. 20.6 for the flow retention times and the initial set time (IST) and final set time (FST) of concrete mix in response to changes in the admixture dosages.

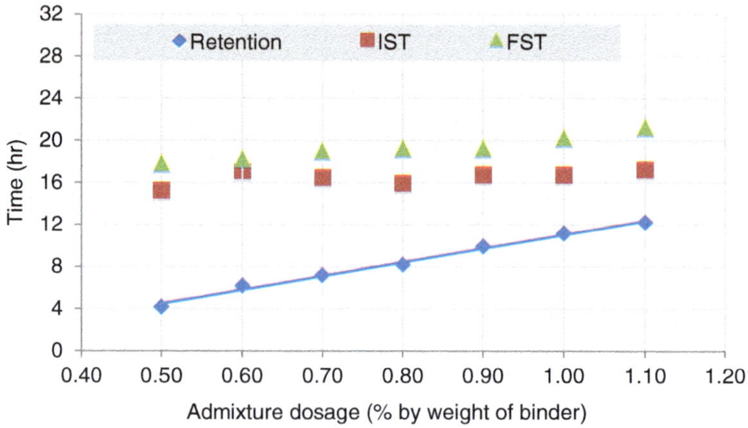

Fig. 20.6 Influence of admixture dosage on workability retention and setting times

20.9 Practical Aspects

20.9.1 Minimizing Variability

20.9.1.1 Multiple Quarries and Crushers

Typically on hydropower or underground construction projects, controlling the aggregate quality is tougher than in a typically quarried rock feed aggregate crusher. This primarily happens because economical production of aggregates necessitates the use of either river boulder material (RBM) or blasted tunnel muck (BTM) as feed materials. While dealing with pumpable concrete, having uniform (or equivalent) settings for all the crushers and screen sizes becomes essential in order to minimize intentional variations. Critical among the produced crushed aggregates is the fines fraction especially 0–300 micron.

20.9.1.2 Supply of Binders

Sticking with a lean supplier base for binders helps in optimizing and restricting the number of admixture(s) and other critical performance indices of concrete like bleeding. This also helps in controlling the properties of resulting concrete and hence the time cycle requirements.

20.9.1.3 Chemical Admixture(s)

Restricting the sources of binders helps optimize admixture to a greater extent – often restricting it to a single component addition. Many a times it becomes essential to use an additional admixture component (e.g., a viscosity modifier or a retarder) to be able to use two or more sources of cements or combinations with supplementary cementitious materials. Hence the real quality control begins by having proper control over the supply chain.

20.9.1.4 Dealing with Weather

Any construction activity that spans over several seasons should account for the climatic variations in its work plan. There is definite impact of changes in the climatic conditions on the properties and performance of concrete and extra care is required in case of LDP. The workability (rheology), setting behavior, and strength gain of concrete change with changes in the ambient temperature and humidity. Critical to understand are the differences and the variations between the external (concrete production point) and internal (concrete placement point, gantry) climates, since the concrete goes through changes as it moves through long-distance pipe. For example, concrete could be produced at 5–7 °C and relative humidity of 30 % and placed at 35–37 °C and relative humidity of 95 %. Such transitions need to be carefully monitored along with specific responses of concrete, viz., workability, retention, setting behavior, etc. To regulate this, setting time was measured regularly under these ambient climatic conditions at both the ends.

20.9.2 Pumping Methodology

There are multiple reasons why pumping blockages could happen. Figure 20.7 summarizes salient ones. These are primarily categorized as blockages due to:

(i) Mix deficiencies – including lack of fines, inadequate binder and paste content, lack of consistency in supplied concrete, proper selection of admixture system, adequate lubrication provided during priming of pipeline, and responsive adjustments to concrete w.r.t. climatic changes. Most of these are described in earlier sections.
(ii) Batching plant and pipeline – selection of adequate mixing time so that the admixture could disperse and react, proper selection, installation, and maintenance of pipeline system. These are described in the following sections.
(iii) Operating errors could be catastrophic at times; hence it is essential to list and control these at the planning stage through proper communications, training, and mock-up trials.

Fig. 20.7 Causes of pumping blockages

20.9.2.1 Pipeline: Installation and Quality

Careful, safe, and fast pipeline installation for such long distances is an essential aspect of overall time management. Important aspect of pipeline installation includes anchor clamping to a leveled concrete base. Adequate clamps of suitable strength should be used while fixing the pipeline. The number of bends, curvatures, and reducers should be reduced to a minimum through careful planning and deliberation over the pipeline layout. Keeping logs of pipes, joints, bends, and reducers and swapping these periodically help keep uniformity in wear and tear and enhance life and performance.

The checking of quality of pipeline is essential for the continuity and safety of the LDP operation. Common things to be periodically checked include cleanliness of the pipeline, wear and tear of clamps, rubber seals, bends, and tapers. The thickness of the pipeline was checked on a two-month basis. This should be balanced with volume of concrete pumped through the pipeline and benchmarked with the rated life expectancy of the pipe.

20.9.2.2 Water Cleaning Doesn't Mean "OK" for Pumping

For LDP applications, water cleaning is usually done on routine basis while ensuring smooth ball passing before and after pumping. Experience suggests that even though water cleaning is done twice, there could be some water and/or chunks of hardened concrete left behind at various stages of cleaning. Hence numbering of pipelines, clamps, elbows, and bends and maintaining a periodic log of cleaning and QC checks greatly help in saving time and improving pumping efficiency. Additionally frequent checking of rubber seals and gaskets is critical. It is important to note that even a minor air leakage could jeopardize safety and continuity of LDP for hours.

20.9.2.3 Priming: A Critical Transitional Step

Priming or lubricating of pipes is a critical transitional step, wherein empty pipes gradually get fully filled with concrete. Priming keeps the concrete-steel interface coated with a film of grout, preventing pumped concrete from drying and helping plug formation. If not properly understood and executed, blockages at priming stage are quite common even during normal pumping. In case of long-distance pumping, this step becomes more critical – primarily because tracing blockages are tedious and demoralizing and can cause disproportionate damage.

Fig. 20.8 General setup for full-scale field trials

20.9.3 Full-Scale Trials

Before commencing actual pumping, it is essential to run full-scale field trials (FSFT). These are undertaken with the following objectives:

(i) To investigate pump behavior and performance
(ii) To investigate, troubleshoot, and simplify the construction procedure and fix control points
(iii) Train the teams on procedures and safety measures

Figure 20.8 shows a general setup for one of the full-scale trials.

20.10 Sustainability Impact

The use of this method saved the burning of fuel required for transporting concrete in transit mixers. Additionally, the use of fly ash saved substantial cement consumption. If OPC mix was to be used, the mixture was not pumpable with up to 450 kg/m^3 of cement and was giving frequent disruption of operations. Comparatively, on an average, adopting fly ash-based mixture saved 110 kg/m^3 of cement.

Assuming emission of CO_2 at a rate of 1 kg/1 kg production of cement, the savings in CO_2 generation amounts to 110 kg/m^3.

20.11 Summary

(i) Long-distance pumping, a sustainable and viable method of conveying and placing concrete, can be made successful by having proper quality assurance and quality control of raw materials.
(ii) Inclusion of fly ash helps improve the rheology and pumpability of the mixture, besides bringing the inherent benefits of fly ash usage in concrete.
(iii) Admixture selection and evaluation are critical to the success of LDP.
(iv) In addition to proper pump maintenance, pipeline layout design, installation, alignment, periodic quality checks, and maintenance of are pivotal in avoiding blockages and unnecessary delays. Suitable priming, pumping, and closing methodologies (ball passing, water cleaning, etc.) should be carefully evaluated and pretested before undertaking LDP.
(v) Full-scale trials are very helpful in evaluating concrete, plant, and equipment performances.

References

BIS IS 383 (2002) 1970 Specifications for coarse and fine aggregates from natural sources for concrete. Bureau of Indian Standards, New Delhi
BIS IS 3812 (2003) 2003 Pulverized fuel ash – specifications. Bureau of Indian Standards, New Delhi
BIS IS 8112 (2005) 1989 43 grade ordinary portland cement specifications. Bureau of Indian Standards, New Delhi
Browne R, Bamforth P (1977) Tests to establish concrete pumpability. [Journal]ACI J Proc. – [s. l.]: ACI, May 1977. – 5:74: 193–203
Gray J (1962) Laboratory procedure for comparing pumpability of concrete mixtures. In: The 65th annual meeting of national crushed stone association. [s.n.], Washington D.C. pp 964–971
Kaplan D, de Larrard F, Sedran T (2005a) Avoidance of blockages in concrete pumping process. ACI Mater J 102-M21:183–191. [s.l.]: ACI International
Kaplan D, de Larrard F, Sedran T (2005b) Design of concrete pumping circuit. ACI Mater J 102:110–117. [s.l.]: ACI International
Rio O et al (2011) Pumping quality control method based on online concrete pumpability assessment. ACI Mater J 108:423–431. [s.l.]: ACI International

Printed by Printforce, the Netherlands